WELDING PROCESSES

AND

POWER SOURCES

By

Edward R. Pierre Enterprises
P.O.B. 14021, Spokane, Wash. 99214

~~~~~~~~~~~~~~

Formerly associated with:

LOCKHEED MISSILES AND SPACE COMPANY
ALUMINUM COMPANY OF AMERICA
AIR REDUCTION PACIFIC COMPANY
UNITED STATES NAVY
MILLER ELECTRIC MFG. CO.
POWER PUBLICATIONS

First Edition

First Printing—February, 1967
Second Printing—March, 1968
Third Printing—March, 1972
Fourth Printing—March, 1973

Second Edition

First Printing—January, 1974
Third Printing—August, 1978
Fifth Printing—April, 1981

Printed in United States of America

Distributed by
Burgess Publishing Company
7108 Ohms Lane
Minneapolis, Minnesota 55435

ISBN 0-8087-1645-X

Library of Congress Number 73-86974

Dedicated to my children

ACKNOWLEDGEMENTS

The author gratefully acknowledges the assistance of many people who have contributed to the preparation of this book. In particular, a hearty "Thank You!" to the many Community College and Vocational-Technical Institute Welding Instructors who contributed ideas and advised on subject matter and arrangement of chapters. The following organizations provided assistance with photographs and technical data:

Air Products & Chemicals, Inc.; Air Reduction Company; American Optical Company; American Welding Society; Arcair, Inc.; Arc Rods Company; Combustion Engineering Corporation; Hobart Brothers Company; The International Nickel Company, Inc., (Dr. Norman Kenyon); Miller Electric Mfg. Co.; National Cylinder Gas Division of Chemetron Corporation; North Carolina Department of Community Colleges; The Simpson Company; Smith Welding Equipment Division of Tescom Corporation; Thermal Dynamics Corporation; University of Wisconsin—Engineering Department, Metallurgy (Dr. Richard Moll).

To all the individuals and Companies who have encouraged me through the years of writing and editing this book I say a most grateful "Thank You!"

— E. R. Pierre

Chapter Index

Page

SOME BASIC CONCEPTS OF WELDING xvii

Three states of matter — related to specific material — solid to liquid to gas — reversal of process — gaseous to liquid to solid — principle applied to metal — positioning of joint — heat applied — metal heats — metal melts at edges of joint — fuses at joint edges — cools to ambient temperature — meaning of fusion welding.

Chapter 1 1

OXYGEN-ACETYLENE WELDING PROCESS FUNDAMENTALS

Brief history — use of fire — meaning of "blacksmith" — first oxygen and acetylene manufactured — process fundamentals — hottest fuel-gas and oxygen flame is oxyacetylene — safety rules — torch bodies — injector or low pressure type — medium or equal pressure type — torch care and maintenance — torch mixers — torch tips — gas pressure regulators — how to set up an equal pressure torch for welding — setting up an injector welding torch — fluxes used — practical gas welding data — heat transfer comparisons in metals — brazing concepts — filler rods — summary.

Chapter 2 20

COMMUNICATION WITH COMMON TERMS

Definitions and simplified meanings of terms that are in everyday use in the welding industry. Electrical, metallurgical and welding terms are discussed. This portion of the work alphabetized for easy reference.

Page

Chapter 3 73

WELDING METALLURGY FUNDAMENTALS

Meaning of metallurgy — welding metallurgy — two types of metallurgists — metals welded — ferrous metals — non-ferrous metals — basic metal structures — grains — crystals — pure metal and alloy melting temperatures — crystal formation — crystal structure types — phase diagram for iron — carbon content importance — cast iron — carbon types present — methods of welding cast iron — steels and certain alloying elements — their effects — controlling grain size — the effect of cooling rates on steel — types of microstructures — isothermal transformation (TTT) data — martensite and hardenability — metallurgy and the weld joint — effects in heat affected zone — weld metal deposits in carbon steel — stainless steels — martensitic stainless steel — ferritic stainless steel — austenitic stainless steel — stainless filler alloys — sigma phase — carbide precipitation — summary.

Chapter 4 98

ELECTRICAL FUNDAMENTALS

Introductory data — static electricity — basic theory of matter — protons and electrons — conductors and insulators — section summary — the electron theory of current flow — direction of current flow — measurement of electrical current — atomic movement in matter — the Kinetic theory — electrical conductors — volts, amperes and ohms — electron movements in conductors — electrical energy terms and their meanings — watts and watt-hours — electric energy for welding — resistance losses — electric power generation — hydro-electric — nuclear — fossil fuel generating plants — KVA —— KW — power transmission — summary.

Chapter 5 124

WELDING POWER SOURCE TRANSFORMERS

Basic functions of transformers — characteristics of transformers — welding transformer fundamentals — voltage-turns ratios and what they mean — iron cores — coil characteristics — power losses — eddy currents, what they arc — how minimized — hysteresis losses explained — iron core losses — iron core permeability — power source efficiency — magnetic coupling concepts — calculation of power source efficiency.

Chapter 6 136

Power Factor

Power factor terms defined — primary KW — primary KVA — vector — kilovars — what is power factor? — step by step analysis — power factor determination — specific power source example — normal industry practice with regard power factor correction — vector diagram — summary.

Chapter 7 142

SHIELDED METAL ARC WELDING

Fundamental welding equipment — safety rules — historical development — principles of the process — proper arc length — functions of the electrode flux — electrode core wire considerations — polarity selection — current density and electrode melt rates — arc voltage, arc length correlation — metallurgically affected areas — weld deposit — heat affected zone (HAZ) — base metal — power sources used — alternating current, NEMA Class 1, Class 2, Class 3 — advantages of alternating current — direct current — development — motor-generator type — static rectifier — engine driven — safety equipment.

Chapter 8 160

AC TRANSFORMER WELDING POWER SOURCES

AC transformer concepts — the step-down welding transformer — component illustrations — explanation of functions — iron core — primary coil — secondary coil — simple two winding transformer — circuit diagram for simple step-down transformer with illustrated voltage and amperage values — the calculation of electrical turns and voltage ratios — current and voltage ratios inversely proportional — electric field — magnetic field — electric field concept — duty cycle — methods of output current con-

trol — mechanical — movable coil — movable shunt —
electrical — troubleshooting ac power sources —summary.

Chapter 9 179

RECTIFIERS FOR WELDING POWER SOURCES

General development of dc welding power — function of
any rectifier — battery banks — engine driven generators — electric motor-generators — transformer-rectifier
power sources — rectifier development — vapor arc rectifiers — metallic rectifiers — silicon rectifiers — three
phase primary power for dc welding — MG power sources
— explanation of components — stator assembly — rotor
assembly — commutator and brush assembly — comparison of motor-generator and transformer-rectifier power sources — selenium rectifiers — method of manufacture — selenium cell components — completed selenium
stack — silicon rectifiers — diodes — basic element —
semi-conductor and what it is — methods of manufacture
— "growing" the silicon crystal — making the diode —
silicon rectifier components — installation of silicon rectifiers — testing diodes — summary.

Chapter 10 198

AC/DC and DC TRANSFORMER-RECTIFIER POWER SOURCES

Basically an ac power source — rectifier and dc circuitry added — ac/dc power sources use single phase primary power — output current control either mechanical
or electrical — typical volt-ampere curves illustrated —
reasons for dc welding power — historical development —
rectifier type power source — operate from either three
phase or single phase primary power — illustration of rectified single phase and three phase dc power — engine
driven power source with rectifiers — rectifier stacks illustrated.

Chapter 11 206

KEY CIRCUITS USED IN SOME DC AND AC/DC POWER SOURCES

Term "key" circuits defined — saturable reactor —
illustrations of components — component function ex-

plained — step-by-step circuit buildup — self-inductance — electro motive force — reactor concept — methods of varying current in the welding circuit — tapped reactor — dc control circuit concept — infinite current control — small amount of dc controls large amount of ac — saturation explained — ac and dc circuits isolated from each other — reactor core saturation only common factor.

Current feed-back circuit — in dc portion of circuit — reactor core saturation graph — current feed-back permits super-saturation of core — illustration and explanation of current feed-back concept — stable short circuit current — magnetic flux lines increased in reactor core — maximum current output in all ranges.

Control circuit — simplified drawing — voltage check points — ac portion of circuit — dc portion of circuit — other control voltages possible — half-wave rectification in control circuit — control circuit is warning system.

Chapter 12 224

TROUBLESHOOTING SOME DC POWER SOURCES

Introduction — best tools are the perceptive senses — volt-ohm meter — maintenance and operating manual use — primary voltage and phase — line fuse testing — power source terminal voltage linkage — results of improper primary voltage linkage — checking open circuit voltage — the control panel — remote current control switch — control circuit fuse — current selector rheostat — use of the rheostat to limit remote current maximum value — starting current rheostat (some models) — effect of low current starts — effect of high current starts — summary.

Chapter 13 236

GAS TUNGSTEN ARC WELDING

Principles of the process — basic equipment needed— compared to other welding processes — typical tungsten arc electrode holder — shielding gases used — argon — helium — typical deposit characteristics — gas flow requirements — current density — gas ionization—arc starting — thermal placement in the weld — types of power sources used — deposit characteristics with each type power source — tungsten electrode chart — comparison of electrode size at same current with DCSP and DCRP — cleaning action — reverse polarity — theory of how oxides are removed — refractory oxides — summary.

Chapter 14 255
TUNGSTEN ELECTRODES

Powder metallurgy — methods of manufacture — sintered material — thermal treatment — swagging the ingot — rods for electrodes are drawn — melting points — boiling points — comparison of carbon and tungsten — the term sublime defined — electron emission characteristics — thoriated electrodes — zirconium — better current carrying capacity — better arc initiation — grinding methods for tungsten electrodes — results of improper grinding methods — weld inclusions — shattered electrodes — types of tungstens and their uses — electrical pointing of tungsten electrodes — correct dimensions for grinding tungsten electrodes — summary.

Chapter 15 264
HIGH FREQUENCY SYSTEMS

The term high frequency defined — used for arc stabilization — partial shielding gas ionization — provides safe high voltage at the electrode tip — assists arc initiation — promotes electron emission — spark gap type oscillators — high frequency circuit diagram and explanation — the high frequency "skin effect" — why the high frequency circuit functions as it does — the tank circuit components — how high frequency is impressed on the welding circuit — how frequency is determined — problems with high frequency — trouble shooting high frequency problems — lack of high frequency at the electrode — voltmeters and high frequency — direct radiation from the power source — direct radiation from welding leads — direct feed-back to primary power lines — pick-up and re-radiation from power lines — summary.

Chapter 16 274
POWER SOURCES FOR GAS TUNGSTEN ARC WELDING

Either ac or dc power sources used — some units designed specifically for the process — power sources operating from single phase power — power sources operating from three phase power — single phase ac rectified illustration — three phase ac rectified illustration — standard equipment for TIG power sources — dc component — de-rating ac power sources for TIG welding — chart for de-rating power sources — transformer heating results — uses of ac, dcsp and dcrp for TIG welding.

Chapter 17 281

GAS METAL ARC WELDING

Brief explanation of process — basic concepts and equipment — metal transfer type by deposition speed — terminology used — GMAW — MIG — process uses dcrp — ac tried — effect of dcsp — basic equipment required — power source development — types of metal transfer — open arc — short circuit transfer — definitions of metal transfer types — spray transfer — buried arc transfer — pulsed current transfer — globular transfer — short circuit transfer — typical voltage ranges — slope settings for short circuit transfer — amperage control for the process and equipment — how the constant potential power source automatically sets amperage — summary.

Chapter 18 297

SHIELDING GASES AND ELECTRODES

Argon — chemically inert — ionization potential — definition of ionization potential — argon characteristics — concentrated arc column — result of heat concentration — cleaning action — gas purity — illustration of argon shielded weld in aluminum.

Helium — chemically inert — ionization potential — heat carrying characteristics — results in voltage gradient — how helium is obtained — illustration of helium shielded weld in aluminum — where helium is used.

Carbon Dioxide (CO_2) — a compound gas — not inert — dissociates into oxygen and carbon monoxide — comparative arc column shapes — open arc welding with CO_2 — used for mild steel — summary.

Argon-Oxygen — reasons for adding oxygen — superheat molten weld metal — retard cooling rate — flatten weld bead profile — eliminate undercut.

Argon-Helium — reasons for use — methods of combining the gases — selecting the proper proportion.

Argon-CO_2 — where used — methods of mixing — percentages of each gas — the mixing chamber — gas cost — metals welded.

Helium-Argon-CO_2 — reason for development — stainless steel, short circuit transfer — low crown profile — pipe welding — carbide precipitation — summary.

Other gases — chlorine addition — reasons for adding chlorine to argon — danger in use — nitrogen — where used — summary.

Welding wire — alloys and deoxidizers — how they function — what they are — combined functions of some materials — aluminum, a versatile material — the effect of shielding gas selection — CO_2 and mild steel — element transfer efficiencies — deposition rates — flux cored electrodes — some historical data — some of the benefits — the method of manufacture — how flux cored electrodes are sized — the effect of shielding gas — applications — specifications — summary.

Chapter 19 311

WELDING POWER SOURCES FOR GAS METAL ARC WELDING

Welding power used is dcrp — conventional power sources used first — response times explained for MG set and transformer-rectifier units — first constant potential power sources developed — discussion of conventional power sources — magnetic coupling explained — how achieved — open circuit voltage requirements — arc length adjustment with conventional power sources — used for aluminum — requires limited maximum short circuit current — excessive spatter, poor welds if too high — use of conventional power sources for gas metal arc welding process — spray or globular transfer only — not normally recommended for the process — summary.

Constant potential power sources — define constant potential and constant voltage — flat volt-ampere curve — considerable amperage change with very little voltage change — transformer design different — tight magnetic coupling — primary-secondary coils close together — method of controlling open circuit voltage — volt-ampere

characteristics — no amperage control mechanism on constant potential units — illustration of current control by
power source — low open circuit voltage — CP power
sources preferred for gas metal arc welding — types of
voltages in welding — open circuit voltage — load voltage
— arc voltage — variable voltage settings — parallel volt-
ampere curve development — voltage drop per hundred
amperes of output remains essentially constant — slope —
either resistance or reactor controlled — resistance slope
modifies output volt-ampere curve — reactor slope uses
impedance — circuit reactance — limits maximum short
circuit current — slows response — impedance defined —
slope reactor concepts shown and explained.

Lenz's Law applied — results of impedance — where
slope is required and used — the inductor or stabilizer —
what it is — where in circuit — what it does — gas metal
arc spot welding — applications. Air carbon-arc gouging
with constant potential power sources — volt-ampere
curve illustration and explanation — carbon electrode
chart with ampere ratings each diameter.

Chapter 20 337

WELDING PROCESSES COMPARISON AND USE

Object of comparison — lowest net welding costs —
comparing shielded metal arc and gas metal arc welding —
joint preparation — joint groove angles — reason for difference — economical use of filler metal — cost of
scrap metal removed — welding electrodes — purchased
pound cost — deposited pound cost — losses with shielded metal arc welding — losses with gas metal arc welding
— results of comparison.

Weld cleaning — shielded metal arc — costs more than
actual arc time — gas metal arc — very little cleanup.

Operator training — long term with shielded metal arc
— minimal with gas metal arc — equipment is semi-automatic.

Fast controlled welding — less stop and start of welds
— less possibility of failures — gas metal arc welding
conditions pre-set — welding operator concentrates on
weld.

Current density — what it is — how calculated —

minimum and maximum current density — what occurs if either limit is exceeded — penetration patterns — comparison of welding speeds and deposits — arc columns compared — heat input per linear inch of weld — bead width-to-depth ratio — other benefits.

Gas metal arc welding applications — 80% steel welding — spray arc — short circuit transfer — shielding gases used.

Porosity — reasons why it occurs — how to prevent porosity — silicate residue — what it is — how formed — methods of removal.

Aluminum — light weight — good electrical and thermal conductivity — cleanliness requirements — methods of cleaning — sheared aluminum plate — foreign particle entrapment — proper cleaning methods — spray transfer.

Welding procedures — some electrode specifications available — some code data — responsibility for certifying welding operators and the process — charts and tables.

Chapter 21 349

ROTATING TYPE WELDING POWER SOURCES

All rotating type power sources electro-mechanical — generators produce a maximum KW — open circuit voltage control — correlation of weld amperage and electrode diameters — types of welding generators — types of motive power used — electric motors — fuel powered engines — dc generator design concept — generating sequence — armature — commutator and brushes — ac generator design concept — power generation — from armature to slip rings — ac alternator design concept — power generation — armature coils in stator — comparison: rotating armature and rotating magnetic field designs and efficiencies. Paralleling rotating type welding power sources — auxiliary power plant operation — summary.

Chapter 22 363

SPECIAL WELDING PROCESSES

Special use processes and equipment — submerged arc welding — high amperage power sources — either ac or dc — fundamental process equipment required —

process operation explained — can be either portable manual operation or fully automatic — explanation of flux — advantages and disadvantages.

Electron beam welding — system illustrated and explained — principle of operation — metals welded with electron beam — precautions suggested.

Ultrasonic welding — no specific heat application — use transducers to change energy from electrical to vibratory — transducer explained — uses of ultrasonic welding — metals and metal thicknesses suggested.

Plasma arc welding — plasma defined — plasma production either transferred arc or non-transferred arc — methods explained — special controls and torch equipment required — special nozzle with constricted orifice — process fundamentals — types of arcs used — power source requirements — summary.

Laser welding — basic concepts of process — generation of the laser beam — directed by optical focusing equipment — several types available — others in development — light does not diffuse with distance — summary.

Epilogue 376

Appendix 377

Data Charts 378

Index 402

Preface

This Second Edition of Welding Processes And Power Sources has been made possible because of the continued interest and support of thousands of welding people around the world.

The five new chapters were added at the specific request of many Welding Instructors in both Industry and the academic schools. As before, the information gathered within these pages is not necessarily new. It is correlated and written in such a manner as to promote knowledge of the tools, processes and theory of welding.

Every page of the First Edition has been re-written and brought up to date using the latest terms and technology.

Those who read this text will inevitably gain from their labor. Not necessarily because of the contents of the book but because they had the desire and will to study and learn. —EP—

"Education is the progressive discovery of our own ignorance".
—Will Durant

SOME BASIC CONCEPTS OF WELDING

Over the centuries Man has joined many different types of materials using various methods. Some very old manuscripts used the word "welding" to describe the union of thoughts, materials or ideas. Yet, long before Man conceived the idea of welding two or more objects into one entity, Nature had already done so.

Consider the three known states of matter. They are solid, liquid and gaseous. Relating the welding concept to a specific material such as water will assist in making the explanation more clear.

In solid form water is called "ice". It has a melting point of 32° F. At this temperature the substance may be in either solid or liquid form. If it is a solid it is called ice. If it is in liquid form it is called water.

Assume that a block of solid ice is set into a container into which a regulated amount of heat energy is induced. The result is that the solid ice begins to melt. But it does not "melt" away. Instead, the solid material changes to another form, or state, which is termed liquid. The interesting point is that the water will remain at a constant temperature of 32° F. until all the ice has disappeared. This occurs even though the heat energy input to the mass is held at a constant rate.

As soon as the last bit of ice has disappeared the temperature of the liquid (water) will begin to rise. The **rate of temperature rise** will depend on the **rate of heat energy input** to the liquid mass. It is assumed that the rate of heat energy input exceeds the heat losses of the material.

Assuming the heat energy input is sufficient to raise the water temperature to its boiling point (212° F.) we find that another transition takes place. At this temperature the liquid water is transformed into gas in the form of steam. If the steam is contained within a vessel or flask the temperature and pressure of the substance will continue to rise as the water is changed into steam. This again presumes the rate of heat energy input has remained constant and that it exceeds the heat losses of the substance.

In this manner the steam is super-heated and, as long as it is contained within the flask, will expand and build pressure. Reversing the procedure creates a situation where heat is removed from the mass. For this example we will consider the rate of heat removal to be constant although it is not necessary to accomplish the end result.

The super-heated steam cools to 212° F. Since the temperature will no longer sustain the gaseous state of the substance, it reverts (condenses) back to the liquid state. In this transition the temperature of the liquid will remain at 212° F. until all the gas (steam) has changed to the liquid state.

The liquid (water) will then decrease in temperature at the prescribed rate of heat energy removal. The actual **rate** may be fast or slow depending on the specific conditions prevailing. When the liquid reaches the 32° F. temperature a solid will begin to form. The substance has again reached the **transition point** between liquid and solid. The material substance will remain at the 32° F. temperature until the total mass has changed from liquid to solid.

To carry the water analogy a bit further suppose two ice cubes are placed in a shallow metal container. If they are allowed to melt to a completely liquid state there would be one liquid mass where there had been two solid pieces before. If the liquid is placed in a freezing compartment it will solidify into one solid piece of ice. In effect, we have combined, or "welded", the two separate pieces into one homogenous mass.

The next step is to apply the information we have obtained to welding metals. As we will see, the basic concepts are very similar.

The same melting and solidification concept may be applied when welding two pieces of metal together. With metals only the abutting edges are joined by melting. The melting points of various metals are shown in the appendix of this book.

Now let us talk a bit about welding metals. Iron is one of the most commonly welded metals in industry today. Combined with carbon, and usually some fractional percentages of other elements, it makes an alloy called steel. Depending on the percentage of carbon content of commercial steel, the melting point will vary from approximately 2,500-2,700 degrees F. The exact melting point temperature is not important to this phase of the discussion.

The two pieces of carbon steel to be joined should be positioned so the weld can be easily made in the joint. The method used to join the steel may be any of the commonly used welding processes. In this example, the joint will be a simple corner weld. The actual weld will be accomplished by fusing the metal edges together with no filler metal added. The only requirement for welding is a source of sufficient heat energy to melt the metal edges. This can be done with an oxy-acetylene torch flame, a gas tungsten arc welding torch or any other competent heat energy source.

Upon first application of the heat energy to the joint to be welded nothing happens that is apparent to the naked eye. As a matter of fact, the metal does not even appear to change color. As heat energy input continues, however, the metal edges do turn color. First there is a dull red, then a bright red, orange and finally a brilliant white. At this time a small molten pool appears at each metal edge. If the torch operator is skilled in the oxy-acetylene welding techniques, he will cause the melted edges of the metal parts to flow together by manipulation of the torch. Once the first union of the two edges is accomplished it is a simple matter to complete the weld joint. As the heat source is moved along the weld joint the completed weld will rapidly change color from the welding heat to its natural color.

Examining the sequence of events discloses that the steel followed exactly the same pattern of behavior as the ice. In both cases there was a solid material substance to begin with. By applying heat energy the solid reached its melting point at which it changed from a solid to a liquid. Of course, the melting of the steel pieces was localized to the weld joint because the metal would not transfer heat readily to other areas of the metal pieces. While in the liquid state both the water and the steel were fused. The water was then cooled and frozen to produce one solid piece of ice. The steel weld was permitted to cool to room, or ambient, temperature as one solid steel weldment. In both cases the result was a single object where there had been two separate pieces.

Fusion welding, with or without the addition of filler metal, is the act of bringing metals to the liquid state at the point of juncture (the joint), combining the molten mass, and allowing it to cool to the solid state. While this is over-simplifying the process considerably it is the procedure that is followed each time a fusion weld is made.

Some of the various methods used when applying heat energy to a joint for welding will be discussed in this book. The oxy-acetylene welding process will be briefly discussed as one heat source. Most of the text materials will be concerned with the use of electrical power as the heat source for welding operations in industry. To this end, there will be a chapter on Electrical Fundamentals so the reader has some background when he progresses to the sections concerning welding power sources and the welding processes.

Chapter 1

OXYGEN-ACETYLENE WELDING PROCESS FUNDAMENTALS

Flame! A mixture of oxygen and some type of combustible material which, when ignited, produces heat and light. A controlled flame for welding metals is what this chapter is all about.

Since before recorded history, Man has known fire—for warmth, for cooking and later for heating and shaping metals. A basic tool handed down from antiquity is the blacksmith's forge. The forge is the container in which the fire is shaped and controlled by the blacksmith.

The welder of today, with his relatively sophisticated welding equipment, has a rich heritage from the blacksmiths of yesteryear. The very term "Blacksmith" comes from the work he did and the metal he used. The black iron (wrought iron) he worked with provides part of the name. His actions in "smiting" (striking) the iron with a hammer gives us the rest of the name. A blacksmith smites black iron to shape it into usable objects.

It was not uncommon for blacksmiths to manufacture all of their own tools including their anvil. Unfortunately the art and craft of blacksmithing has almost disappeared from the American scene. It is still an honored profession in many parts of the world, including Europe.

It was about 1900 when the first practical methods were developed for manufacturing both oxygen and acetylene. Gas welding and cutting torches were developed and a new welding process was born.

The manufacture of iron and steel has progressed very greatly in recent years. As a matter of fact, approximately 80% of all welding accomplished with all welding processes is done on some type of iron or steel alloy.

In the oxygen-fuel gas welding processes, the flame heats the metal to be welded where the two pieces join or abut. This area is called the **weld joint**. When the weld joint reaches the proper temperature the edges of the base metal melt. The molten metal of the edges is caused to flow together by proper manipulation of the welding torch.

As the welder moves the welding torch and flame along the metal joint the molten weld metal cools and solidifies. The result is a weld joining two pieces of metal into one homogenous mass through the use of controlled heat input.

Most gas welding is done with oxygen and acetylene gases. It is normally referred to as **oxy-acetylene** welding. The gas used for fuel will be selected for its heat output when mixed with oxygen. For example, oxy-acetylene is used for metals having relatively high melting points. These include the carbon steels, cast irons, stainless steels, copper and copper alloys.

Other fuel gases, having lower flame temperatures, may be used for metals and alloys which have relatively low melting points. For example, hydrogen, natural gas and several manufactured gases are separately combined with oxygen for welding aluminum, magnesium, lead, zinc and some precious metals.

The hottest flame produced by any of the oxygen-fuel gas mixtures is oxy-acetylene. The three basic flames used for oxy-acetylene gas welding are illustrated in Figure 1. Each flame type has its specific use with various metals and alloys. For example, a carburizing flame is normally used when welding aluminum; a neutral flame when welding carbon steels; and an oxidizing flame is used for brazing with bronze alloys.

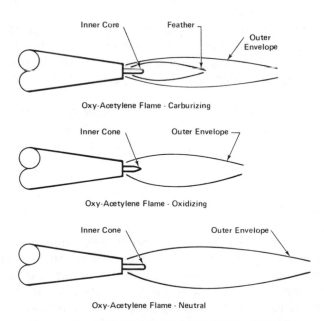

Oxy-Acetylene Flame - Carburizing

Oxy-Acetylene Flame - Oxidizing

Oxy-Acetylene Flame - Neutral

Figure 1. TYPICAL OXY-ACETYLENE FLAMES.

Safety With Oxy-Acetylene Equipment.

In the welding business there are basically two kinds of people—the "quick" and the "dead".

The term "quick" as used here means the live, SAFE worker. Only a fool takes chances with welding equipment.

The term "dead" as used here means the UNSAFE workers who are often injured or worse through careless and ignorant use of welding equipment.

Welding is a safe occupation for anyone who practices a few common sense rules for safe operation. This part of the safety discussion will be concerned only with gas welding safety practices. Electric welding safety will be considered under the proper chapter titles.

When working with oxygen and acetylene gases and welding equipment, there are some natural hazards which must be watched carefully.

Some safety rules, facts and comments are listed in the following paragraphs. Read them, understand them and follow them!

Rule 1.

Oxygen is a tasteless, odorless and colorless gas. It will not burn or explode by itself but it will support combustion in all materials. OXYGEN SHOULD NEVER BE USED TO BLOW DIRT OR DUST FROM CLOTHING, WORKBENCHES OR ANY OTHER OBJECT! Even so-called non-inflammable and fire retardant clothing and materials will burn if saturated with oxygen. All that is needed is a slight spark to set it off.

Rule 2.

Oxygen should never be used with, or around, grease and oil. Even the slightest trace of grease mixed with oxygen will cause violent explosions to occur. Therefore, NEVER USE GREASE OR OIL ON ANY OXYGEN OR ACETYLENE WELDING EQUIPMENT INCLUDING TORCHES, VALVES, REGULATORS AND HOSE FITTINGS. Oxygen is so sensitive to grease and oil that even a slightly oily rag, which might be used to wipe off cylinder valve threads before attaching the regulator, would leave enough oil to cause an explosion.

Rule 3.

Acetylene is a fuel gas that becomes unstable at operating pressures above 15 p.s.i. The acetylene gas cannot evolve out of the acetone in the cylinder and it is possible to have acetone pulled from the cylinder in liquid form. ACETYLENE SHOULD NEVER BE USED AT OPERATING PRESSURES ABOVE 15 P.S.I.

Rule 4.

Storage of fuel gases and oxygen, particularly in cylinders, should be in accordance with OSHA regulations. All fuel gases should be stored in a clean, dry area away from other combustible materials, open flames, etc. Oxygen should be stored in a separate area away from any combustible materials, grease, oil, or open flames. All gas cylinders should be stored in the upright position and should be held in place by restraining chains or cables. THERE ARE NO EXCEPTIONS TO THIS RULE! !

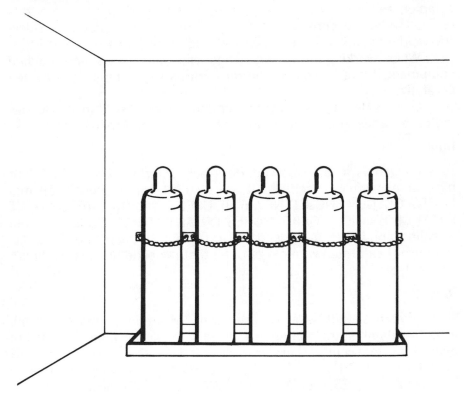

Figure 2. SAFE STORAGE OF CYLINDERS.

Rule 5.

ACETYLENE PIPING SHALL NEVER BE CONSTRUCTED OF COP- PER PIPE OR TUBING. Copper and copper alloys with more than approximately 65% copper content will form acetylides with acety- lene gas. Acetylides are violently explosive and can be set off by heat or a very slight shock of any kind. Again, DO NOT USE COPPER PIPE OR TUBING WITH ACETYLENE.

Rule 6.

Protective SAFETY GLASSES with properly shaded lenses shall be worn when welding with oxy-acetylene. Molten weld metal and other spatter can seriously damage, or destroy, eyesight. Regular hardened safety glasses should always be worn when working in any metal shop. You only get one pair of eyes. If you lose them, there is no replacement. Wear safety eye protection when you work!

Rule 7.

THERE SHALL BE NO HORSEPLAY, SKYLARKING OR GENERAL JACKASSING AROUND IN ANY WORKSHOP AREA! Serious injury or death can be the result of just "fooling around". Don't do it!

Rule 8.

Never pick up a piece of "cold metal" without first testing it to be sure it is cold. The blacksmith who was making horseshoes and tossing them onto the ground as he finished each one watched in amusement when a young apprentice walked over to the pile of 'shoes, picked one up and immediately dropped it. "What's the matter, son? Was it hot?", he asked. The young fellow looked ruefully at his burned hand and replied, "Not particularly so. It just doesn't take me long to inspect a horseshoe!" Save your hands by using your head.

Rule 9.

Never put a mixture of oxygen and acetylene into a container such as a paper bag with the thought of exploding it to scare someone. Death can result. Besides, the other guy might not think it is funny. Some short tempered fellow just might stomp a mudhole in your middle section and then walk it dry!

Rule 10.

Be one of the "quick, safe workers", not one of the unsafe, possibly "dead" individuals. SHOP SAFETY IS EVERYBODY'S RESPONSIBILITY!!

Oxy-Acetylene Welding Torches (Body, Mixer, Tips).

The oxy-acetylene torch is a very important tool of your trade and extremely important to your job. If it is damaged through careless or improper use or through poor maintenance it cannot function properly. It could, in fact, become a dangerous instrument. A SKILLED CRAFTSMAN TAKES CARE OF HIS TOOLS—and they will serve him well for years to come.

Torch Bodies.

There are two basic types of oxy-acetylene torches for welding. They are the **Injector (low pressure) type** and the **Equal Pressure (medium pressure) type.**

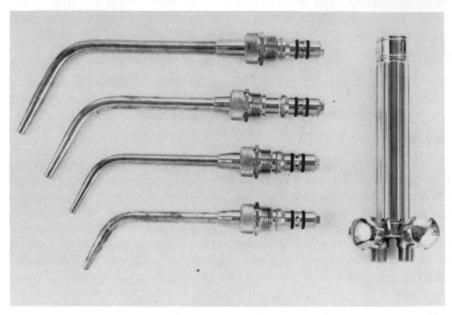

Figure 3. OXY-ACETYLENE WELDING TORCH, TYPICAL.

The **injector, or low pressure, torch** operates with very low acetylene pressure (usually 1-3 psi) but with relatively high oxygen pressure (usually 10-35 psi).

The **medium, or equal pressure, torch** operates with oxygen and acetylene pressure above one (1) psi but not greater than 15 psi on the acetylene regulator. Actual flow rates are balanced for both oxygen and acetylene. The actual pressures used will depend on the torch type and the tip size used for welding. (See Oxy-Acetylene Welding Techniques).

The welding torch body is equipped with two needle valves at the gas input end. One valve is for oxygen and the other is for acetylene. The gas is supplied from either separate cylinders or a manifold system.

Both torch needle valves have packing glands and packing nuts to seal gas leaks. The packing nuts should ALWAYS be checked for tightness before using the torch for welding. Gas leaks are dangerous!

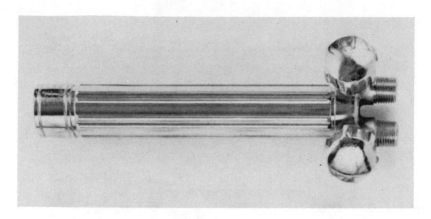

Figure 4. OXY-ACETYLENE NEEDLE VALVES.

Some common sense facts concerning torch body care and maintenance are listed for your information.

1. USE WRENCHES GENTLY TO TIGHTEN BRASS FITTINGS ON ANY WELDING EQUIPMENT. Excessive wrench pressure will possibly strip the threads or otherwise damage the fittings. Some torch tips and mixers are so designed that only hand pressure is needed for adequate tightening of the threaded joint.

2. **Never** place the torch body in a vise. The body could be crushed and the internal parts severely damaged.

3. KEEP ALL OIL AND GREASE AWAY FROM TORCH FITTINGS.

4. If the torch needle valves do not close properly and completely, remove the valves and clean the seat connections and surfaces. If necessary, remove and replace the packing in the packing glands.

5. NEVER USE ANY OXY-ACETYLENE WELDING TORCH UNLESS YOU ARE SURE IT IS IN PERFECT WORKING CONDITION! It's your life.

Torch Mixers.

The torch mixer is a device, or fitting, between the torch body and the torch welding tip. It is in the mixer that the oxygen and acetylene come together to form a gas compound usable for welding.

Some torch designs have one mixer which accomodates a range of welding tip sizes. Other torch designs employ a separate gas mixer for each torch tip size. In this design, the mixers are usually an integral part of the welding torch tip structure. The basic function of a gas mixer is to combine the oxygen and acetylene thoroughly for proper combustion at the torch tip.

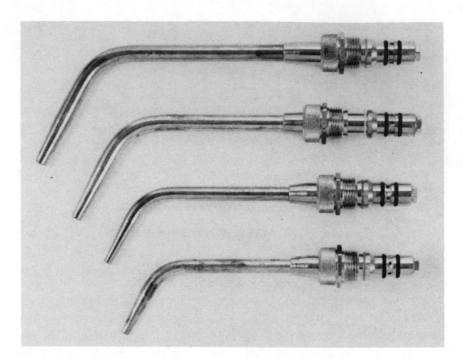

Figure 5. WELDING TORCH TIPS WITH MIXERS.

Torch Tips.

Welding torch tips are manufactured in a variety of orifice (output hole) sizes for industrial use. For a given torch body and mixer design the tip sizes may range from "0" to "12" inclusive. For small flame, delicate work, there are small "Aircraft" type torches which use only small tip sizes. Such torches may be used for light sheet metal work, lead welding, jewelry manufacturing and repair, and other low heat input applications.

Torch tips are normally made of copper or high copper alloys. Care should always be taken when fitting a torch tip to the torch mixer. Cross threading the tip will seriously damage the tip and mixer.

Gas Pressure Regulators.

Gas pressure regulators, as the name implies, control and regulate the flow of gas from the cylinder or manifold system. Separate oxygen and acetylene regulators are used for the two gases.

Acetylene pressure regulators always have **left hand threads** on the connection nut that attaches to the cylinder valve. The acetylene nut has a groove around the outside periphery.

Oxygen pressure regulators will always have a **right hand threaded** connection nut which attaches to the cylinder valve. The oxygen connection nut is smooth on the outside.

The two common types of oxygen and acetylene pressure regulators are the **single stage** and the **two stage** units. Single stage regulators are less expensive to buy but do not have the close control of gas flow over the entire cylinder pressure range. Two stage pressure regulators are more expensive than the single stage units but they do provide full gas flow control and regulation over the entire cylinder pressure range. Two stage regulators have one pressure gauge to show the remaining cylinder pressure (contents) and a second gauge to set the operating gas pressure at the welding torch.

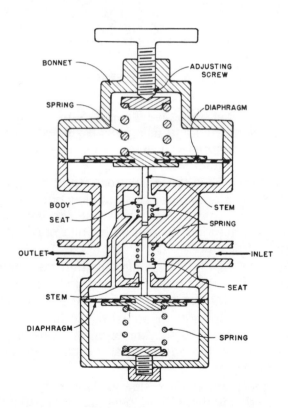

Figure 6 (A). TYPICAL TWO STAGE GAS REGULATOR

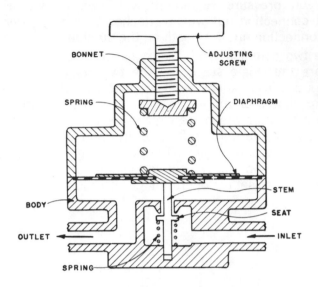

BONNET

ADJUSTING
SCREW

SPRING

DIAPHRAGM

STEM

BODY

SEAT

OUTLET

INLET

SPRING

Figure 6 (B). TYPICAL SINGLE STAGE GAS REGULATOR

Some basic common sense shop practices which have been found useful in connecting and using gas regulators are listed below. Following these procedures will help insure safe, sensible operation of your gas welding equipment.

1. Before attaching the regulator, stand to one side of the cylinder and "crack" the cylinder valve to blow out dust. ("Cracking" the cylinder valve means to open and close the valve quickly to let some gas escape. Standing to one side while cracking the cylinder valve is a safety precaution).

2. Attach the regulator, taking care that the correct regulator is being used and that the threads of the cylinder connection nut are not cross-threaded. Tighten the cylinder nut gently but firmly.

3. Back out the regulator adjusting screw to the no-pressure position BEFORE opening the cylinder valve. Opening the cylinder valve with pressure on the adjusting screw can damage, or destroy, the regulator diaphragm.

4. Open the cylinder valve slowly. The acetylene cylinder valve should be opened not more than one-quarter turn of the cylinder wrench. This is a safety precaution in case you have to shut down fast in an emergency. The oxygen cylinder valve should be opened all the way since it is a double seat valve. It must be seated at the top, or full open, position.

5. The regulator adjusting screw should be turned in slowly to prevent sudden high pressure surges on the diaphragm.

6. NEVER USE OIL OR GREASE ON ANY PART OF THE REGULATOR ASSEMBLY.

7. All applicable OSHA and American Welding Society Safety and Operating Standards must be observed for safe operation of gas welding equipment.

8. Check total gas system for leaks using approved and safe techniques. Liquid soap and water are generally used for this purpose if leaks are suspected.

Oxy-Acetylene Welding Techniques.

Before any welding can be done with the gas welding process the proper torch tip size must be selected. The tip size will depend on the metal type to be welded and the thickness of the metal. Thin gauge materials require less heat volume, and therefore a smaller tip size, than heavier metals. The thermal conductivity of the metal being welded must be considered also. For example, aluminum will require more heat input for a given thickness than will steel because aluminum will transfer heat very quickly through its entire mass. Steel, on the other hand, transfers heat comparatively slowly through its mass.

Keep in mind that, although there is more **flame volume** with a larger size tip, **the flame temperature will remain the same for a given type of flame.** For example, a neutral flame has a temperature of approximately 5,800 degrees F. regardless of the tip size used for welding.

After selecting the torch tip size, and having properly connected all fittings and other equipment, you are ready to bring the gases to the torch. Be sure the cylinders are correctly and safely chained to upright posts or stanchions in accordance with approved safety standards. Make sure all hoses are so arranged that they are not subject to cuts or other damage due to falling metals, lift trucks or other vehicles running over them, etc.

Equal Pressure (Medium Pressure) Torches.

In setting the equal pressure torch for welding it is necessary to follow certain steps for obtaining the best operational characteristics.

The gas regulators should have the adjusting screws backed out all the way to the "no pressure" setting. Open the cylinder valves as previously instructed. **Remember:** the acetylene valve is only opened one-quarter turn of the acetylene valve wrench. The oxygen valve is opened to full capacity so the valve will seat at the top of the screw valve. There should be no pressure on the gas hoses at this time.

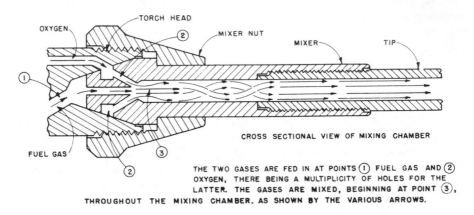

OXYGEN
TORCH HEAD
MIXER NUT
MIXER
TIP
FUEL GAS

CROSS SECTIONAL VIEW OF MIXING CHAMBER

THE TWO GASES ARE FED IN AT POINTS ① FUEL GAS AND ②
OXYGEN, THERE BEING A MULTIPLICITY OF HOLES FOR THE
LATTER. THE GASES ARE MIXED, BEGINNING AT POINT ③,
THROUGHOUT THE MIXING CHAMBER. AS SHOWN BY THE VARIOUS ARROWS.

Figure 7. TYPICAL EQUAL PRESSURE WELDING TORCH.

The next step is to open BOTH needle valves on the torch to full on. You are now ready to ignite the acetylene gas.

Place your hand on the acetylene pressure regulator adjustment screw. Slowly open the adjusting screw until you hear a hissing sound as the acetylene gas flows from the torch tip. You will certainly smell the pungent odor of acetylene.

Ignite the torch flame with a spark lighter designed for the job. Do not use matches! ALWAYS BE CAREFUL WHERE YOU AIM THE TORCH TIP AND FLAME! Never point the flame in the direction of a fellow workman.

Increase the acetylene regulator pressure until the flame is dancing off the end of the torch tip about one-half inch. Remove your hand from the acetylene regulator adjusting screw. All further acetylene regulation, for this torch tip size, will be accomplished at the torch acetylene needle valve. Using the torch acetylene needle valve, decrease the acetylene gas flow until the flame returns to the end of the torch tip. Remove the hand from the torch needle valve.

Place your hand on the oxygen pressure regulator adjusting screw. Slowly turn the adjusting screw in until the torch flame shows a white "feather" and a darker blue inner cone. **Slowly** increase the oxygen pressure until the white feather disappears.

Remove your hand from the oxygen regulator adjusting screw. Move your hand to the acetylene needle valve on the **torch.** Increase the acetylene flow rate with the torch needle valve until the white feather again appears, indicating an excess of fuel gas. The white feather should be six to eight inches in length.

Again place your hand on the oxygen regulator adjusting screw and slowly increase the oxygen pressure at the cylinder. The white feather will again disappear as more oxygen is added to the fuel mixture.

Moving your hand to the acetylene needle valve on the torch, repeat the increase of acetylene as before. At this time, the acetylene torch needle valve can probably be turned full on.

Move the hand to the oxygen regulator adjusting screw and slowly increase the oxygen pressure until the white feather disappears. At this point, both needle valves are fully open and you should have a neutral flame at the torch tip. The inner cone of the flame should have a medium blue color. This is where most of the welding heat is produced. The outer flame, which is called the outer envelope, is a combination of almost transparent white with some blue, yellow and orange flame. The outer envelope serves the purpose of excluding atmospheric air from the weld area. In effect, it is a shielding gas for the weld metal.

The torch flame is now "balanced" for the type and size of welding tip being used. By "balanced", we mean that you have set **equal pressure** on both the oxygen and acetylene regulators, despite the fact that the gauges probably have different output readings. Gauge readings are probably no more accurate than plus or minus about 5%.

You will note that both the acetylene and oxygen gauge pressures read very low on the working pressure gauges. **Do not increase either oxygen or acetylene working pressure further at the cylinders**. The gas pressures you have set at each regulator are the maximum usable for the welding torch tip size you have selected. The equal pressures insure that there can be no "creep" in pressure of either oxygen or acetylene at the torch tip. The flame at the torch tip is a NEUTRAL flame.

When shutting down the equal pressure torch and flame, it is recommended that the **oxygen** needle valve be closed first. When the oxygen valve is completely closed, shut down (close) the acetylene needle valve on the torch. At this point, there should be no flame and no gas issuing from the torch.

Safety is the issue here. It DOES make a difference which needle valve is closed first. The method described is considered the safest because acetylene without pure oxygen has a much lower flame temperature plus it must then depend on atmospheric oxygen for combustion. The positive pressure from the acetylene cylinder, through the gas hose, will prevent atmospheric oxygen from coming into the torch and will virtually eliminate what is known as a "flashback" in the torch and hose assembly.

If, on the other hand, the acetylene needle valve is shut off first, there is the possibility of a flashback in the torch and hose assembly since oxygen supports combustion in even very small amounts of flammable material. The higher pressure of the oxygen could literally suck the torch flame back into the torch and even into the acetylene hose. This could cause an explosion at the acetylene cylinder. This, of course, is contrary to OSHA and most other safety regulations.

The rule is: SHUT OFF THE OXYGEN NEEDLE VALVE FIRST!

The reason for setting the pressure regulators, adjusting screws and torch needle valves as described is very simple.

The two welding gases, oxygen and acetylene, are at exactly equal pressures. Neither gas has an excess pressure that would permit forcing itself into the other gas hose, possibly causing pre-combustion. It is virtually impossible to create a flame—that is, a flashback—beyond the torch mixer into the torch body and hoses.

In all cases the torch flame may now be set at the tip solely by adjusting the needle valves on the welding torch. Remember: the reason for balancing the welding torch flame as described is the inability of either gas (oxygen or acetylene) to cause "creep" at the welding flame. "Creep" may be defined as the slow increase of one gas pressure at the welding flame. Excess acetylene would be apparent as the white "feather" extending through the outer envelope of protective flame. Excess oxygen would be apparent as a darker blue inner cone having a rather pointed flame end and a harsh, turbulent affect on the weld molten puddle.

It must be pointed out that changing the torch tip size requires a change in the oxygen and acetylene pressure regulator pressures. For each welding torch tip size the regulator adjusting screws and gas working pressures should be re-set as previously outlined.

Welding is essentially a common sense craft and skill. If the equal pressure torch is balanced as directed, and all other safety precautions have been observed, the gas welder should have no problem with gases or welding equipment.

Injector (Low Pressure) Torches.

Injector, or low pressure, torches operate on a different principle than equal pressure torches. They are so constructed that oxygen, flowing through a center orifice in the torch body, pulls in the acetylene (or other fuel gas) by suction. The action is called the "venturi" principle. The movement of relatively high pressure oxygen causes the acetylene to flow through a metered orifice in the welding torch body in proper amounts to create a good welding flame condition. (A metered orifice is a hole of specific diameter which will allow only so much gas to flow through).

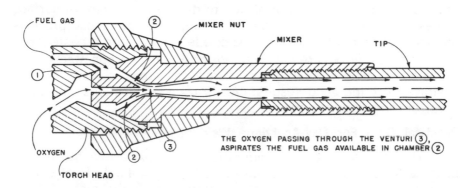

THE OXYGEN PASSING THROUGH THE VENTURI ③, ASPIRATES THE FUEL GAS AVAILABLE IN CHAMBER ②

Figure 8. TYPICAL LOW PRESSURE TORCH.

To setup prior to initiating the injector type torch flame, the welding torch needle valves are left closed. All pre-flame regulation is done at the pressure regulator adjusting screws.

The adjusting screws of both the oxygen and acetylene regulators are screwed out to the no-pressure position. The cylinder valves are opened as previously described with the oxygen valve full open and the acetylene valve only one-quarter turn open.

The acetylene regulator adjusting screw is then turned in until about two (2) psi shows on the gauge. The oxygen regulator adjusting screw is turned in until approximately 15-35 psi shows on the gauge. The exact amount depends on the torch tip size used.

Moving your hand now to the torch needle valves, turn on the acetylene valve and ignite the gas with a spark lighter as before. BE CAREFUL WHERE THE TORCH TIP IS POINTED SO THAT NO ONE IS

INJURED. Next, turn on the oxygen needle valve and adjust for the proper type of welding flame desired. The movement of the oxygen through the torch will pull the required amount of acetylene gas into the mixer and torch tip.

In shutting down the injector type welding torch, shut off the oxygen first for the safety reasons previously explained. Then shut off the acetylene needle valve. Be sure both welding torch needle valves are completely closed.

Th injector type welding torch is used a great deal where natural gas is the fuel gas. For oxy-acetylene welding, the equal pressure torch is usually preferred by experienced welders.

Fluxes For Gas Welding.

Gas welding of non-ferrous metals (containing no iron) normally requires some type of chemical flux to dissolve and remove surface oxides. Exceptions to this rule are lead, zinc and most precious metals. Fluxes may be furnished as powders, liquids or in paste form.

Chemical fluxes are not required for welding carbon steels. Gas welding of stainless steel and cast iron does require some type of chemical flux for good results. All brazing applications, including silver brazing, requires flux.

Practical Data For Oxy-Acetylene Welding.

The most important consideration in the oxy-acetylene welding process is heat control in the base metal being welded. Heat energy input will cause expansion in the heated area of the metal. When the base metal cools after welding, it will contract, often with considerable distortion of the welded part.

The distortion of base metal welded structures that often accompanies oxy-acetylene welding is not necessarily caused by the heat input to the weld although this is a factor. Basically, the distortion is caused by the contraction that takes place in the metal as it cools from welding temperature. **The problem is that metals will contract approximately 10% more on cooling than they expanded on heating.** This is a "rule of thumb" since the expansion and contraction coefficients will differ for various specific metals. Locked-in stresses, held and constrained by the rigid, non-expanding cold base metal not heated during welding, create both tension and compression forces that can cause distortion in the welded end product.

The whole idea of oxy-acetylene welding is to put as little heat as possible into the base metal and still accomplish the weld job. There are some metals which have relatively poor heat transfer characteristics. Such metals are said to have poor "thermal conductivity". The term "poor thermal conductivity" means that such metals will not permit heat to spread through their mass very much at all and then only at a very slow rate. Other metals have good to excellent heat transfer abilities (good termal conductivity). In these metals, heat will disperse very rapidly throughout the mass of the base metal. The temperature of metals having good heat transfer qualities tends to equalize throughout the metal structure very rapidly.

Aluminum, magnesium, copper and their alloys are examples of metals which have excellent heat, or thermal, conductivity. Heat input from welding will flow to all parts of the metal mass very quickly.

Some of the commonly welded metals that have relatively poor heat conductivity are carbon steel, nickel and nickel alloys, and the stainless steel family of alloys. With these metals, the heat remains localized in the immediate area where it is introduced. We could expect more distortion in the metals with poor heat conductivity than would be the case with metals having good heat conductivity.

It makes sense that it will take less total heat energy input to gas weld carbon steel or stainless steel than it takes to weld copper or aluminum. The reason is that copper and aluminum will conduct the heat away from the weld joint area faster than steel or stainless steel. More heat energy input is therefore necessary for copper and aluminum to overcome the heat losses; the heat which has spread throughout the base metal structure.

The metals listed in the table below show the relative heat transfer rates through some typical materials. They are not listed in any particular order of ability to transfer heat.

Fast Heat Transfer	Slow Heat Transfer
Aluminum	Carbon Steels
Aluminum Alloys	Low Alloy Steels
Copper	Nickel
Copper Alloys	Nickel Alloys
Magnesium	Stainless Steels
Magnesium Alloys	Titanium

Figure 9. RELATIVE HEAT TRANSFER RATES OF SOME METALS.

Oxy-acetylene welding is not suitable for all metals. For example, gas welding should not be attempted on refractory metals such as tungsten, columbium, molybdenum and tantalum. Gas welding techniques are not applicable to reactive metals such as zirconium and titanium. These metals are normally welded with one of the arc welding techniques such as gas tungsten arc welding, electron beam, etc.

Torch Handling Techniques.

Probably the most important welding skill necessary for oxy-acetylene welding is the welding operator's ability to manipulate the torch for heat control in the weld. A steady hand, good eyesight and good judgment concerning what is seen in the weld puddle area are all necessary factors for performing top grade oxy-acetylene welding.

Through proper torch manipulation heat energy input to the weld joint can be closely controlled. Slight movements of the torch tip, through the welding operator's wrist and hand movements, can cause the molten weld metal in the puddle to "wet out" and flow smoothly into the weld joint. Correct torch handling techniques are most important to successful oxy-acetylene welding. Good operator torch handling techniques require practice, Practice, PRACTICE!

Some Brazing Concepts.

The oxy-acetylene welding process is also used for joining hard-to-weld or dissimilar metals by a brazing process. In general the filler metal used may be a copper or brass alloy, a silver alloy or, in the case of aluminum, an aluminum alloy is used.

To define: **Brazing** is a metal joining process where the metal parts are cleaned thoroughly at the joint edges, brought to a suitable temperature and joined with a filler metal which has a melting point in excess of 800° F. but lower than the melting point of either of the metal pieces being joined. Chemical fluxes are normally used for brazing operations to remove the surface oxides from the metal being joined and to promote "wetting out" of the filler metal.

A unique factor in brazing is that the **base metal does not melt.** The part is heated to the appropriate temperature and filler metal is then melted by the torch flame. The filler metal flows over the solid surface of the base metal, aided in its wetting action by the chemical flux used. For brazing, using brass filler rod, a borax-based flux is often the choice of the welding operator.

In most brazing operations, the filler metal diffuses at the base metal surface and flows into the grain boundaries of the base metal by capillary attraction. The brazing action may be compared to grass roots spreading into surface soil in a lawn.

— 18 —

Brazing Filler Rods.

Copper alloys of brass and bronze are normally used for brazing cast irons, iron and steel alloys. Silver brazing alloys and techniques differ from regular brazing since this technique is usually used for lap or sleeve joints of very close tolerances. The nominal radial clearance in a typical sleeve joint to be silver brazed is 0.002"-0.003". It is interesting to note that silver filler alloys flow **to** heat rather than **away** from heat as do most brazing alloys.

For brazing aluminum and aluminum alloys, some form of aluminum brazing filler alloy is used. Special fluoride fluxes are normally used when gas welding aluminum and aluminum alloys.

Summary.

Oxy-acetylene welding is considered a repair and maintenance process in most welding shops today. It is not normally employed in production welding applications. There is some application of oxy-acetylene welding for small diameter pressure piping where local operating codes require the process.

It is well for the professional welder to have the knowledge of gas welding as part of his abilities. Much can be learned from the oxy-acetylene welding process in methods of controlling heat input to metals.

Chapter 2

COMMUNICATION WITH COMMON TERMS

Introduction.

Whenever we are called upon to explain something to someone else we are faced with the choice of words. Selecting the correct words and phrases to tell our story can be a real problem. Very often we use terms that are perfectly clear in their meaning **to us** but which our listeners do not seem to understand. We have, in such cases, a communication problem. The listeners do not comprehend the meaning of our words, thoughts or ideas. As a matter of fact, we might as well be talking to a brick wall for all the good it is doing.

We are all familiar with the term "Touchdown!" No one questions the fact that some football team must have scored six points. To fishermen, the term "bite" can mean any one of several things depending on how it is used. It can mean a fish on the line, the result of a bothersome mosquito or simply a mouthful of sandwich.

In our everyday work life terms are used to discuss various jobs, tools, materials, etc. The unfortunate thing is that many people either hear or use terms with which they are not familiar. Rather than appear unknowing and uninformed, they nod their heads and say nothing. Their intention is to find out what it means later but "later" usually never comes. Of course, many times common sense will help you figure out the approximate meaning of the word or phrase.

A glossary of terms is usually found in the back of a book such as this. Very often it is printed in such small type that it is difficult to read and so nobody uses it much.

In discussing common terms for the welding industry, however, it is necessary for the reader-student to have some understanding **before** he studies the course. Based on experience, most instructors would agree that almost any student will pass up understanding a word rather than look up its meaning in a glossary or dictionary.

With this thought well in mind, and because of the prime necessity of knowing what you are talking about when discussing welding, this chapter of the text is devoted to practical definitions of common terms used in metal working shops. The work is alphabetized for easy reference by the reader when he encounters an unknown word or phrase in the body of the text.

The terms defined are based on their general use in electricity, metallurgy, testing methods and welding in general. Study them well for they are the short-hand of industry.

1. Acetylene

Acetylene is a colorless fuel gas which is created when appropriate amounts of water and calcium carbide are mixed together. Acetylene gas has a rather sharp and pungent odor. It is classed as a hydro-carbon fuel. When mixed with oxygen in the proper proportions and ignited, the mixed gases are capable of producing the hottest welding flame of any of the commercial oxygen-fuel gas mixtures. Acetylene should never be used at working pressures exceeding 15 psi.

2. Alternating Current (AC;ac).

Alternating current is an electrical current that has both positive and negative half-cycles alternately. Current flows in the same direction during the total half-cycle, stops at the instant of time it passes through the "zero" line, then reverses direction of flow for the next half-cycle. The term "alternating current" is derived from the alternation of direction of current flow.

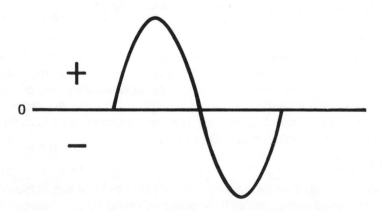

Figure 10. ONE COMPLETE AC SINE WAVE FORM.

3. Ampere (I).

The ampere is another name for electric current. It is the unit of electrical rate measurement. This means that a certain amount of current is flowing past a given point in an electrical conductor every second. Amperage, or current, flows in an electrical conductor.

4. Ampere-Turn.

An ampere-turn is a current of one ampere flowing in a coil of one turn. Its primary use is in coil design for electrical apparatus.

5. Angstrom.

An angstrom is a unit of length based on the metric system. It has a mathematical value of 1×10^{-8} cm.

6. Anode (+).

The positive pole in a welding arc is called the anode. In dc welding the work is the anode when using straight polarity. For dc reverse polarity the electrode is the anode. The positive anode is symbolized by the (+) sign.

7. Arc (Welding).

A welding arc can be created when there is sufficient voltage and amperage at the electrode tip to overcome the natural resistance to the flow of current. The resistance may be caused by the resistance of the air space between the electrode tip and the base metal. Some resistance to current flow is inherent in the electrode and the base metal. To have enough force, or electrical pressure, to overcome the resistance to arc initiation requires adequate open circuit voltage at the output terminals of the welding power source.

8. Arc Blow.

The deflection or distortion of a welding arc from its normal path because of magnetic forces in the arc is termed **magnetic arc blow.** Another type of arc blow is called **atmospheric arc blow.** It is caused by vagrant winds or breezes blowing in the arc area. Magnetic arc blow is most prevalent with dc welding. Atmospheric arc blow occurs with both ac and dc welding.

9. Arc Time.

Arc time is considered to be the actual time the arc is maintained when making an arc weld. It is often referred to as "operator duty cycle."

10. Arc Voltage.

Arc voltage is the actual voltage force measured across the welding arc between the electrode tip and the surface of the weld puddle. Arc length and arc voltage are normally correlated. The greater the arc length, the higher the arc voltage. In gas shielded welding processes, arc voltage may vary for a specific arc length due to the difference in densities of the different shielding gases. Arc voltage should not be confused with load voltage which is measured at the output terminals of the welding power source.

11. Atomic Hydrogen Process.

Atomic hydrogen welding is an arc welding process in which the arc is maintained between two tungsten electrodes in an atmosphere of hydrogen. The circuit is completed between the electrodes and no electrical grounding of the workpiece is necessary. The hydrogen atoms separate into sub-atomic particles in the heat of the welding arc. In so doing they absorb heat energy. When the sub-atomic particles strike the relatively cold workpiece they re-combine and release the stored heat energy which is transmitted to the workpiece.

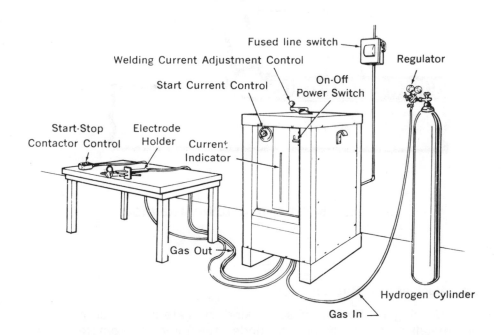

Figure 11. ATOMIC HYDROGEN WELDING EQUIPMENT.

12. Backfire.

A backfire occurs when the tip of an oxy-acetylene torch over-heats, usually from being held too close to the molten weld puddle. The backfire is apparent as a momentary disappearance of the welding flame within the torch tip followed immediately by either the re-appearance of the flame or the total extinction of the flame. The backfire gets its name from the "popping" sound made when it occurs. In almost all cases, a backfire is the result of poor welding techniques.

13. Backhand Welding Technique.

The welding technique where the electrode, or torch, is directed away from the line of travel and toward the welding puddle is termed the backhand welding technique.

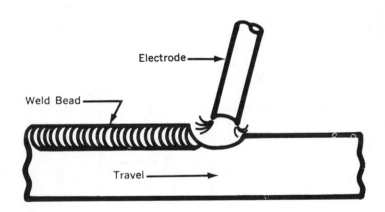

Figure 12. BACKHAND WELDING TECHNIQUE.

14. Backing, or Backup.

Backing is the use of some kind of material (either metallic, non-metallic or gas) behind or under the joint to promote better quality in the root of the weld. Backup bars may be used to regulate weld deposit cross sectional shapes and configurations.

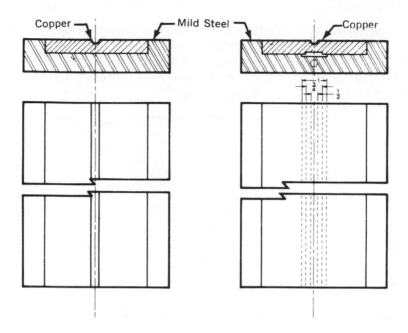

Copper ——   ——— Mild Steel ——   ——Copper

Figure 13. SOME TYPICAL BACKUP BAR SHAPES.

15. Bare Electrode.

A bare electrode is a solid wire which may be either a cut "stick" electrode or a continuous wire as used for Gas Metal Arc Welding. It has no flux coating other than that occurring incidental to the wire drawing operation.

16. Base Metal.

a) In welding applications, "base metal" refers to the metal to be welded. It is sometimes referred to as the parent metal.

b) In metallurgy, the "base metal" in an alloy is that metal having the highest percentage of content in the alloy. For example, brass is a copper-based alloy with copper the dominant, or base, metal.

17. Bead.

A bead is one pass of a deposited weld. "Top" bead refers to the portion of the weld above the surface level of the base metal. "Under" bead refers to the portion of the weld extending below the bottom surface of the base metal joint.

18. Buttering.

Depositing weld metal on the surface of a joint to improve weldability of the base metal is called "buttering". This welding technique is usually employed for cast iron to reduce dilution of the weld metal deposit with the base metal. The term buttering is considered obsolete by the American Welding Society.

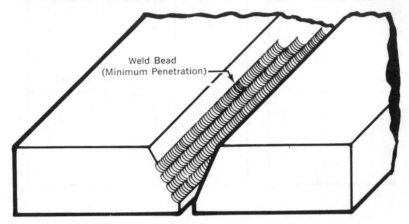

Weld Bead
(Minimum Penetration)

Figure 14. PARTIAL BUTTERING OF A JOINT.

19. Capacitor.

A capacitor is an electrical device which, when connected to an ac circuit, causes current to lead voltage by 90 electrical degrees in time. It is a device which can store some electrical energy momentarily while power is on the circuit. The peak of the **current** wave trace will reach maximum amplitude, or strength, 90 electrical degrees before the voltage wave trace. This is the result of the storage of, and discharge of, electrical energy.

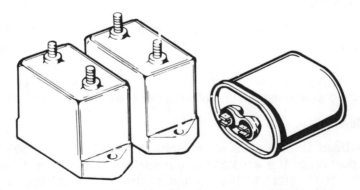

Figure 15. SOME CAPACITOR CONFIGURATIONS.

20. Capacitance.

The properties in a system of dielectrics and conductors that results in the storage of an electrical charge is called capacitance. It is the result of the capacitors ability to store electrical energy. **Capacitance is always an electrostatic effect and is voltage induced.** The unit of measure of capacitance is the farad. The farad is too great a value for most applications. The micro-farad (one millionth of a farad) is more commonly used in industry.

21. Capillary Attraction.

The combination of adhesion and cohesion forces which causes molten metal to flow between closely spaced **solid metal surfaces,** even against the force of gravity, is called capillary attraction. The term is normally used in brazing, especially silver brazing.

22. Cathode (−).

The negative pole in an arc is termed the cathode. In dc welding the electrode is the cathode when using straight polarity. For dc reverse polarity the workpiece base metal is the cathode.

23. CFH (Cubic Feet per Hour).

The various shielding gases used for Gas Tungsten Arc Welding and Gas Metal Arc Welding processes are normally measured in flow rates of cubic feet per hour (CFH).

24. Charpy Test.

The Charpy Test is a pendulum-type single impact test in which the specimen is usually notched to control the point of fracture. The specimen is supported at **both ends** as a simple beam and broken by the falling pendulum. The energy absorbed, as indicated by the subsequent rise of the pendulum, is a measure of the impact strength (or notch toughness) of the material.

25. Chipping.

a) Chipping is the mechanical removal of metal from a weld joint. It is also used in the preparation of some types of joint edges, particularly on relatively thin plates. Chipping is normally done with an air operated impact type gun and cold chisels.

b) In welding, chipping is the term used to describe the removal of slag residues from welds made with flux-covered electrodes. Chipping may be done manually or with semi-automatic equipment.

26. Circuit.

Any system of electrical conductors that is designed to carry current is termed an electrical circuit. A circuit functions when the proper voltage and amperage is impressed on the electrical system.

27. Coil

A coil is usually made from insulated copper or aluminum wire. Copper is used most because it has better current carrying capabilities with less electrical resistance. A coil is wound on a "coil form" which determines the inside dimensions of the coil. The design of an electrical coil determines how many electrical "turns" the coil will have in order to perform its proper function. An electrical "turn" is one complete wrap of the coil wire around the outer edge, or periphery, of the coil.

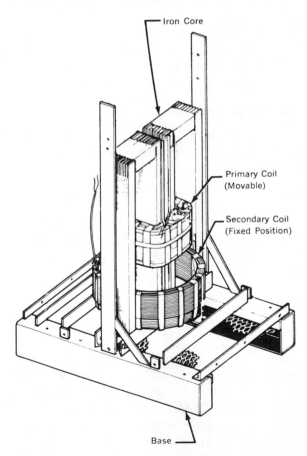

Iron Core

Primary Coil
(Movable)

Secondary Coil
(Fixed Position)

Base

Figure 16. PRIMARY AND SECONDARY COILS.

28. Conductor

Any electrical path is considered to be an electrical conductor. Good conductors are normally thought of as materials that offer the least electrical resistance to current flow. For example, most metals are considered good electrical conductors. A few, such as copper and aluminum, are used specifically for that purpose.

29. Constant Current Power Source.

Constant current welding power sources are those units which have a substantial negative volt-ampere output characteristic. Sometimes referred to as "droopers" because of their volt-ampere output curves, **constant current welding power sources have limited maximum short circuit current.** According to NEMA standards (NEMA EW-1), the maximum short circuit current for this type of welding power source should be not more than approximately 150% of rated amperage. Constant current welding power sources are normally used for Shielded Metal Arc, and Gas Tungsten Arc, welding processes.

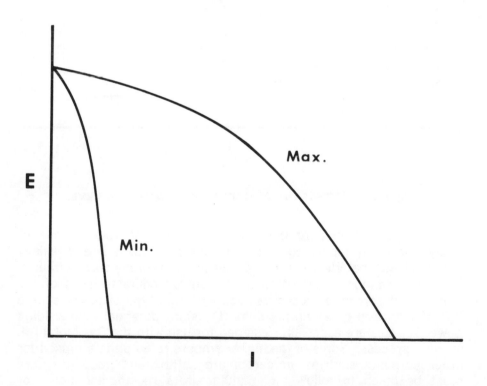

Figure 17. TYPICAL VOLT-AMPERE CURVE, CONSTANT CURRENT.

30. Constant Potential Power Source.

In welding terminology, the terms "potential" and "voltage" are synonymous in their meaning. Constant potential means a stable, consistent voltage regardless of the amperage output of the welding power source. The actual design of most constant potential type welding power sources provides a relatively flat, slightly negative volt-ampere output characteristic.

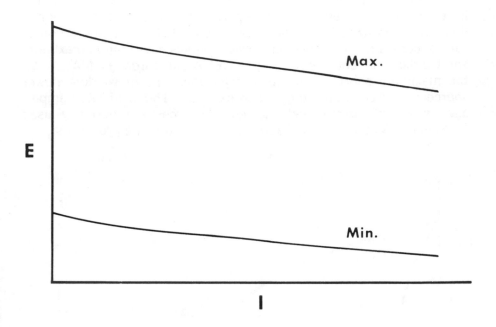

Figure 18. CONSTANT POTENTIAL VOLT-AMPERE CURVE.

The term "constant potential" or "constant voltage" is not really true when applied to welding power sources. There is usually sufficient internal electrical resistance in the power source circuitry to cause a minor voltage drop in the output characteristics of the unit. For the welding processes used with this type of power source the voltage drop poses no problem. Constant potential type welding power sources are especially designed for use with the Gas Metal Arc welding process. Some of the higher ampere rated units are used for other purposes such as air carbon-arc cutting and gouging. Care must be taken to follow the manufacturers recommendations for carbon electrode size when using this type of welding power source.

31. Contactor.

A contactor is simply a type of switch. A contactor may be located in either the primary or secondary circuit of a welding power source.

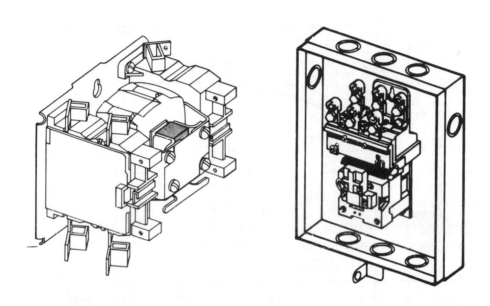

Figure 19. PRIMARY AND SECONDARY CONTACTORS.

A **primary contactor** is located in the relatively high voltage, low amperage primary circuit of the power source. When the contactor points are in the "open circuit" position, there is no primary power to the main transformer. When the contactor coil is energized it closes the contactor points and the electrical circuit is energized. Power now goes to the main transformer, the welding power source is energized and open circuit voltage is apparent at the output terminals of the power source.

A **secondary contactor** may be located internally in the power source (often done with engine driven equipment) or it may be in the welding circuit external to the power source. It is a high amperage, low voltage electrical device.

32. Cooling Curve.

The cooling curve shows the time and temperature relationship during the cooling of a metal. It may also show the transformation characteristics of the metal if the information is required. This type of curve is very useful in calculating the freeze, or solidification, rate of the various metals when they are welded.

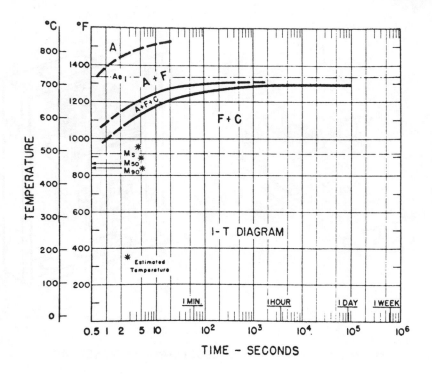

Figure 20. TYPICAL COOLING CURVE FOR CARBON STEEL.

33. Core, Iron.

In transformer type welding power sources, the iron core is the magnetic link between the primary coil and the secondary coil. Iron core material is made of laminated, insulated sheet steel stock of relatively thin gauge. Approximately 0.018"-0.020" thickness is typical for core material. The steel is **insulated on both sides** so each lamination is an electrically isolated entity. The steel used in most present day manufacture of transformers is rolled at the steel mill in such a manner as to have preferred grain orientation, shape and size. Special electrical steel is used because it has better magnetic permeability with less heating due to hysteresis losses or eddy currents. Some steel core materials have relatively high silicon content. The silicon performs the same function of decreasing heat losses as the specially rolled electrical steel but not as effectively.

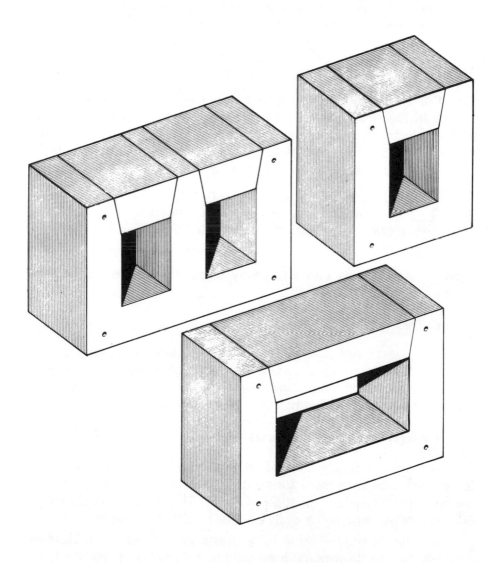

Figure 21. SOME IRON CORE CONFIGURATIONS.

34. Counter Electro-Motive Force.

Electro-motive force is another name for voltage. Counter electro-motive force (counter voltage) opposes the impressed ac voltage in an electrical circuit. For example, it is apparent in an ac reactor circuit as the induced voltage created by the ac sine wave form cutting the magnetic field flux lines. **The induced voltage opposes the alternating current flow in the circuit and tends to limit the amount of current flow at the output terminals of the welding power source.** It is the electrical principle on which Lenz's Law is based.

35. Current Density.

Current density is the amount of current per unit area of electrode. In the United States, it is the amount of current **per square inch** of electrode cross-sectional area. To calculate current density for any electrode diameter, divide the current value by the electrode cross sectional area in square inches (inches²). For example:

$$0.030'' \text{ dia. electrode} = 0.00071 \text{ inches}^2.$$
$$\text{welding current} = 100 \text{ amperes.}$$
$$0.00071 \overline{) 100} = 141,000 \text{ amperes per inch}^2$$
$$\text{approx. current density.}$$

A data chart showing current density calculation figures is included in the appendix of this book.

36. Current Flow.

The question of which direction current flows is academic until the subject of electronic circuits and tubes is considered. It has been proven that electrons **must** flow from **negative to positive** in electron tubes. This concept is the basis for the electron theory.

37. Cycle(s) per Second (Hertz).

A cycle is a double alternation of ac power. As shown in Figure 22 the ac sine wave trace forms 360 electrical degrees. The horizontal "zero" line is a function of time. The example is based on 60 cycle power where one cycle equals 1/60th of a second.

As illustrated, the wave form starts at "0" electrical degrees and ascends to its maximum amplitude (strength) at 90 electrical degrees. **Note that electrical degrees are definitely a function of time.**

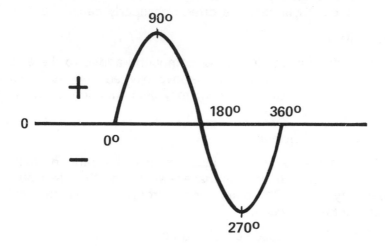

Figure 22. ONE ALTERNATING CURRENT CYCLE.

At the 90 degree mark one quarter cycle has been completed. The curve then descends from maximum amplitude to the zero line. The 180 degree point shown is the completion point for the first half cycle. At the precise instant of time the sine wave curve crosses the zero line there is no power. Current flow has been in **one direction only** during the entire first half cycle.

At the beginning of the third quarter cycle there is a reversal of current flow direction. The sine wave trace again starts at "0" but this time it is below the zero line. The wave trace builds up to its maximum strength, or amplitude, at 270 degrees as illustrated. The fourth quarter cycle shows the curve decreasing to the zero line at point 360 degrees. At this instant of time one full cycle (hertz) has been completed. The total cycling process is repeated 60 times per second as long as current is flowing in the ac circuit.

38. Cylinder Gas.

All gases used in the various welding processes and operations must be contained within some type of holding or storage device. Gas cylinders are made in a variety of sizes and shapes for this purpose. All gas cylinders are required by law to have the name of the gas it contains plainly marked on the outside of the cylinder.

Most gas cylinders are capable of storing gases at very high pressures. For example, a standard oxygen cylinder contains oxygen at approximately 2,200 psi when full. Controlled release of the gas is achieved through the cylinder valve which regulates the amount of gas released. Cylinders are often improperly called "bottles".

39 Deoxidizer.

A deoxidizer is an element or compound added to the electrode flux, or core wire chemistry, to remove oxygen and its derivatives from the weld. The same element may also function as an alloy in the deposited weld metal.

40. Deposition Efficiency.

The term deposition efficiency means the ratio of metal deposited, by weight, to the purchased weight of the electrode. For example, a typical 5/32" diameter electrode will have the following approximate loss by weight:

Rod stub loss, based on 2" stub ends $=$ 17%
Flux and spatter losses, by weight $=$ 27%
Total loss, by weight $\overline{44\%}$ approx.

41. Dielectric.

A dielectric material will not conduct direct current (dc). It is used in the manufacture of capacitors to insulate the electrical conductors from one another.

42. Direct Current (DC;dc).

Direct current is an electrical current that flows in **one direction only** and which has either a positive or negative polarity. There is no change of current flow direction as there is with alternating current (ac).

43. Ductility.

Ductility is the metallurgical property of a metal that permits it to deform under stress without rupturing. Metals with good ductility may be stretched, formed or drawn without tearing or cracking. Gold, silver, copper and some iron alloys are metals that exhibit good ductility. A ductile metal is not necessarily a soft metal.

44. Duty Cycle.

Duty cycle is based on a ten minute period of time for most welding power sources. Every legitimately manufactured welding power source is rated at a certain amperage and load voltage for a given period of time within the ten minute span. The National Electrical Manufacturers Association (NEMA) Standard EW-1 sets the time period at ten minutes.

Duty cycle is not accumulative. A welding power source rated at 300 amperes, 32 load volts, 60% duty cycle is designed to supply the rated amperage, at the rated load voltage, for **six minutes out of every ten minutes.** The welding power source must **idle and cool** the other **four minutes** to complete the ten minute time cycle. This will prevent damage to the insulation materials on the various power source components such as the coils and transformer core materials.

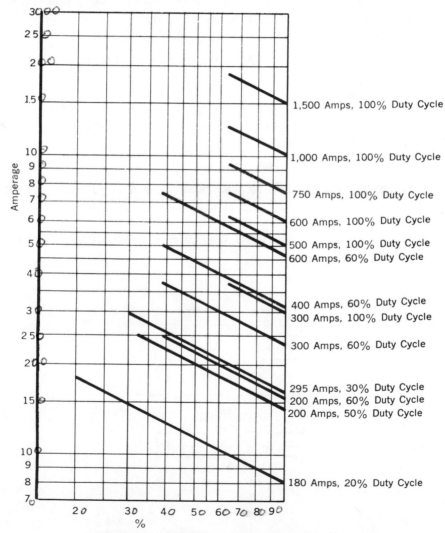

Figure 23. TYPICAL DUTY CYCLE CHART.

45. Dynamic Electricity.

The term dynamic electricity means electrical power at work. An excellent example of electrical dynamics is the welding arc. The current and voltage in the arc are constantly fluctuating and there is constant change in the power ratio of arc volts to amperes.

46. Effective Value (AC).

Alternating current is constantly changing its direction of current flow so that the total **net current flow** is zero. In practice, measured values of ac volts and ac amperes are taken as **effective values** unless otherwise specified. Often the effective value is referred to as the "Root Mean Square", or RMS, value.

RMS value = 0.707 × (Maximum, or peak, value).

47. Elasticity.

In metals, elasticity is the ability of a metal to return to its original shape and dimensions after being deformed in some manner and then having the load removed.

48. Electrical Charges (Voltage).

Electrical charges are termed **positive and negative.** A basic electrical law is, "Unlike charges attract and like charges repel". For example, two negative electrical charges would repel each other. A positive and a negative electrical charge would attract each other.

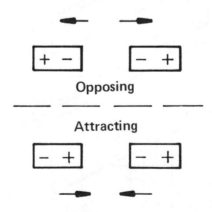

Figure 24. **ELECTRICAL CHARGES ATTRACT AND REPEL**

49. Electrical Conductivity.

Materials that promote the flow of electrical current are considered to have good electrical conductivity. Metals, such as aluminum and copper, are used as electrical conductors because they have the ability to release great quantities of free electrons when voltage is impressed on them. It is the movement of free electrons in an electrical conductor that constitutes the flow of electrical current.

50. Electrical Degree.

An electrical degree is applied as a unit of measure of time for alternating current. One cycle (one hertz) equals 360 electrical degrees.

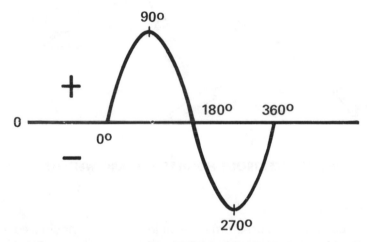

Figure 25. ELECTRICAL DEGREES IN ONE AC CYCLE.

51. Electrode.

The term electrode is employed in arc welding to describe the conducting material between the electrode holder and the welding arc. An electrode carries the welding current. The two basic types of electrodes are "consumable" and "non-consumable". Consumable electrodes are used with the Shielded Metal Arc welding process and the Gas Metal Arc welding process. Non-consumable electrodes are used with the Gas Tungsten Arc welding process. (Tungsten is considered a non-consumable electrode because it is not designed to be part of the weld deposit. Consumable electrodes are designed to be an integral part of the weld deposit).

52. Electrode Holder.

The electrical device used for making contact with an electrode when arc welding is called an electrode holder. It may be a simple tong-type clamping device, or a twist-lock unit, for holding the cut consumable electrode employed for Shielded Metal Arc welding. It can be a fairly complex torch or gun which not only holds the electrode material but also acts to provide shielding gas and possibly cooling water for the operation. Gas Tungsten Arc and Gas Metal Arc welding processes normally use such sophisticated types of electrode holders. The shielding gas is for the protection of the weld area and the electrode. The cooling water, if used, flows in the gun or torch to remove heat in the electrode holder body.

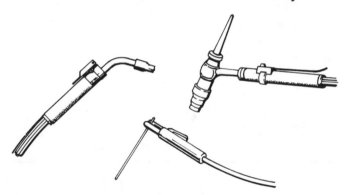

Figure 26. ELECTRODE HOLDERS FOR ARC WELDING.

53. Electron (—).

The electron is a sub-atomic particle which is considered to be the fundamental unit of negative electricity. It is the smallest particle of matter known to man that carries a negative electrical charge. Electrons are capable of moving from one place to another within atomic structures. It is the electrons that move when electrical current flows in an electrical conductor.

54. Electron Bond.

Electron bond is the term used to describe the ability of electrons to move within the atomic structure of a material. Materials that have a "loose" electron bond allow electrons to leave the atom easier, and with less potential force being applied, than do materials having a "tight" electron bond.

All of the metals have a relatively loose electron bond while wood and rubber, for instance, have tight electron bonds. As you might surmise, the metals are relatively good electrical conductors while wood and rubber are classed as poor electrical conductors.

55. Embrittlement

Embrittlement describes the reduction or loss of normal ductility in a metal due to physical or chemical changes. In welding, hydrogen embrittlement can occur if improper welding procedures are used. It is for this reason that low hydrogen welding electrodes were developed. Such electrodes have closely controlled flux chemical analysis and moisture content to prevent the formation of hydrogen gas while welding.

56. Flashback.

A pre-ignition of the oxy-acetylene gas mixture in the mixing chamber of a welding or cutting torch is called a flashback. Its occurance can be extremely dangerous since the flame could follow the fuel gas back to the cylinder. The resulting explosion could cause severe damage or injury. If a flashback occurs in the torch, the gas cylinders should be shut off immediately, fuel gas first, to prevent further combustion.

A flashback is usually caused by a clogged or overheated torch tip. If a flashback occurs, the torch should be examined for damage and proper mechanical operation before it is used again. In particular, check the tip and the seating of the tip in the mixer. The mixer and torch body connection should be examined for damage. It would be a good idea to check the torch needle valves for burned packing or other damage before proceeding with the welding job.

57. Flowmeter.

A gas flowmeter is a device for measuring the flow of shielding gases when welding. The measurement is usually made in cubic feet per hour (CFH) or liters per hour. The flowmeter may be attached to a gas cylinder valve or a manifold piping system.

58. Flux.

a)   In electrical terms, flux is another name for magnetic lines of force. Magnetic lines of force are located in a magnetic field.

b)   For gas welding, various types of chemical fluxes are employed to assist in breaking down and removing metal oxides from the welding area. Fluxes also promote wetting action in the weld deposit.

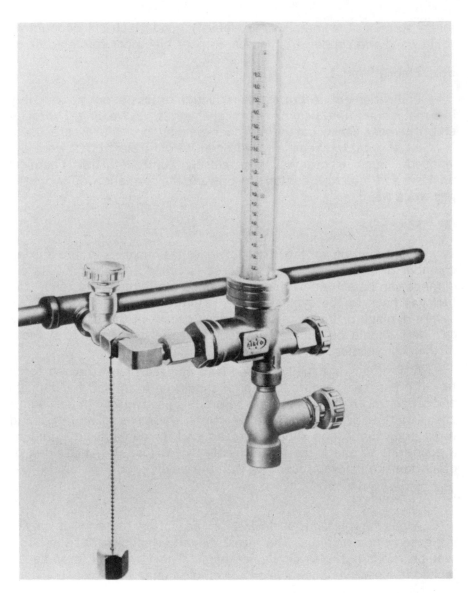

Figure 27. TYPICAL GAS FLOWMETER.

59. Forehand Welding.

Forehand welding is a technique in which the electrode or torch is directed **towards** the line of travel in the weld. Although it is used primarily with the oxy-acetylene gas welding process, the forehand technique may be used for any of the arc welding processes.

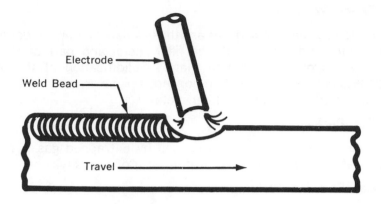

Electrode

Weld Bead

Travel

Figure 28. THE FOREHAND WELDING TECHNIQUE.

60. Frequency.

Frequency is the rate at which alternating current makes a complete cycle of reversals. It is expressed as "hertz" or "cycles per second". Most electrical power in the United States is 60 hertz frequency. In other parts of the world it is common to find 50 hertz power.

61. Fuse.

A fuse is basically an over-current protective device for electrical circuits. Fuses normally have a circuit-opening fusible link that may be directly heated and destroyed by a short time surge of excessive amperage. Fuses may be either the cartridge type or the plug type.

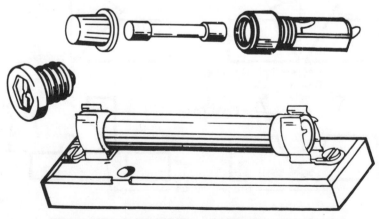

Figure 29. FUSE TYPES, CARTRIDGE AND PLUG.

62. Fusion Welding.

Fusion welding procedures are those where welding is done without pressure and where both the filler metal and part of the base metal fuse together as molten metal. Solidification of the molten fused mass produces a single homogenous piece or weldment. Fusion welding may be done without filler metal.

63. Gas Pocket.

A cavity in the weld metal caused by entrapped gas is called a gas pocket. It is a weld defect in the form of porosity.

64. Gas Shielded Welding.

Arc welding processes that employ some type of gas shielding around the electrode tip and weld metal area to protect the area from surrounding atmosphere are called "gas shielded welding" processes. Some shielding gases may be inert to the products of the weld zone while others are either carburizing or oxidizing. The type of shielding gas used depends on the metal to be welded and the welding process employed. In oxy-acetylene welding, the outer envelope flame is actually the shielding gas for the molten metal in the weld.

65. Groove Angle.

The term "groove angle" describes the total included angle of the prepared joint groove between two parts to be joined by welding. It is sometimes referred to as the included angle of the joint.

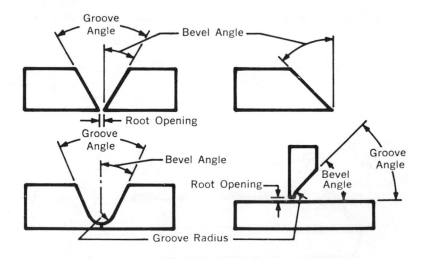

Figure 30. TYPICAL GROOVE ANGLES OF A JOINT.

66. Heat Affected Zone (HAZ).

The heat affected zone is that area of the base metal not melted during the welding operation but whose physical characteristics and properties were altered by the heat induced from the weld joint. The heat affected zone begins at the interface of the weld deposit and the base metal. It ends in the base metal where no physical changes in structure or properties has occurred.

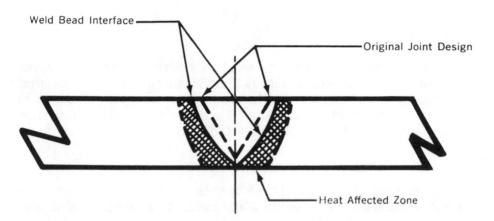

Figure 31. WELD WITH HEAT AFFECTED ZONE.

67. Hertz (Hz;hz).

The term **hertz** has replaced the term **cycle(s)** per second in present electrical terminology. The word is used in honor of Heinrich R. Hertz, a German physicist, who was a pioneer in the development of electrical concepts. The term "hertz" is abbreviated "Hz".

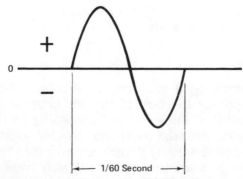

Figure 32. ONE HERTZ EQUALS 1/60 SECOND FOR 60 CYCLE PER SECOND POWER.

68. High Frequency.

High frequency covers the entire frequency spectrum above approximately 50,000 hertz per second. The numerically high frequency rate causes the current to flow on the surface of the electrical conductor. This is known as the "skin effect" of high frequency power transfer. High frequency is used primarily as a means of bringing relatively safe high voltage to the Gas Tungsten Arc electrode tip. Improved arc initiation and improved arc stabilization with ac welding are other benefits.

69. Hold Time.

Hold time is a term employed in resistance welding. It describes the time when pressure is maintained on the workpiece after the resistance welding current has ceased to flow. Its purpose is to maintain a constant pressure on the weld joint until solidification of the weld nugget has been completed.

70. Horsepower.

The measure of rate of work may be made in horsepower. One horsepower equals 746 watts, electrically. One horsepower is also equal to lifting a 33,000 pound weight to a height of one foot in one minute, mechanically.

71. Hysteresis.

Hysteresis is the resistance, or reluctance, of magnetic particles to polar orientation when subjected to a magnetic field. As the direction of current flow changes each half cycle with alternating current, so must the molecules in the iron core material of a transformer change polarity each half-cycle. Since this is work being done, energy is being expended. The energy is expended as heat in the iron core material, causing it to lose electrical efficiency.

72. Impedance.

Electrical impedance is a combination of resistance and reactance which opposes the flow of current in an alternating current circuit. The resistance value is fixed according to the type and diameter of conductor material used and may be construed as "real". The reactance value is entirely dependent on other factors in the circuit. It may have a maximum value or it may have no value at all. Reactance values are construed as "apparent" values.

73. Inclusions.

Inclusions are usually non-metallic particles that appear in a weld deposit. They may be slag residue that is trapped in the weld because of fast solidification characteristics inherent in the weld metal. Inclusions are normally considered as weld deposit contaminants or defects and should be removed.

74. Induced Current.

Current that is caused to flow in an electrical conductor by magnetic action is termed induced current. The conductor is subjected to a magnetic field of varying strength or intensity, usually caused by the ac wave form characteristics, and current is induced to flow. Current flows in the circuit only when the circuit is complete. Induced current is part of the total KW induced in the secondary coil of a welding transformer.

75. Inductance.

Inductance is the electrical phenomenon that causes voltage and amperage to be apparent in the secondary circuit of a welding transformer power source. It is the electrical influence exerted by current flow in a conductor, through a magnetic field, on adjacent conductors. There is no physical contact between the two conductors. **Inductance is always a magnetic effect and is current induced.**

For example, current is brought to the primary coil of a welding transformer. The primary coil is located around one leg of the iron transformer core. A magnetic field is created when current is caused to flow in the primary coil. The energy stored in the magnetic field is induced into the secondary coil of the transformer without physical contact between the two coils.

76. Inert Gas.

An inert gas will not combine with any known element. In welding, argon (Ar) and helium (He) are the two inert gases used. They are used with the Gas Tungsten Arc, and Gas Metal Arc, welding processes. There are a total of six known inert gases on the Earth. Besides argon and helium, there are neon, xenon, krypton and radon. These last four gases are not economically feasible for welding use.

77. Inert Gas Shielded Arc Cutting.

Inert gas shielded arc cutting is a technique used for severing metals. It employs an electric arc within an inert gas atmosphere.

High current densities are used with the process. All metals including those classed as refractory may be cut with this process.

78. Inert Gas Shielded Arc Welding.

Several arc welding processes are involved when discussing inert gas shielded arc welding. The various processes are used when joining metals with an electric arc shielded by an inert gas such as argon or helium. High current densities are a feature of the processes used with this type of gas shield. Gas Tungsten Arc, and Gas Metal Arc, welding processes fall into this general category of arc welding.

79. Ion.

An ion is an atom of matter that has gained or lost one or more electrons and which, therefore, carries an electrical charge. **Positive ions,** called cations, are deficient in outer electrons. **Negative ions,** called anions, have an excess of outer electrons. In most welding applications, it is the positive ions with which we are concerned.

80. Ionization Potential.

Ionization potential is the energy necessary to remove an electron from an atom thereby making it an ion. The potential energy (voltage) necessary will depend on the material being ionized. The term ionization potential is usually related to the shielding gases used with some welding processes.

81. Izod Test.

The Izod test is a pendulum-type single impact test in which the specimen is usually notched, **held at one end,** and broken by the falling pendulum. The energy absorbed, as measured by the subsequent rise of the pendulum, is a measure of impact strength or notch toughness of the material.

82. Joint.

A "joint", as used in welding terminology, is the junction of two or more pieces of metal that are to be, or have been, welded. Joint design is critical to many types of weldments. Some of the more common groove joint designs are illustrated.

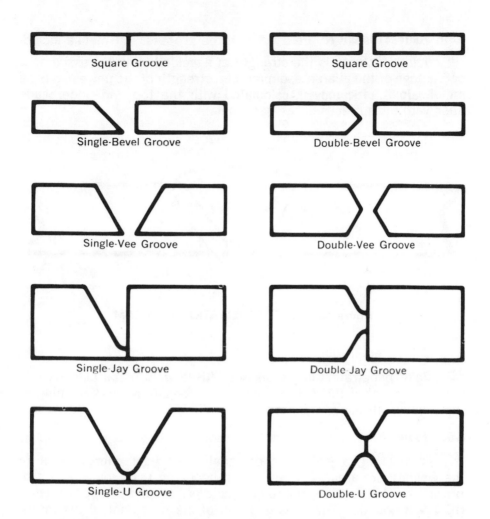

Square Groove   Square Groove

Single-Bevel Groove   Double-Bevel Groove

Single-Vee Groove   Double-Vee Groove

Single-Jay Groove   Double-Jay Groove

Single-U Groove   Double-U Groove

Figure 33. SOME COMMON GROOVE JOINT DESIGNS.

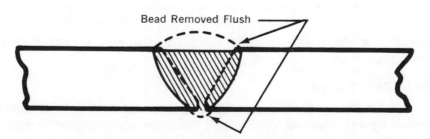

Bead Removed Flush

Figure 34. SPECIMEN PREPARED FOR JOINT EFFICIENCY TEST.

83. Joint Efficiency.

Joint efficiency is the strength of a welded joint expressed as a percentage of the guaranteed minimum strength of the unwelded base metal. Joint efficiency is calculated with the top and underbeads flush with the surfaces of the base metal.

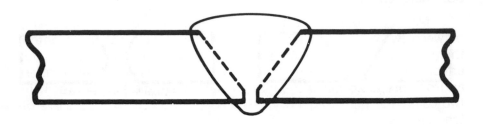

Figure 35. JOINT PENETRATION DIAGRAM.

84. Joint Penetration.

Joint penetration is the distance fused weld metal extends into the base metal of the welded joint. It is considered as beginning at the original surface of the weld joint.

85. Joule.

In welding, a **joule** is a designation of heat energy input to the welding arc area. The term is in honor of J. P. Joule, a British physicist. The watt is equal to one joule per second or approximately 0.74 foot-pounds. The joule is a unit of measurement of the metric system. The equation for determining joules of energy input is as follows:

$$H \text{ (joules per inches)} = \frac{E \text{ (volts)} \times I \text{ (amperes)} \times 60}{S \text{ (speed in inches per minute)}}$$

Specifically, joule energy input is the heat energy imparted to the weld puddle by the welding arc.

86. Load Voltage.

Load voltage is measured at the output terminals of the welding power source **while welding.** It is the total voltage load, including arc voltage and the voltage drop through the welding cables and connections, which the power source senses.

87. Locked Rotor Current.

The locked rotor current of an electric motor is the steady-state current taken from the primary power system with the rotor locked in one position. Calculation of locked rotor current is made with rated voltage, and rated frequency in the case of alternating current motors, applied to the motor.

88. Magnetic Coupling.

Magnetic coupling is the term used to describe the relationship between two coils so situated that a magnetic field set up in one of them will interact on the other one. Another term that describes this effect is inductive coupling. In power circuits for welding transformers an iron core is normally the intermediate magnetic link between the two coils. For high frequency systems air core coupling coils are often used.

Some welding power sources have "loose" magnetic coupling in the primary-secondary coil arrangement of the main transformer. These are the conventional, or constant current, type power sources which have a drooping volt-ampere output characteristic.

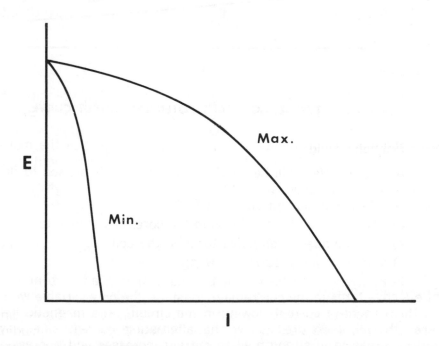

Figure 36. A TYPICAL CONSTANT CURRENT VOLT-AMPERE CURVE.

Other types of welding power sources have a "tight" magnetic coupling between the primary and secondary coils. This type of welding power source is called a constant potential, or constant voltage, unit. They have a relatively flat volt-ampere output characteristic curve.

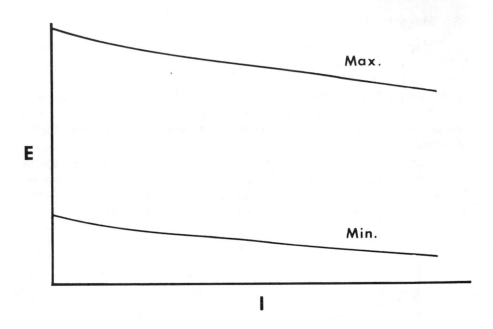

Figure 37. A TYPICAL CONSTANT POTENTIAL OUTPUT CURVE.

89. Magnetic Field.

A magnetic field will be created when current is caused to flow through a coil wrapped around an iron core. The strength of the magnetic field will depend on three factors:

    a) The mass and type of iron in the core.

    b) The number of effective turns in the coil.

    c) The amount of current flowing in the coil.

In a welding transformer, if the mass of iron and the number of effective turns in the coil remain constant, the only variable would be the amount of current flowing in the circuit. The magnetic field strength will, therefore, follow the alternating current sine wave form, increasing in strength when current increases and decreasing in strength when current decreases.

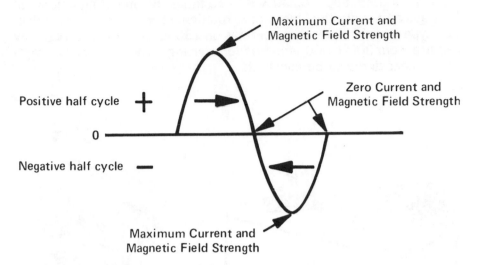

Figure 38. ALTERNATING CURRENT AND MAGNETIC FIELD STRENGTH.

90. Malleability.

Malleability is the characteristic of metals and other materials that permits plastic deformation of the material without rupturing. It indicates the pliability of a metal in forming operations. A good example of malleability in metals is the deep-drawing of household utensils such as pans.

91. MIG.

The term MIG is the abbreviation for **M**etal **I**nert **G**as. This is the term used to describe the Gas Metal Arc Welding process when it was first developed. It is applicable only when using an inert gas, such as argon or helium, for weld shielding.

92. Mild Steel.

Mild steel is another name for a type of low carbon steel. It has a maximum carbon content of approximately 0.25% carbon. Mild steel is considered to be non-heat treatable for hardness. It may be normalized or stress relieved after welding with no problem.

93. Motor, Electric.

An electric motor literally converts electrical energy to mechanical energy. Most electric motors are comprised of a rotor (usually the armature) and a stator. The primary power from the utility company causes the motor to run by turning the rotor. Depending on the type motor used, approximately 3 to 10 times the operating amperage is required for starting the motor rotation. For example, an induction type electric motor that operates on 230 volts, 100 amperes, may require from 300-1,000 amperes for starting. The primary line voltage is considered to be constant.

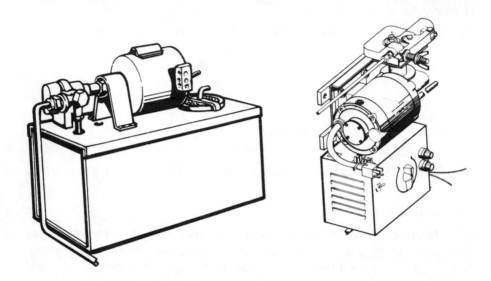

Figure 39. ELECTRIC MOTOR APPLICATIONS.

94. Negative Charge (−).

Negative charge is the term used to indicate the type of electrical charge carried by an electron. It is illustrated by the negative, or minus, sign (−).

95. NEMA.

NEMA is the abbreviation for the National Electrical Manufacturers Association. It is a self-regulating group of electrical manufacturers whose purpose is to promote common, safe standards of manufacture for electrical apparatus. It is entirely supported by member companies. The NEMA rating on a welding power source means that it has passed all the required tests and is certified to produce the rated amperage at the rated load voltage and duty cycle. This information is normally shown on the data plate or front panel of the welding power source.

96. **Nucleus.**

In physics the nucleus is the heavy central core of an atom. Most of the atomic mass, and all of the positive electrical charge, is contained in the nucleus of an atom. This fact is most important when considering the gas shielded welding processes. A diagrammatic view of a helium atom is shown in Figure 40. Note that the nucleus is not shown in proportional scale to the electron. Its mass is many times greater than the mass of the orbiting electrons.

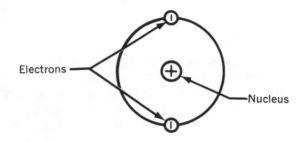

Figure 40. HELIUM ATOM—DIAGRAMMATIC VIEW.

97. **Nugget.**

A nugget is the fused metal in a resistance spot, seam or projection weld. It is sometimes mistakenly used to describe a fusion weld deposit. A nugget is considered to be a solid lump of metal rather than a continuous weld seam.

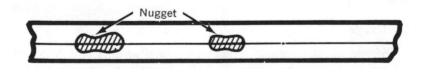

Figure 41. RESISTANCE SPOT WELD NUGGET.

98. **Ohm.**

The ohm is the practical unit of electrical resistance. There is one ohm resistance when a pressure of one volt causes a current of one ampere to flow in a circuit or conductor. Ohm's Law is given as follows:

$$E = IR \text{ or, } I = \frac{E}{R} \text{ or, } R = \frac{E}{I} \text{ where } E = \text{Volts}$$

$I = $ Amperes

$R = $ Resistance in Ohms

A simple way to remember Ohm's Law is with the timeless pyramid:

99. Open Circuit Voltage.

As the name implies, open circuit voltage means that no current is flowing in the circuit because the circuit is not complete. The voltage is impressed on the circuit so if, and when, the circuit is completed current will flow immediately. For example, a welding power source that is energized but under no welding load will have open circuit voltage measurable at the output terminals. When measuring open circuit voltage always measure across the two output terminals of the welding power source. Voltage is the potential difference in value between two conductors such as the terminals.

100. Overheating.

Overheating means raising the temperature of a metal to such a degree that its properties are impaired. When the original properties cannot be regained by heat-treating or some other method of working, it is known as burning the metal. Burning may occur in arc welding because of excess heat input to the weld deposit. The term **overheated** may refer to damaged insulating material in welding transformers.

101. Oxygen.

Oxygen is an odorless, colorless, invisible gas which comprises approximately 21% by volume of the Earth's atmosphere. Oxygen is necessary to, and apparent in, all living things including human life. It has the ability to support combustion in all burnable materials and, in concentrated quantities, it can cause theoretically non-flammable materials to burn.

102. Permeability.

The word "permeability" comes from the word **permeate** meaning to "go through", fill up or saturate something. In electric welding power sources, permeability is a measure of ease with which a material, such as electrical steel for welding power source transformer cores, can accept magnetic lines of force. Silicon steels and steels with specially controlled grain shape and orientation are used because of their excellent magnetic permeability.

103. Phase.

Phase is an electrical term which indicates the space relationship of electrical windings (coils) and the changing values of the recurring cycles of ac voltage and current. Due to the positioning (or phase relationship) of the windings the various voltages and amperages will not be similar in all respects at any given instant of time. Each winding will lead, or lag, another in **relative position.** Each voltage will lead, or lag, another in **time.** Phase angle plays an important role when considering the power factor of a welding power source. The illustration shows the relative phase angle of a purely inductive circuit, a capacitive circuit and a totally resistive circuit. The use of these drawings will be further explained in a subsequent chapter entitled "Power Factor".

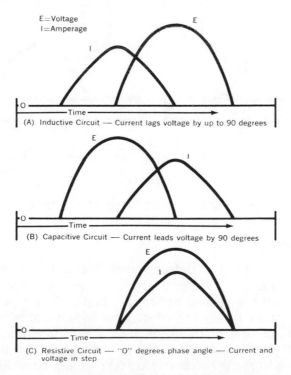

E = Voltage
I = Amperage

(A) Inductive Circuit — Current lags voltage by up to 90 degrees

(B) Capacitive Circuit — Current leads voltage by 90 degrees

(C) Resistive Circuit — "O" degrees phase angle — Current and voltage in step

Figure 42. RELATIVE PHASE ANGLES FOR CIRCUITS.

104. Polarity.

Electrical polarity primarily concerns dc welding. Since direct current (dc) flows in one direction only at any given time it is said to have polarity. The term means that the current has stabilized directional flow. Straight polarity means that the electrode is negative and the workpiece is positive. Reverse polarity means the electrode is positive and the workpiece is negative. With ac welding power there is no polarity because of the change of direction of current flow each half-cycle.

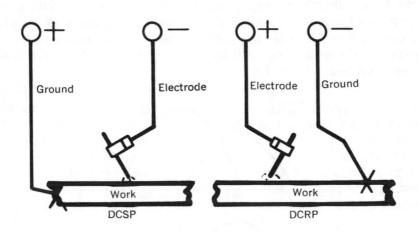

Figure 43. POLARITY CONNECTIONS FOR DIRECT CURRENT.

105. Porosity.

Porosity is a form of defect sometimes found in weld deposits. It is usually caused by trapped gases in the deposited weld metal. Porosity will always have a spherical shape. The type of porosity normally indicates its source of origin. In-line porosity usually indicates a lack of heat input to the weld area. Random porosity may indicate that welding speed of travel is too fast. In this case the gases evolving from the molten weld metal do not have sufficient time to reach the surface of the weld metal before it solidifies.

106. **Positioned Weld.**

A positioned weld is one that has been placed in the best position for downhand welding. Usually a welding positioner is used although the weldment may lend itself to positioning by manual methods. The reason for using the weld positioning technique is speed of fabrication and economy of welding process.

107. **Positive Charge.**

"Positive charge" is the term used to indicate the type of electrical charge carried by a proton. The symbol for a positive charge is the plus sign ($+$).

108. **Power Factor.**

Power factor is the measure of time phase difference between the voltage and the current in an alternating current circuit. It is expressed as a percentage of power used (primary KW) to total power drawn (KVA) from the primary power system.

The ratio of used power in KW to the total power in KVA is called power factor. Dividing the primary KW (used power) by KVA (the total power drawn) will give the power factor percentage of a welding power source. Power factor may be improved by the addition of capacitors to the primary power circuit. The addition of capacitors to an inductive electrical circuit improves power factor by **demanding less primary amperage** from the primary power lines. The addition of capacitors is called **power factor correction.**

109. **Power Source, Welding.**

A welding power source is an electrical, or electro-mechanical, device for changing high primary voltage, low amperage, into usable welding voltage and amperage. Power sources are classed as rotating equipment or static equipment. Transformer type welding power sources are called static power sources because they have no moving parts concerned with the development of welding power output. Motor-generators are rotating equipment because they have a rotor assembly turning within the stator thus generating welding power.

Welding power sources are generally designated either constant current or constant potential type power sources. Other names which are used for constant current power sources are conventional, drooper, MG set or rectifier unit. A constant current welding power source has a negative, or drooping, volt-ampere output characteristic.

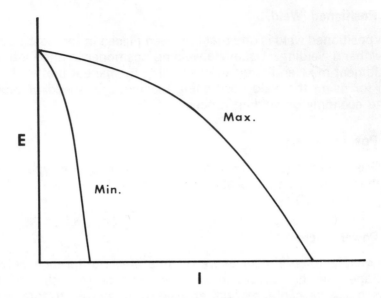

Figure 44. CONVENTIONAL VOLT-AMPERE CURVE.

Such power sources are normally used for Shielded Metal Arc welding and Gas Tungsten Arc welding processes.

Constant potential type welding power sources produce a slightly negative, relatively flat volt-ampere output characteristic curve. This type of welding power source is used primarily for the various techniques of Gas Metal Arc welding.

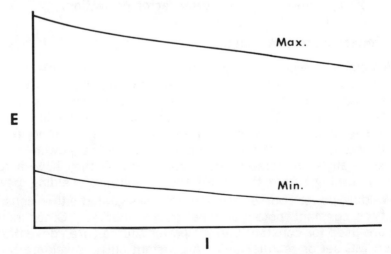

Figure 45. CONSTANT POTENTIAL VOLT-AMPERE CURVE.

110. Primary Coil.

The primary coil of a welding transformer is where the primary line voltage is impressed and where the primary current flows. The primary coil is located on one leg of the main transformer core. The current flowing in the coil causes a magnetic field to form. The primary coil of a step-down type welding transformer will have smaller diameter conductor wire than the secondary coil. The reason is that the primary coil carries relatively low amperage. The secondary coil carries higher level welding amperage.

111. Proton (+).

The proton is the fundamental unit of positive electricity. It has approximately 1,800 times the mass and weight of an electron. The electrical charge of a proton is equal, but opposite, to the electrical charge of an electron. If an atom has an equal number of protons and electrons it is called a neutral atom.

112. Reactance, Inductive.

The characteristic of a current conducting coil wrapped around an iron core which causes the current to lag voltage in time phase peak is called reactance. The current wave form reaches its peak strength, or amplitude, later in time than the voltage wave trace. The time lag is between "0" and 90 electrical degrees.

113. Reactor.

A reactor is an electrical device consisting of at least one current conducting coil wrapped around an iron core. Normally located in the secondary ac circuit of a welding power source, a reactor's function is to change the effective reactance of an ac circuit. **Reactance** may be either inductive or capacitive. A **reactor** is always an inductive device. Capacitive reactance would normally be a part of the power factor correction circuit in some power sources.

114. Rectifier.

A rectifier is an electrical device that permits the flow of electrical current in essentially one direction only. Its function is to change alternating current to direct current. The two types of static rectifiers used in welding power sources are silicon diode **rectifiers** and selenium rectifiers.

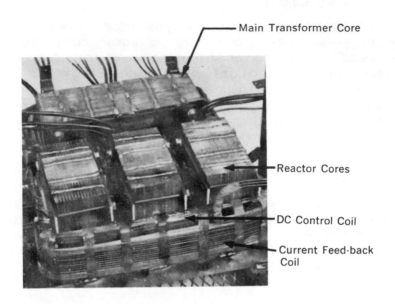

Figure 46. REACTOR ASSEMBLY FOR CURRENT CONTROL.

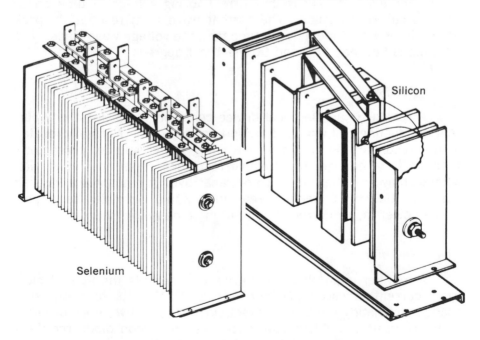

Figure 47. TWO TYPES OF RECTIFIER STACKS.

115. Refractory Material.

A refractory material is considered to be any material that has a melting point in excess of 3,600° F. For example, tungsten is considered a refractory material since it has a melting point of 6,170° F. The known refractory elements are listed in the illustration.

Material	M.P./°F.		Material	M.P./°F.	
Boron	3690	(Approx.)	Osmium	4900	(Approx.)
Carbon	6740	''	Rhenium	5755	''
Columbium	4474	''	Ruthenium	4530	''
Hafnium	4032	''	Tantalum	5425	''
Iridium	4449	''	Technetium	3870	''
Molybdenum	4730	''	Tungsten	6170	

Figure 48. KNOWN REFRACTORY ELEMENTS.

116. Regulator.

A gas regulator is a mechanical device used to control the flow of gases for welding operations. The two basic types are **single stage** regulators and **two stage** regulators. The function of any regulator is to furnish a reasonably constant flow of gas at accurate rate and pressure regardless of the actual pressure at the source of supply. The source of supply may be a cylinder of gas or a manifold system. When cylinder pressure reaches approximately 25 psi the cylinder is, for all practical purposes, empty. It should be replaced with a full cylinder of gas for best welding results.

117. Relay.

A relay is a type of switch operated by electro-mechanical force. It is usually used as a control mechanism in electric circuits.

118. Reluctance.

The term reluctance refers to that characteristic of a magnetic path, or material, which resists the flow of magnetic lines of force through its body. Metals and other materials with high reluctance values have poor magnetic permeability.

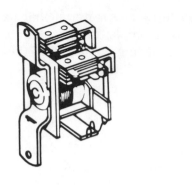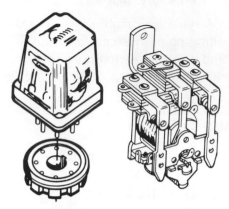

Figure 49. SEALED AND UNSEALED RELAYS.

119. Residual Magnetic Field.

A residual magnetic field is the magnetic field remaining in a ferrous metal part after the source of the magnetic field has been removed. It is sometimes referred to as latent magnetism.

120. Resistance.

The properties in an electrical conductor that oppose the passage of current are called electrical resistance. The **ohm** is the practical unit of measure of electrical resistance. The energy dissipated in overcoming electrical resistance is apparent as heat in the conductor.

121. Resistance Welding.

Resistance welding is a method of electric welding which employs resistance heating and pressure. The lap joint of the workpiece is an integral part of the welding circuit. The resistance of the metal at the interface of the joint where current flows creates a localized heating effect. This factor, combined with the applied pressure, causes a fusion bond to be made. Sometimes called "spot" welding, resistance welding may be applied to most of the commonly welded metals.

122. Reverse Polarity.

Reverse polarity is a term used in dc welding. For welding with reverse polarity, the electrode is connected to the positive terminal of the welding power source. The workpiece is connected to the negative terminal. A reverse polarity connection is illustrated in Figure 50.

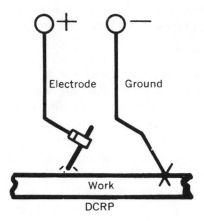

Figure 50. CONNECTIONS FOR REVERSE POLARITY.

123. Root Pass.

The root pass is the first, and most important, pass made in a weld joint. In a multi-pass weld the root pass must provide full penetration and excellent quality because the balance of the weld is essentially fill-in of the joint.

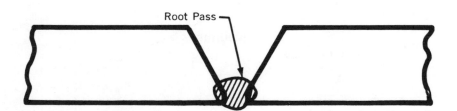

Figure 51. PROPERLY APPLIED ROOT PASS.

124. Seal Weld.

A seal weld is any weld that is made in a vessel to prevent leakage. It may, or may not, be a strength weld.

125. Secondary Coil.

The secondary coil of a welding transformer is located on the main transformer core. The energy created in the magnetic field, by the primary coil, is induced into the secondary coil. Although there is no direct connection between the two coils there is voltage apparent in the secondary coil. It can be measured as open circuit voltage at the output terminals of the power source.

126. Slope.

The term slope has been used for a number of years by the welding industry. When gas tungsten arc welding, the term "sloping off" is used to indicate a decrease in welding current at the end of the weld to permit crater filling and to eliminate crater cracking.

When discussing slope control for the gas metal arc welding process, it is **the shape of the volt-ampere curve** to which we refer. The higher the slope number, the lower the maximum short circuit current of the welding power source.

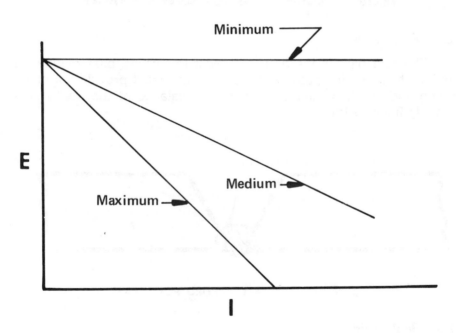

Figure 52. **TYPICAL SLOPE CURVES.**

127. Spatter.

Spatter is the metal expelled from the weld area during the actual welding operation. Spatter should be kept to a minimum since it involves cost due to lost filler metal, weldment cleaning and grinding.

128. Spot Welding.

A spot weld is normally made in materials having some type of overlapping joint design. Although the term is usually referred to in resistance welding, it is also applicable to gas tungsten arc welding and gas metal arc welding processes. Note that resistance spot welds must have electrodes on both sides of the joint. Gas tungsten arc, and gas metal arc, spot welds are made from one side of the joint only.

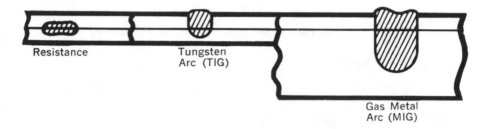

Resistance   Tungsten
              Arc (TIG)

Gas Metal
Arc (MIG)

Figure 53. TYPICAL SPOT WELDS.

129. Squeeze Time.

Squeeze time is the time that elapses between the intial application of pressure and the start of current flow when resistance welding. It permits the positioning of materials to be welded before the actual weld is made.

130. Static Electricity.

Static electricity is electricity without movement. It is the opposite of dynamic electricity.

131. Straight Polarity.

The term straight polarity is associated with dc welding. The electrode is connected to the negative terminal of the welding power source for dc straight polarity welding. The workpiece is connected to the positive terminal of the power source. A straight polarity connection is illustrated.

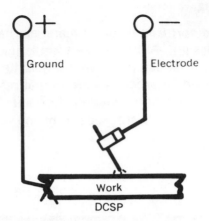

Figure 54. **CONNECTION FOR STRAIGHT POLARITY.**

132. Stringer Bead.

A stringer bead is a narrow, single pass bead made with little, if any, oscillation of the electrode. It is usually used for strength welds and for maintaining control of the fluid weld puddle. Stringer beads are normally applied to vertical welds. There is normally much less possibility of slag inclusions, porosity and other weld defects when a weld is made with stringer beads.

133. Switch.

A switch is an electrical device having points of contact that can complete an electrical circuit. When the contact points are open no current can flow. When the switch points are closed, current can flow because the circuit is complete.

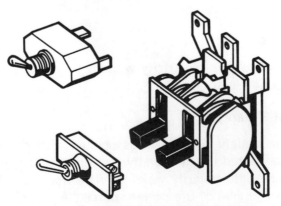

Figure 55. **VARIOUS SWITCH TYPES.**

134. Thermal Conductivity.

All of the metals have some measure of thermal conductivity. Thermal means "heat" and conductivity means to "transfer". Thermal conductivity means heat transfer and, in metallurgy, it means heat transfer through metals and metal structures.

Some of the metals that have good thermal conductivity are copper, aluminum and magnesium. It is interesting to note that these metals also have good electrical conductivity. The basic reason is that all three metals have relatively loose electron bonds in their atomic structures. This means that the metal atoms release electrons with very little force applied to the atoms.

135. TIG.

In welding, the initials TIG are the abbreviation for Tungsten Inert Gas. This is a non-technical shop term often used to describe the Gas Tungsten Arc Welding process.

136. Timer.

A timer is an electrical device that performs a pre-set function after a period of time has elapsed. It is also used to regulate and control welding time in many applications.

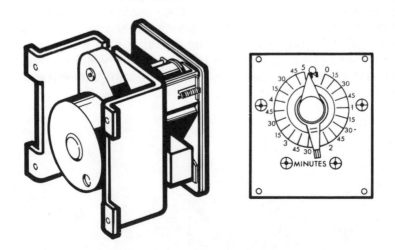

Figure 56. FRONT AND REAR VIEWS OF A TIMER.

137. Ultimate Strength.

The ultimate strength of a material is the maximum stress it can stand in tensile, shear or compression loading. When a metal reaches its ultimate strength it ruptures.

138. Unit Charge.

The electrical energy of one particle of matter is called a unit charge. For example, an electron is one negative unit charge. A proton is one positive unit charge. A neutron is neutral and has no unit charge.

139. Volt (E).

The electrical term volt means the electrical pressure, or force, that causes current to flow in an electrical circuit. Voltage is often termed "emf" which means **electro-motive force.** The electrical symbol for voltage is "E". **Voltage does not flow but it is the motive force that causes current to flow.**

140. Volt-Ampere Curve.

Volt-ampere curves are graphs which indicate the output characteristics of a welding power source. The co-ordinates are output voltage and output amperage. The vertical axis (ordinate) always indicates voltage. The horizontal axis, properly called the abscissa, is the amperage base line. When the curve is plotted it shows **open circuit voltage** where it touches the vertical axis. At the point where the curve touches the horizonal axis the reading is **maximum short circuit current** for the setting of the welding power source. Volt-ampere curves are plotted under static loading conditions. It would be virtually impossible to plot output curves under dynamic welding conditions. Volt-ampere curves are used to show the current and voltage capabilities of a weld power source.

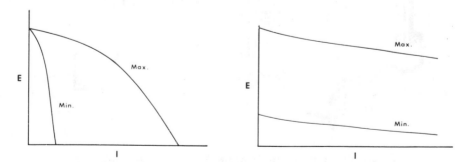

Figure 57. TYPICAL VOLT-AMPERE CURVES.

141. Volt Meter.

A volt meter (voltmeter) is an electrical measuring device for measuring voltage in a circuit. Voltage is the potential difference in value between two electrical conductors. Most voltmeters are battery operated units.

142. Watt.

The watt is a unit of electrical power measurement. In dc power watts are calculated by multiplying volts times amperes. A watt is the amount of power required to maintain a current of one ampere flowing in a circuit where the electrical pressure is one volt. One horsepower equals 746 watts.

143. Watt-Hour.

The watt-hour is the unit of electrical energy equal to the power of one watt being continuously used for one hour. It is watt-hours that is charged for on your electric bill each month.

144. Weave Bead.

A weave bead is made by oscillating the electrode tip from side to side of the weld transverse to the direction of travel. It is usually used as a top, or cap, pass. Weave beads, if not carefully made, may have slag inclusions within the weld deposit as well as lack of fusion at the weld edges.

145. Weld.

In metals, a weld is the uniting of two separate pieces by forging, melting and fusing or through friction of the abutting surfaces. There are other methods of welding but these are the basic concepts. The purpose of making a weld is to unite two or more pieces of metal into one homogenous mass with sufficient strength to serve whatever purpose is intended.

146. Weldability.

Weldability is the characteristic of a metal that permits it to be welded under a specific set of conditions. All of the common metals are weldable with one or more of the known welding processes.

147. Welder.

A welder is a person qualified to perform a manual or semi-automatic welding operation. The word "welder" is often erroneously used to describe a **welding power source.**

148. Welding Procedure.

A welding procedure details the methods and practices used in making a completed weld. Welding procedures should be written for each welding application. The welding operator should follow the written procedure to insure quality weldments.

149. Weldment.

A weldment is an assembly of pieces whose component parts are joined by welding. A weldment may be a single part or it may be a sub-assembly which, with other parts, is joined into a total welded assembly. A weldment is the finished product of welding.

150. Yield Strength.

When testing metals, there is a definite proportion between the stress applied and the strain resulting from such stress application. Yield strength is the value at which the metal shows a marked change from the normal proportionality of stress and strain. At the yield point, the metal deforms with no increase in stress applied. This is the yield strength.

Chapter 3

WELDING METALLURGY FUNDAMENTALS

Metallurgy is the study of metals in both their pure and alloyed forms. How they react to other metals and elements, to heat and cold, to both hot working and cold working, is all in the realm of metallurgy.

People who work with metals as engineers are usually called Metallurgists or Metallurgical Engineers. Associated with Metallurgists are Metallographers, Metallurgical Technicians and testing laboratory personnel. All of these people are working right along with the welder to make sure the weld metal and the base metal structure go together in the best possible manner for strength and performance.

There are two broad classifications of Metallurgists in the metals industries. The **Extractive Metallurgist** is employed in the mining, smelting and refining of metal ores. The **Process Metallurgist** is concerned with the application of metals in the manufacturing industries. It is the process metallurgist who is most concerned with welding metallurgy. In this area he may be the one to specify the welding procedure used, the welding process, electrode filler metal type and classification, any preheat and postheat required and any other data necessary to assure a good sound weld joint.

Welding metallurgy is concerned with metals and how welding affects their physical and mechanical properties and chemical composition. Even slight changes in chemical composition can cause substantial changes in the physical and mechanical properties of metals.

Metals Welded.

There are two basic types of metals in use today by the welding industry. They are the **ferrous metals** containing substantial amounts of iron and the **non-ferrous metals** which have essentially no iron at all in their chemical content.

Non-ferrous metals may be hardenable by cold working such as rolling, bending, hammering, etc. Such cold work reduces the ductility, and increases the hardness, of the metal. When heated after cold working, non-ferrous metals soften to the annealed state, regaining ductility but losing hardness and strength.

Some alloys of non-ferrous metals may be hardened by heat treatment. For example, certain aluminum alloys are classed as heat-treatable. The aluminum alloy 6061 falls in this category. This will create some loss of strength in the heat-affected-zone of the welded part due to overaging. There is no known fusion welding process that can weld heat treated and aged aluminum alloys and obtain an "as welded" part that has 100% joint efficiency. It is preferable to weld heat-treatable aluminum alloys in the annealed condition and then heat treat them subsequent to welding.

There is a substantial loss of ductility when welding heat treatable aluminum alloys. Heat treatment after welding will restore much of the strength of the metal but will not improve the ductility lost in welding.

Ferrous metals are basically iron and steel alloys. An alloy is a combination of two or more elements which may make a metal with better physical or chemical properties than either of the two, or more, elements basic to the alloy. In steel it is carbon that is the primary alloying element.

Most heat treating of iron and steel is done between the temperatures of 1400°-1850° F. When welding iron or steel, a welder actually melts the surface of the joint in the localized arc area. Since iron has a nominal melting range of 2500°-2700° in the commercial grades, it is logical that a temperature gradient exists between the molten weld metal and the unheated base metal. This gradient is from a high temperature at the melting point of the weld metal, through the heat-affected-zone, to the base metal still at room temperature.

Somewhere between the melting point of the weld metal and the room temperature of the base metal, part of the heat-affected-zone is in the range of the 1400°-1850° F. heat treating temperature. It becomes apparent that the welder heat treats the heat-affected-zone each time he makes a weld simply because the heat from the weld travels through the heat-affected-zone on its way to dissipating in the mass of the base metal. A very good rule to know is that heat will transfer faster through cold metal than it will through hot metal. This means that the heat of the weld will disperse, or flow, quickly to the colder base metal being welded. This is why a weld will lose its red heat very fast as the welder moves along the weld joint.

All of the carbon steels have some alloying elements in their chemical composition. Not all of these elements are really desirable.

For example, all plain carbon steels contain iron (Fe), carbon (C), manganese (Mn), sulphur (S), silicon (Si), and phosphorous (P). Both the sulphur and the phosphorous are held to very minimum percentages, usually not more than 0.05% maximum.

In addition to the carbon steels, there are a group of metals called alloy steels. It may seem a little odd that certain steels are called "alloy steels" and carbon steels are just that—carbon steels—when in fact **all steels are an alloy of two or more elements.** Alloy steels are so named because one or more elements are added for a specific purpose. Some of the alloying elements found in **low alloy steels** are nickel (Ni), molybdenum (Mo), chromium (Cr), tungsten (W), vanadium (V), titanium (Ti) and columbium (Cb).

Alloying elements are added to steels to increase their physical properties including strength, ductility and toughness. Some elements, including non-metallic ones, are added to steel for other reasons. Silicon (Si), for example, is added to promote fluidity in the weld puddle and to achieve better wetting action at the toes of the weld. It is also used as a deoxidizer—an oxygen scavenger—in steel. This means that the silicon scavenges the oxygen out of the steel when it is being manufactured at the steel mill. Such steels are called "killed", or "semi-killed", steels and are usually easily welded without porosity problems. Some welding electrodes have deoxidizers such as silicon as part of their chemistry for the purpose of removing oxygen from the weld puddle. This action is similar to the steel makers use of silicon except on a much smaller scale.

Basic Metal Structures.

Any substance that occupies space is called matter. Everything is some sort of matter, even the air we breathe. There are three known **states of matter.** They have been defined as **solid, liquid** and **gas.** There has been some thought that superheated gas plasma might be a fourth state of matter but no conclusion has been reached on this concept by the academic community as a whole. In welding we are basically concerned with metals in either the liquid or solid state.

Metals are normally solid state materials at room temperature. When a metal is heated to its specific melting point it changes from solid to liquid form. If it is a pure, unalloyed metal, the solid-liquid transformation occurs at a specific temperature.

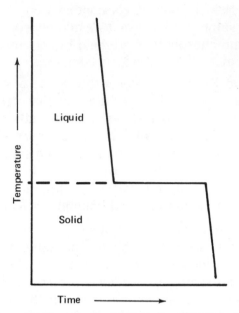

Figure 58. SOLID-LIQUID TRANSFORMATION, PURE METAL.

If the metal is an alloy of two or more elements the solid-liquid transformation **usually** occurs over a temperature range.

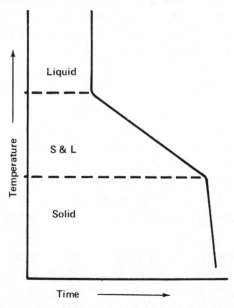

Figure 59. SOLID-LIQUID TRANSFORMATION, ALLOY METAL.

When weld metal is liquid it is a collection of atoms having no definite atomic pattern or arrangement. As a liquid, metal has no crystalline shape or form. When the liquid metal cools to a temperature below its melting point, it solidifies into crystals (or grains). As used here, the terms "crystal" and "grain" have the same meaning. A grain is a crystal formation in metal that has an irregular shape and, therefore, irregular boundaries.

As molten liquid metal cools it reaches a temperature where it begins to solidify. Particles of matter called grain (crystal) **nuclei** begin to form. Naturally, the crystal nuclei forms at the coolest spot in the weld. This would normally be at the interface between the molten metal and the unmelted base metal. The nuclei tend to attach themselves to existing grains of the base metal at the weld interface.

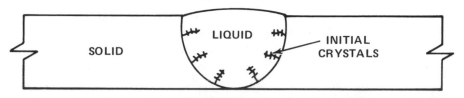

A-INITIAL CRYSTAL FORMATION

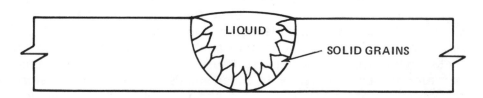

B-CONTINUED SOLIDIFICATION

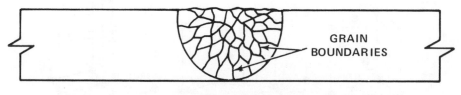

C-COMPLETE SOLIDIFICATION

Figure 60. NUCLEI AND CRYSTAL FORMATION IN WELDS.

As the molten weld metal continues to cool the solidification of metal progresses with the nuclei growing into full size, irregularly shaped grains. As illustrated in Figure 60, the grains grow until they meet an abutting surface of another grain. The grain, or crystal, shapes are of no specific design or pattern in this case.

At point (C) of the illustration the solidification of the liquid weld metal has been completed. Note that the grains in the weld deposit center are smaller and finer in texture than the grains at the weld and heat-affected zone interface. This is explained by the fact that the grains at the heat-affected zone and weld interface solidified first and were, therefore, at higher temperatures longer while in the solid state. This provided more time at elevated temperatures for grain growth in the metal.

Remember: As the weld metal cools from liquid to solid form, the heat flow out of the liquid metal **must** go through the grains that solidified first, then through the heat-affected zone of the base metal until the heat dissipates into the base metal mass.

Formation Of Crystal Structures In Metals.

We have found that liquid metal has no crystal (grain) structure since the atoms are in random location and arrangement. As the liquid metal cools and solidifies, crystal solids form irregular grain shapes. The **solid grains** have specific atomic patterns or arrangements which are typical of whatever metal is being welded. It is the atomic arrangement in the crystals that determines the type of metal.

The crystal structures that form in metals may be classed in three main categories. They are **Body Centered Cubic (BCC), Face Centered Cubic (FCC)** and **Hexagonal Close Packed (HCP).** These structures are found, in one form or another, in most metals. Some metals retain the same crystal structure from room (ambient) temperature to their melting point. Other metals have specific transformation temperatures where they change from one crystal structure to another.

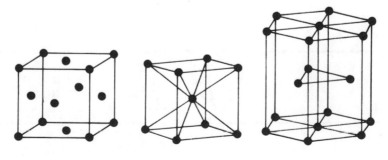

The chart in Figure 61 shows some of the metals and their crystal structures.

STRUCTURE	METALS
Body Centered Cubic (BCC)	Iron (Alpha & Delta phases); Chromium; Columbium; Tungsten; Vanadium; Molybdenum.
Face Centered Cubic (FCC)	Iron (Gamma phase); Copper; Gold; Lead; Nickel; Silver; Aluminum.
Hexagonal Close Packed (HCP)	Cobalt; Magnesium; Tin; Titanium; Zinc; Zirconium.

Figure 61. CRYSTAL STRUCTURE IN SOME METALS.

Phase Transformation.

There are many different metal, and metal alloy, phase diagrams presently published. In this brief discussion of Welding Metallurgy Fundamentals we will, therefore, concern ourselves basically with iron-carbon alloys and the iron-iron carbide phase diagram. Additional information is readily available from the American Welding Society and the American Society for Metals.

A **phase transformation** occurs in some metals when they are subjected to elevated temperatures above room temperatures. By "phase transformation" we mean the changing of a metal from one crystalline structure to another. For example, in **pure iron** the liquid metal solidifies at 2,795° F. The crystal structure that forms is body centered cubic (BCC) and is called **delta iron.** As the iron slowly cools, the body centered cubic delta structure changes (transforms) to face centered cubic (FCC) at a temperature of 2,535° F. The face centered cubic structure is called gamma iron.
Gamma iron is also known as "austenite". The iron is non-magnetic in this form and structure.

The face centered cubic structure of the austenite is retained by the pure iron as the metal continues to cool.

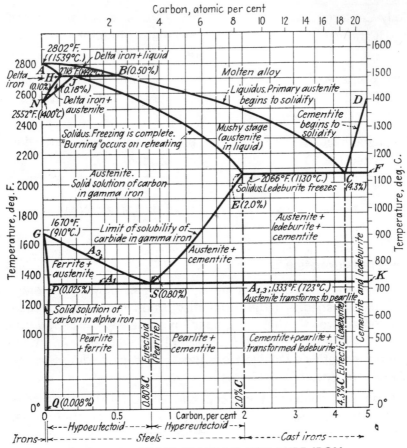

Figure 62. PHASE DIAGRAM FOR PURE IRON.

When the pure iron reaches a temperature of 1,670° F., the face centered cubic structure transforms back to the body centered cubic crystalline form. Below this temperature the metal is known as alpha iron. It is interesting to note that both alpha and delta iron have the body centered cubic crystalline structure. They are given different names to identify **the high temperature phase (delta)** and **the low temperature phase (alpha).** The ability of a metal to change phase through transformation is called **allotropic.**

To define: **Allotropic** = "The ability of atoms to change their orderly arrangement in metal and exist in two or more crystalline structures **at different temperatures**". Iron and steels are allotropic metals. Other metals that undergo allotropic transformation are titanium, cobalt and zirconium.

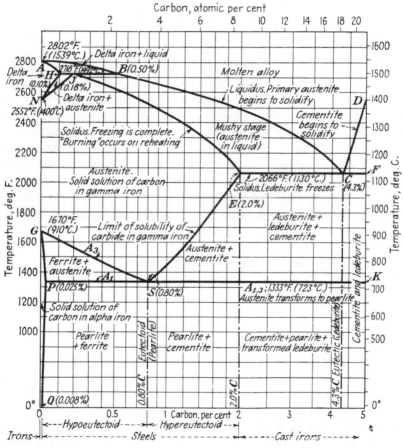

Figure 63. THE IRON-IRON CARBIDE PHASE DIAGRAM.

In looking at the iron-iron carbide phase diagram it is easy to see that, **as the carbon content goes up,** there is a marked change in the type of iron present. In addition, it is evident that phase transformation temperatures increase or decrease with varying carbon content in the iron.

Note that the alloy steel may have carbon content up to 2%. It is difficult to weld steels with more than 1% carbon content and special techniques are required. Above 2% carbon the metal is considered to be cast iron.

Commercial steels are normally classed by carbon content. For example, **low carbon steel** is considered to have an approximate maximum of **0.15% carbon content.** The **mild steels** range up to a maximum of about **0.30% carbon content**. Above 0.30% carbon, and up to about **0.60% carbon,** the steel is called **medium carbon steel**. The medium carbon steels are heat treatable for hardness.

Steels having a range of 0.60% carbon up to approximately **1.5% carbon** are classed as **high carbon steels.** Very often such steels are used for tool and spring steel applications. They are very hard materials and heat treat easily but they have poor weldability. Cooling rates for high carbon steels must be slower than the cooling rates used for low carbon steels. The reason is that a fast cooling rate for high carbon steel would cause it to become extremely brittle. High carbon steel is normally oil quenched where low carbon steel is normally quenched in water. Water has about a 6-7 times faster cooling rate than oil. Water is, therefore, used for carbon steels up to 0.30% carbon. Higher carbon steels normally require oil quenches to slow the cooling rate and cooling time.

Another method of slowing the cooling rate of medium and high carbon steels is with preheat. Usually 300°—400° F. is sufficient to protect most steel weldments. Preheat slows the cooling rate because heat will move more slowly through hot metal than it will through cold metal. The heat from the weld is not abruptly quenched as it would be in cold base metal.

Cast Iron.

In the iron-iron carbide phase diagram you see that above 2% carbon content the metal is considered to be cast iron. There is a substantial difference between the steels and cast iron. True, the percentage of carbon in cast iron is high but that is not the key factor. It is the **form of the carbon** that makes such a difference in the metals.

For example, most carbon steels have their carbon contents in either dissolved form or as iron carbides. Not so with cast iron. Although some carbon is in the dissolved form with cast iron, most of it is in the form of elemental carbon or graphite. There is no particular arrangement of the flake graphite in the cast iron. Rather, it is in random disarray at all angles. When stressed, it isn't the cast iron metal that breaks; it is the flakes of graphite which propagate the crack and fracture. It is logical that the smaller the graphite flakes, the less opportunity for fracture and the stronger the cast iron.

Cast iron may be welded by several welding processes. One of the best is with preheat, an oxy-acetylene torch and cast iron rod. It is hot, dirty work because the cast iron filler metal must be stirred vigorously into the weld. Arc welding is normally accomplished with high nickel electrodes. In this case, the preferred technique is to bring the work-piece to room temperature and weld only a little bit at a time, peening lightly after each short pass.

Carbon Steels.

The chemical combining of iron and carbon in carbon steels produces an iron-carbide molecule having **three atoms of iron and one atom of carbon to the molecule (Fe_3C)**. It is carbon that primarily controls the hardenability of carbon steel. The higher the carbon content, the easier it is to harden the steel.

Some of the other elements found in low carbon steel are manganese, phosphorous, sulphur and silicon. Let us take a brief look at what effect these elements have on the steel product.

Manganese is used in steels basically to combine with sulphur to form manganese sulphides. The intent is to prevent or minimize the formation of iron-sulphides which tend to cause a "hot-shortness", or crack sensitivity, in iron when it is heated. Iron-sulphides form coatings around the individual grains of metal. Since they have a low metallurgical melting point, they are in liquid form at the rolling temperatures of the steel. If they are allowed to form due to insufficient manganese in the steel composition, iron-sulphides cause loss of intergranular adhesion during the rolling process at the steel mill and the metal literally breaks up. The addition of 2-3 times the sulphur weight in manganese creates the manganese-sulphides, which are spherical in form, within the grains of metal and cause no problems in rolling the metal.

Manganese is also added to steel for its effect of increasing the hardenability of steel. It does cost more than carbon, however, so it is used sparingly. The normal maximum content is 1.5% manganese. Higher percentages of manganese will change the steel from water hardening to oil hardening. This indicates that a much slower cooling rate is necessary for steels with manganese in excess of 1.5% content.

Phosphorous is present in all steels but not by choice. Phosphorous cannot be totally removed from steel in smelting and refining but its level can be controlled to very fractional percentages. Usually steel has a maximum of 0.05% phosphorous. This element has a tendency to increase the brittleness of steel thereby making it crack sensitive. Phosphorous does add strength to the base metal but the brittleness it also adds makes any welding of the steel difficult due to cracking.

Some low alloy, high strength steels have no phosphorous percentage limit in their compositions. They are structural steels which meet the ASTM A-242 specification. Steel makers are using phosphorous as a low cost strengthener to meet this specification. Welding people should check the steel specifications carefully to be sure they do not try to weld the unlimited phosphorous version of ASTM A-242 low alloy steel.

The effect of sulphur in combination with manganese has been mentioned. However, certain free machining steels use sulphur in the form of manganese-sulphide to improve machinability. This would be most evident in bar steels. Although the bar stock is often referred to as "cold rolled" it is actually "cold drawn" through a die to obtain both a **clean finished surface** and precise dimensional sizing. During the drawing operation the manganese-sulphides, which are present in the iron grains as spherical shapes, are elongated to needle-like inclusions for better machinability. Such high sulphur steels are difficult to weld. The best technique would be a braze type weld with minimum dilution of the base metal with the filler metal. The preferred electrode would probably be an E-7018 low hydrogen type used with amperage on the low side of the operating range.

One fallacy we would like to dispel now is, "you can't weld cold rolled steel successfully!" That is not a true statement. As a matter of fact, the finished surface condition of the metal has little to do with its weldability. **As long as the metal surfaces are clean of dirt and oxidation, only the composition of the steel primarily controls its weldability.** Certainly the **external finish** of the steel has nothing to do with its weldability.

All metals are granular; that is, all metals have grains or crystals as their basic structure. The grain size may vary from quite small to very large in metals. Grain size does have some effect on the weldability of a steel. For example, metals having small grain size are tougher, stronger and normally easier to weld.

Large grain size may make welding some metals more difficult. The large grains promote crack sensitivity since they have less ductility and less toughness. This is because large grains are susceptible to cross-grain fracture. As an example, it is much easier to break a peppermint stick of candy than it is to break a peppermint candy ball. The analogy relates to large, elongated grains as compared to small, spherical grains.

Grain size can be controlled to some extent at the rolling mill. This is done by making the last hot rolling pass at as low a temperature as possible to produce the finest grain size. Another method of controlling grain size in steels is "normalizing" the metal. Normalizing is accomplished by re-heating the steel to a temperature above its lower critical temperature of 1333° F. minimum (the exact temperature will depend on the metal chemical composition), holding it at that temperature for one hour for each inch of cross sectional area, or portion thereof, and then cooling as rapidly as possible in air. The result is smaller grain size, better physical properties in the metal and better weldability.

Alloying Elements In Carbon Steel.

Some of the alloying elements in carbon steel have been discussed and their relative functions described. There are many other alloying elements which have not been brought out and this is a good time to consider them. The following list discusses the elements and their principal functions.

Carbon = The most important alloying element in steel.

Sulphur = An undesirable element in steel; causes brittleness.

Phosphorous = An undesirable impurity in steel.

Silicon = Used mainly as a deoxidizer in steel; promotes fluidity in the weld puddle.

Manganese = Increases hardenability and strength in steel; combines with sulphur as manganese-sulphides.

Chromium = Soluble in iron, Cr is added in amounts to 9% to increase oxidation resistance, hardenability, and elevated temperature strength.

Molybdenum = A carbide former, Mo is normally added in amounts of less than 1%. Mo increases hardenability, and elevated temperature strength.

Nickel = In low alloy steels there is up to 3.5% Ni used to increase toughness and hardenability. For high alloy and stainless steels up to 35% Ni is used.

Aluminum = Aluminum added to steel as a deoxidizer and grain refiner. Very fractional percentages used.

Dissolved

Gases = Hydrogen, oxygen and nitrogen are all soluble in steel. These dissolved gases must be removed for they will embrittle the steel.

Anytime alloying elements are added to carbon steel they tend to make the metal more difficult to weld. More care must be taken with the welding procedures and the filler metal chemical composition to insure strong, sound welds at proper strength.

It is true that small changes in chemical composition in steels will possibly cause considerable change in the physical properties of the metal. Combine that fact with the knowledge that iron and most steels undergo allotropic transformation (the change in crystal structure that occurs on heating and cooling the metal through specific temperatures) and it is logical that a wide range of metallurgical properties are possible.

The Effect Of Cooling Rate On Carbon Steel.

The ability of austenite to transform to certain desirable lower temperature microstructures in steel is essentially the ability of carbon steels to be hardened by heat treatment. The transformation temperatures will vary with the composition of the carbon steel.

The effect of the cooling rate on austenite and the resulting lower temperature microstructure of carbon steel is described in the following paragraphs.

a) **Pearlite.**

When austenite is **slow cooled** there is sufficient time for the extensive diffusion of metal atoms to form ferrite and carbide layers. This is typical **coarse pearlite** and the resulting carbon steel microstructure is soft and ductile.

b) **Fine Pearlite.**

Fine pearlite is formed by using a **slightly faster cooling rate.** The steel is less ductile and is harder than coarse pearlite.

c) **Bainite.**

When **faster cooling rates** are employed, no pearlite appears. **Bainite** appears at lower transformation temperatures, with a feathery arrangement of fine carbide needles in a ferrite matrix. This structure has high strength, low ductility and fairly high hardness characteristics.

d) **Martensite.**

Martensite is achieved by **extremely fast quenching** of the austenite. It is a hard, relatively brittle phase of steel. The martensitic microstructure has an acicular needle-like appearance. Quenched martensite is normally too brittle for industrial use and must be tempered for ductility. Tempering martensite provides a tough, more ductile steel with slightly decreased hardness.

Certain commonly used heat treating terms refer to the **rate of cooling** of austenite to room temperature. In the table following, certain techniques are listed in the order of increased (faster) cooling rates from above the A$_3$ temperature as shown on the iron-iron carbide phase diagram.

1) Furnace annealing = Slow furnace cooling (Slowest).

2) Normalizing = cooling in still air. (Slightly faster).

3) Oil quench = Quenching into an oil bath (Medium fast).

4) Water quench = Quenching into a water bath (Fast).

5) Brine quench = Quenching into a salt brine bath. (Very fast).

If a heat treatable steel of specific type were heat treated and quenched with materials in the order named, beginning with furnace annealing, it would increase in hardness if quenched from a specific temperature using each of the mediums. For example, the metal would be softest after furnace annealing and hardest after brine quenching. Note in the table that rate of cooling (time) is very important to the hardening of steels. The faster a heat treatable steel is cooled, the harder it is.

The Isothermal Transformation (TTT) Diagram.

In metallurgy the TTT diagram is a method of showing graphically the **time, temperature** and **transformation** at which pearlite, bainite and martensite form when austenite is cooled. **It is important to know that every steel composition has its own TTT diagrams.**

The TTT curve shown in the illustration is for a eutectoid steel. A eutectoid steel has a carbon content between 0.80% and 2.00% carbon. In this particular case, the carbon content is 0.80% carbon.

Creating the TTT curve was accomplished by heating samples of the 0.80% carbon steel to their austenitizing temperature of approximately 1550° F., then quenching them to a series of temperatures between 1300° F. and room temperature. The quenching mediums were hot salt pots, hot lead pots, oil quenches and water quenches, all at the proper temperatures.

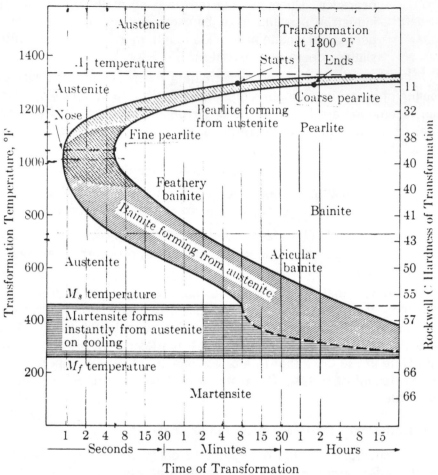

Figure 64. TTT DIAGRAM FOR 0.80% C. EUTECTOID STEEL.

The sample held at 1300° F. **did not begin to transform** for about 500 seconds, or eight minutes, and **did not finish transformation** for about 4,000 seconds (about one hour). The microstructure formed was coarse pearlite and was fairly soft (about Rockwell "C" 15).

When the quench temperature was reduced to 1050° F. the action speeded up. The transformation started in one second and was completed in five seconds. This is the "nose" of the TTT curve. The microstructure of the transformed steel was fine pearlite with an increased hardness to Rockwell "C" 41. (Take the time to inspect the isothermal TTT diagram and plot the difference in quenching **temperature** and **time** to find the **transformation** characteristics).

As shown in the TTT diagram, further decreasing the quenching temperature increases the time to start transformation. The microstructure becomes bainite with a corresponding hardness in the steel specimen.

If the specimen is cooled rapidly enough to get past the "nose" of the TTT curve, which would be cooling to 450° F. or lower instantaneously, the microstructure formation would be martensite. (see MARTENSITE). Martensite only forms at temperatures below the M_s point and is substantially completed at the M_f point. The slowest cooling rate required to form 100% martensite may easily be determined from the TTT curve. It is called the "critical cooling rate" which just misses the nose of the TTT curve. As you can see, the time involved is less than a second.

As carbon content and alloying content increase in steel, the nose of the TTT curve moves to the right. In this situation, the heat treating conditions which produce less rapid cooling can be used and martensite is easily formed. When there is lower carbon content and alloy content in the carbon steel, the nose of the TTT curve moves to the left.

Hardenability And Tempering Martensite.

Hardenability of steel is a measure of the ease with which the formation of non-martensitic microstructures can be prevented and 100% martensitic structures be obtained. In general, the steels with higher carbon content and/or alloy content have higher hardenability characteristics.

To harden steels properly, it is necessary to produce martensite in the base metal by proper chemistry and quenching. As we know, martensite is very hard and brittle and totally unsuitable for engineering applications of steel. However, by tempering martensite it is possible to achieve a microstructure in the metal that is more ductile, increases toughness, and decreases brittleness without significantly decreasing the metal strength.

The heat treatment for tempering martensite is simple. The martensite must be heated to some temperature below the A_1 temperature (approximately 1333° F.). This permits the unstable martensite to change to tempered martensite by allowing the carbon to precipitate in the form of tiny carbide particles of Fe_3C. (Fe_3C is iron carbide).

The desired strength and ductility can be controlled by selecting the proper tempering times and temperature. The use of higher tempering temperatures results in softer metal with more ductile properties and less strength.

Metallurgy and The Weld Joint.

In order to intelligently consider the welding of a steel material it is necessary to know the chemistry of the material and any previous heat treatment that it may have had. This is called the metallurgical and heat treatment history of the metal.

Weld joints have two basic areas that are of interest and concern to the metallurgist and welder. The two areas are the **heat-affected zone (HAZ)** and the **deposited weld metal.** The weld metal is a cast structure which is part deposited weld filler metal and part melted base metal. The heat-affected zone is totally base metal which has undergone possible physical and chemical changes due to the heat of the weld passing by.

We have said that the heat treating range for carbon steels is approximately 1400°-1850° F. The heat-affected zone of a weld is in the temperature range from room temperature at the base metal to just below the melting point of the steel at the weld and heat-affected zone interface. This means that some portion of the heat-affected zone is going to be transformed into austenite by the heat of the weld. It is apparent that the properties of the heat-affected zone, for the particular steel, will be determined to a major extent by the iron-iron carbide phase diagram and the TTT diagram and curve.

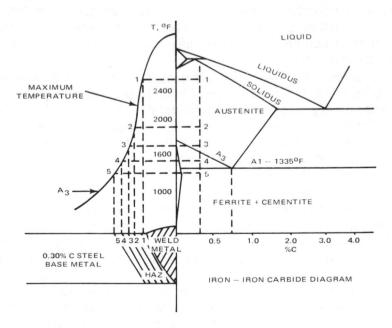

Figure 65. WELD CROSS SECTION AND HAZ PROFILE.

The illustration in Figure 65 shows the temperature profile of the maximum temperatures achieved at various points in the heat-affected zone of a typical 0.30% carbon steel. As you can see, by referring to the iron-iron carbide phase diagram, much of this profiled area has been heated above the A_1 temperature.
For example:

Point 1—has been heated in excess of 2400° F. The austenite that forms will be coarse grained because of the grain growth at this temeprature.

Point 2—has been heated to 1800° F. and fully austenitized. Grain growth has not occurred; some grain refinement may occur.

Point 3—has been heated to just above the A_3 temperature. This is not enough heat to completely homogenize the austenite.

Point 4—this area has been heated to approximately 1400° F. which is between the A_1 and A_3 temperatures. Part of the structure is converted to austenite and the resulting mixture of products during cooling can result in poor notch toughness.

Point 5—this area has been heated to about 1200° F. which is **below** the A_1 temperature and no austenite is formed. The base metal may be spherodized and softened.

The effects of welding on the cooling rate depends on the conditions, such as the heat input in joules per inch, the base metal thickness and the amount of preheat in the base metal.

The greater the amount of heat input to the weld area the slower the cooling rate will be. Heat input is measured in joules per inch of weld bead length. Base metal thickness is important to the cooling rate since thicker metals will cool more quickly than thinner metals. The reason is that thicker metals have more mass and, therefore, act as a "heat sink" for the heat energy put into the weld. **Remember:** Heat flows more readily through cold metal than it does through hot metal. The base metal preheat is important for the reason just given. If the base metal has any temperature above room temperature the cooling rate of the material will be slowed down. The hotter the base metal, the slower the cooling rate of the steel.

Stress Relieving.

It is possible to have both bainite and martensite in the heat-affected zone of a weld. Such microstructures would probably occur in steels of fairly high carbon content or those having alloy additions of some magnitude. As welded, they would create a problem because they would tend to make the heat-affected zone brittle and crack-prone.

Stress relieving weldments in a range of 900°-1200° F. will temper the microstructure and make it more ductile. It will not, of course, remove any cracks that might have occurred due to the stressing of the material while welding and cooling.

Deposited Weld Metal For Carbon And Alloy Steel.

The weld metal used in making a joint must meet the same structural strength requirements as the base metal. There are four main factors that govern the metallurgical characteristics of weld metal deposits. They are:

1. The chemical composition of the consumable electrode.
2. The chemical composition of the base metal.
3. The chemical reaction of (1) and (2) when the metal is molten and mixed together.
4. The cooling cycle of the **solid weld metal.**

Let's examine each of these factors and see what is done to protect the weld metallurgical characteristics.

Most filler electrodes, either cut lengths with flux coatings, or bare continuous wire, have low carbon content. If the carbon content were fairly high there would be the possibility of creating martensite in the weld with attendant hardness and cracking possibilities. This is not normally possible with low carbon filler metal.

The chemical composition of the base metal is a fact of life and there is nothing much you can do about it except to **know what it is**. The selection of filler metal will be primarily based on the type of base metal to be welded.

If the proper electrode chemical composition is selected the chemical reactions between the filler metal elements and the base metal elements will cause no problems. Consider that there is a mixing of base metal with the molten filler metal and the result must be a weld at least as strong as the minimum guaranteed strength of the base metal.

The method of cooling will control the time of cooling to a great extent. How fast the cooling takes place; how thick the base metal is (since this is a form of heat sink); the chemical composition of the base metal; and the chemical composition of the electrode material all have an important bearing on the physical characteristics of the weld deposit.

The Metallurgy Of Stainless Steel.

The "stainless steels" are iron-based high alloy metals with substantial amounts of chromium and nickel and smaller amounts of other alloying elements in their matrix. Such alloys have excellent resistance to oxidation and corrosion. They will normally have improved physical properties at elevated temperatures due to the alloying elements. Chromium is the principle element responsible for the improved properties. The amount of chromium will vary from about 12% to about 30% in the high alloy materials.

There are three basic types of "stainless steels". They are:
1) Martensitic Stainless Steel.
2) Ferritic Stainless Steel.
3) Austenitic Stainless Steel.

Both martensitic and ferritic stainless steels are alloys of iron with varying amounts of chromium and carbon. If nickel is present it will usually range from 1.5-4% by volume.

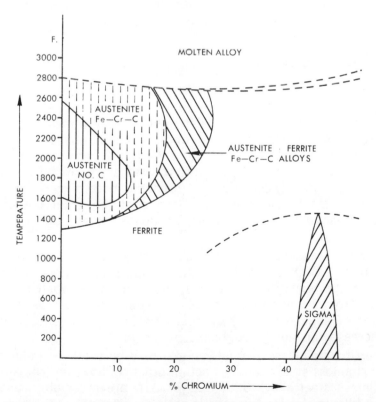

Figure 66. THE IRON-CHROMIUM-CARBON PHASE DIAGRAM.

Austenite can only exist up to about 12% chromium in **carbon-free** iron. This is shown in the iron-chromium diagram. Note the loop at the extreme left of the diagram. With carbon additions, and depending on the amount of carbon added, the austenite extends to approximately 18% chromium in iron. There are certain high chromium, low carbon, iron alloys which are called **ferritic stainless steels**. The ferrite phase of these alloys is stable up to the melting point of the alloy. There is no phase transformation in this type of stainless steel.

Martensitic Stainless Steel.

As the name implies, the martensitic stainless steels are hardenable by heat treatment. The various cutlery grades of stainless steel are included in this group.

Extreme care should be exercised when welding the martensitic stainless steels. The most important reason for using extreme care is that martensite tends to be produced in the heat-affected zone of the weld. Preheat is highly desirable to slow the cooling rate of the metal, thus permitting austenite transformation to take place. Preheat of martensitic stainless steel will also lower the cracking tendency of the metal when welded.

Ferritic Stainless Steel.

The ferritic stainless steels contain 17-27% chromium without significant amounts of nickel, carbon or other austenite-forming elements. The ferrite phase will exist right up to the melting point of the alloy. This is shown in the iron-chromium phase diagram in Figure 66. Due to the fact of retaining the ferrite phase to the melting point there is little, if any, austenite formed as the metal is heated. Although not normally considered hardenable metal alloys, some ferritic stainless steels can develop austenite **at the grain boundaries.** This can change to martensite in the heat-affected zone when the metal is cooled.

Ferritic stainless steels are subject to grain growth at temperatures above 2,000° F. Rather large, coarse grains will probably be evident in the heat-affected zone of the base metal when the metal is cooled from welding temperatures. Fortunately, the grain size can be refined (reduced in size) by post-weld heat treatment. This will also increase the notch toughness of the metal.

Ferritic stainless steel filler metal may be used for welding the ferritic stainless steel alloys. In actual practice, however, the austenitic stainless steel alloys 308, 309 or 310 are most often used as welding filler metal for the ferritic stainless steels.

Austenitic Stainless Steels.

The austenitic stainless steel alloys are basically an iron-chromium-nickel (Fe-Cr-Ni) group. Varying amounts of carbon and other alloying elements are added to provide special properties such as oxidation and corrosion resistance.

The austenite microstructure is stabilized at all temperatures by the addition of nickel and lesser amounts of carbon and manganese. By stabilizing the austenite at all temperatures the austenite-ferrite transformation is eliminated.

The austenitic stainless steels cannot be hardened by heat treatment since there is no transformation to ferrite. This is beneficial because there can be no hardened areas in the heat-affected zone of the base metal. The austenitic group of stainless steels have excellent weldability in all the alloys.

Stainless Steel Weld Filler Metals.

The most important single factor to consider in stainless steel welding is to match the filler metal as closely as possible to the base metal alloy chemical content. Of course, strength requirements, corrosion resistance and oxidation resistance are usually the factors which most influence weld metal electrode selection.

Although there is no alpha ferrite in austenitic stainless steels, there is usually a small amount of delta ferrite in the metal. This is advantageous since it will help to minimize hot-cracking in the weld during the cooling cycle. Ferrite control is highly desirable in austenitic stainless steels. The addition of small amounts of ferrite minimizes the welding problems of hot-cracking in the weld metal.

Sigma Phase.

Although ferrite in the weld metal aids in the actual welding process it can cause physical property problems when subjected to **high temperature service.** Some austenitic stainless steel alloys, particularly those containing ferrite, can develop a brittle phase called the **Sigma phase** when held in a temperature range between 900°-1775° F. At specific temperatures within the range sigma phase formation is very rapid. For example, it will form in one hour at a temperature of 1550° F.

When sigma phase formation occurs in the microstructure of the alloy, corrosion resistance and ductility decrease. Hardness increases markedly while notch toughness is reduced considerably.

While sigma phase may be removed from stainless steel by annealing at temperatures above 1800° F., it is important that careful choice of the alloys be made, especially when the alloys will be used for high temperature service above 1,000° F. The amount of delta ferrite present in the alloy selected should be held to a minimum. Another area that deserves careful consideration is the selection of the electrode alloy employed for specific high temperature applications. Again, the reason is to minimize the amount of delta ferrite present in the weld metal.

Carbide Precipitation.

There is a **critical range** of temperatures to consider when austenitic stainless steels are heated. In the temperature range between 800°-1200° F. chromium precipitates from the metal grains and combines with the carbon at the grain boundaries as chromium carbides. The chromium carbides will cause corrosion attacks to occur preferentially at the grain boundaries as well as in the base metal grains which have been depleted of chromium. Carbide precipitation occurs in those areas of the heat-affected zone which have been heated within the critical range of 800°-1200° F. Of course, the **time** the metal is within this heat range is important. The longer the time at temperature, the greater the carbide precipitation will be.

There are three basic methods of reducing or preventing carbide precipitation in the austenitic stainless steels. They are:

1. Heat treat the metal to above 1800° F. and quench cool rapidly. This will dissolve the carbides and put the chromium back into solution in the metal grains. Rapid cooling prevents carbide precipitation by cooling the metal through the critical range very rapidly.

2. Decrease the carbon content of the stainless steel to a low enough level that there is not sufficient carbon to form carbides. Extra low carbon grades have a suffix of "ELC" or "L".

3. The third method is to add a "stabilizer" element such as titanium (Ti) or columbium (Cb) to the alloy steel. Both of these elements have a greater affinity for carbon than chromium. The formation of TiC (titanium carbide) and CbC (columbium carbide) uses up the available carbon and there is very little carbon left to combine with the chromium. This type of stabilized stainless steel is often used where control of temperatures in the critical range is not possible.

Summary.

1. Strength and ductility of steels, both carbon and alloys, will vary widely with either or both of the following:
    a) Comparatively small changes in chemical composition.
    b) Changes in the heat treatment cycle.
2. The properties of a welded joint are determined by the metallurgical characteristics of the weld metal and the heat-affected zone.
3. Chemical analysis of the weld metal is controlled by the welding process and techniques used, filler metal analysis, and the composition of the electrode flux coating, if used.
4. The strength and ductility of the welded joint are controlled by the following:
    a) Preheat of the weld joint and base metal;
    b) The slower cooling rates caused by preheating;
    c) The heat treatment of the weld and heat-affected zone;
    d) The post-heat used for stress relieving.

Chapter 4

ELECTRICAL FUNDAMENTALS

Introduction.

Electricity! A magnificent servant of Mankind! This chapter will explain some of the things that electricity does, what it is made of, how it is generated and hopefully take some of the mystery out of this most fascinating subject.

The name "electricity" is based on both the Greek and Roman languages. The Greek word for amber is "elektron". The Roman word for amber is "elektrum". From this we can conceive the word electricity. The significance of amber to our society will be explained in the section concerning Static Electricity.

A basic physical law states, "For every action there is an equal and opposite reaction". This is especially true when considering electrical power.

The concepts we discuss in this chapter will start with the very simple beginnings of Man's discovery of static electricity. From that point we will examine the theory of matter, a bit about atoms, protons and electrons, conductors and insulators, and Ohm's Law. Eventually we will discover the methods of generating electrical power. Overall, this discussion of electrical fundamentals is slanted towards electric arc welding but the physical and electrical laws presented are applicable to any use of electrical power.

Static Electricity.

History tells us that Thales, a Greek, first recorded an experiment where he rubbed amber with a silk cloth. He noted that the amber would then attract small pieces of lint and paper due to some force that was imparted to the amber from the silk material. This was the first recorded data concerning static electricity.

An experiment similar in concept to that conducted by Thales will assist you to better understand static electricity. Suspend a table tennis ball by a length of silk thread (about 12 inches of thread is sufficient). Rub a common glass stirring rod briskly with some silk or nylon cloth. Touch the suspended table tennis ball **lightly** with the glass rod, then withdraw the rod. Now bring the glass rod **slowly** into the immediate area of the table tennis ball—and the ball will move **away** from the glass rod! The static electrical charge transferred from the glass rod to the table tennis ball **repels** the ball from the similar static electrical charge on the glass rod.

Another experiment will help us to see another aspect of static electricity. This time, rub a hard rubber comb with a piece of **wool** cloth. The comb will attract pieces of lint and paper as before. If you **slowly** approach the table tennis ball with the comb, the ball will actually move **toward** the comb!

Some conclusions can be deduced from these two rather simple experiments. They are as follows:

1. There must be at least **two different kinds of electrical charges** in existence since in one case there was a **repelling action** and, in the other, there was an **attraction.**
2. It appears, from the action and reaction of the table tennis ball in the two experiments, that **like electrical charges repel and unlike electrical charges attract.**
3. It is evident that some type of **electrical charges can be moved** from one place to another through some kind of electrical conductor.

The examples described are two very simple methods of showing static electricity; that is, electrical energy not in motion as generated power.

It is important to know that there are **two basic types of electrical power.** They are **"static electricity"**, which we have discussed briefly, and **"dynamic electricity"**. The term dynamic electricity refers to current electricity which flows in electrical conductors. Dynamic electricity is electrical power in action. It is the knowledge of dynamic electricity that enables us to have the vast electrical technology that the world presently enjoys.

The Basic Theory Of Matter.

For many years it was thought, and taught, that the atom was the fundamental unit of matter. That theory was literally blown up when the first "atom bomb" was exploded during World War II. Let us take a look at what some of the world's outstanding physicists began to suspect even before 1930!

The term **matter** may be defined as **the physical substance of any object that occupies space.** All matter is made of particles of some substance. **The smallest particle of a substance that contains all the elements of that substance is called a molecule.** For example, a molecule of carbon dioxide (CO_2) contains the elements carbon monoxide (CO) and atomic oxygen (O). The molecule is the particle that is involved in most of the chemical changes that take place in matter.

Molecules may be divided into smaller atomic particles called **atoms.** An **atom** may be defined as **the smallest unit particle of an element that retains all the characteristics of that element.** Over 100 atomic elements have been isolated and classified in a chart known as the Periodic Table of the Elements.

The atom, which for many years was thought to be indivisible, can be divided into smaller sub-atomic particles. Although the sub-atomic particles are smaller than the atom they are very important to its function in matter.

The heavy central core of the atom is called the **nucleus.** Most of the mass of the atom, and all of the positive electrical charge, is contained in the nucleus of the atom. In a broad sense, the nucleus of an atom may be compared to the sun in our solar system. Sub-atomic particles called **electrons** revolve around the nucleus of the atom much as the planets, including Earth, revolve around the sun.

The nucleus of the atom contains **protons** (sub-atomic particles that carry a **positive electrical charge**) and neutrons (sub-atomic particles that carry **no electrical charge** and which are considered electrically neutral). Much smaller sub-atomic particles called **electrons,** which are the fundamental unit of **negative electricity,** are considered to be in orbital path around the nucleus of the atom. This concept was first announced by Dr. Neils Bohr, an eminent physicist, who resided in Copenhagen, Denmark.

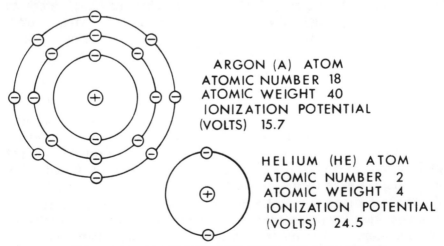

ARGON (A) ATOM
ATOMIC NUMBER 18
ATOMIC WEIGHT 40
IONIZATION POTENTIAL
(VOLTS) 15.7

HELIUM (HE) ATOM
ATOMIC NUMBER 2
ATOMIC WEIGHT 4
IONIZATION POTENTIAL
(VOLTS) 24.5

Figure 67. DIAGRAMMATIC VIEW OF GAS ATOMS.

Usually there is one proton for each electron in the atom. Although the proton is approximately 1800 times greater in mass and weight than the electron, **the two sub-atomic particles have equal but opposite electrical charges.** When there is an equal number of protons and electrons in an atom it is considered to be electrically neutral.

Remember: The proton carries a positive (+) electrical charge and the electron carries a negative (−) electrical charge.

Protons And Electrons.

The number of protons in an atom usually determines the **type of element** present. As we have said, the number of electrons will normally equal the number of protons in the atom. The Periodic Table of the Elements shows the atomic number and the atomic weight of each of the elements known to Man. Atomic number and atomic weight are defined as follows:

Atomic Number = The number of planetary electrons in orbit around the atom nucleus.

Atomic Weight = The total number of protons and neutrons within the nucleus of the atom.

Certain materials have what is known as a "loose electron bond". The term loose electron bond means that electrons may be separated from the positive nuclei of the atom with relative ease. Metals are one type of material that have a loose electron bond. The **electron** is the atomic particle that **must move** since there is no disruption of the integrity of the atomic nucleus and, therefore, **protons cannot move.** It is the electrons that move when electric current flows.

An atom is in the normal condition when the internal energy is at minimum level. This is termed the **normal state.** If the energy of the atom is raised above the **normal state** the atom is said to be **excited.** The act of excitation may occur in a number of different ways. For example, voltage applied to an electrical conductor will excite the atoms of the conductor. Collision of the atoms with high speed sub-atomic particles may cause the energy of the particles to be imparted to the atom. If the energy, or force, is great enough it can cause electrons to leave the parent atom. An atom that has gained or lost one or more electrons is called an **ion.**

If the atom has lost one or more electrons it would be classed as a **positive ion** since it has given up some of its negative electrical charge. **If the atom has gained one or more electrons** it would have a negative electrical charge due to the excess of electrons and would be classed as a **negative ion.** In welding, the primary concern is the ionization of various shielding gases which are normally considered to be positive ions under the influence of the arc heat energy. This is called thermal, or heat, ionization.

Remember: **An ion is an electrically charged atom.**

When electrons are removed from an atom they become free electrons. It is the movement of free electrons in an electrical conductor that constitutes the flow of electric current in an electrical conductor.

Conductors And Insulators.

Materials that allow the movement of large numbers of free electrons are called electrical conductors. For example, **silver** is an excellent electrical conductor because it has substantial amounts of free electrons available when voltage, or electrical pressure, is applied. **Copper** is the next best electrical conductor and, because it is much less expensive than silver, it is used as the most common electrical conductor. **Aluminum,** with approximately 62% the electrical conductivity of copper, is third best as an electrical conductor. Aluminum is considerably less expensive than copper and has been used for increasing numbers of electrical applications in industry.

Free electrons transfer from one atom to another in electrical conductors. As the electrons move the relatively short distance from atom to atom they displace other electrons in orbit around the second atom nucleus. The electrons so displaced move on to other atoms where the action is repeated.

The transfer of electrons continues until the electron flow (as electric current) is apparent all along the length of the conductor. The greater the number of electrons that can be caused to move in a material, at a given level of electrical pressure (voltage), the better the electrical conductivity of the material. **Good conductors** are those materials that **have low electrical resistance to current flow** and the ability to free large quantities of electrons.

Other materials have what is called a "tight electron bond" and therefore few free electrons. Such materials are considered to be poor electrical conductors. In many cases, these poor electrical conductors are actually used in circuits as electrical resistors and insulators. Of course, all materials have some electrical resistance. **There is no known perfect electrical conductor.**

Even resistor and insulator materials have some electron flow, or movement, so **there is no known perfect electrical resistor material.** It makes sense that the best electrical conductors, those having the least resistance to current flow, are used to carry electrical current. Those materials classed as least effective electrical conductors are often used as electrical insulators and resistors.

Summary Of Section.

Some of the terms used in this chapter section are briefly defined here for ready reference.

Matter = The physical substance of any object that occupies space.

Molecule = The smallest particle of a substance that contains all the elements of that substance.

Atom = The smallest unit particle of an element that retains all the characteristics of the element.

Nucleus = The heavy central core of the atom which is electrically charged positive.

Proton = The fundamental unit, or particle, of positive electricity.

Electron = The fundamental unit of negative electricity.

Ion = An ion is an electrically charged atom of matter.

Unit Charge = The smallest unit quantity of an electrical charge. For example, a proton is a positive unit charge; an electron is a negative unit charge.

It is important to know that protons are approximately 1800 times greater in mass and weight than electrons. That is the reason that almost all of the mass of an atom, and all of the positive charge, is contained in the nucleus of the atom. **The protons are all in the nucleus.**

Electrons can be transferred from one place to another within an electrical conductor. It is the movement of free electrons that constitutes the flow of electric current.

The negative electrical charge of the electron is exactly equal in value to the positive electrical charge of the proton.

Conductors are those materials which easily carry electrical current with a minimum of internal electrical resistance. Such materials have loose electron bonds and the ability to free large quantities of electrons. Silver, copper, and aluminum are metals which are considered good conductors.

Resistor materials are those which have high electrical resistance to current flow. Wood, rubber and certain ceramic materials are in this category. Such materials have a tight electron bond.

It is evident that good conductors have the ability to release large amounts of free electrons; resistors do not.

The Electron Theory Of Current Flow.

By now you know that the electron, a sub-atomic particle, is capable of motion within an electrical conductor. As a matter of fact, electrons are capable of motion within all fundamental matter. Knowing that **like electrical charges repel** and **unlike electrical charges attract,** it makes sense that electrons would try to move **away** from an area having a high negative electrical charge; an area with an excess of electrons. It is logical that electrons would be **attracted to** an area of high positive electrical charge; an area where there is an excess of protons and a deficiency of electrons. This concept is the basis for all electric current flow and, most especially, the electron theory of current flow.

There is not enough energy in electrical circuits to cause the protons to leave the nucleus of the atoms. It has to be the electrons, therefore, that move when electric current flows in a circuit conductor. The electron movement is caused by some type of potential energy, called voltage, which has sufficient numerical value to dislodge an electron from the atom. Electrons do move within electrical conductors but only the relatively short distances from one atom to another. As we will discover in this part of the discussion, there are literally billions of billions of electrons moving in an electrical conductor at any given moment of time when electric current is flowing.

Direction Of Current Flow.

In many electrical textbooks, the mathematical formulas and theories are based on the fact that electrical current flows from positive to negative. Most of the modern textbooks for the study of electronics, however, are based on the concept that current flows from negative to positive. The question is, which philosophy is correct?

Actually, it is only of academic interest to know which direction electric current flows. Until you get involved with electronic circuits, that is, where current **must** flow from negative to positive.

It was B. Franklin, Statesman, Editor and Inventor who first determined in which direction electric current flows. This was soon after his famous kite flying experience where he literally pulled lightning from the sky. From his observations of electrical power, Mr. Franklin concluded that electric current must move from positive to negative.

The electron theory of current flow was first recorded by T. Edison during his development of the incandescent light bulb. Before proceeding, it must be remembered that much of what we know about electricity has been developed within the years 1900 to the present time.

Mr. Edison worked with direct current; that is, current which flows in one direction only in an electrical conductor.

During one experiment, he had the problem of removing a sooty, black smoke which appeared on the inside of the glass bulb of the incandescent light. Among the things he tried, to overcome the problem, was the placement of a small metal plate within the sealed glass bulb. The metal plate had two electrical wires leading from the metal plate, through the glass bulb wall, to a galvanometer (a type of electric current metering device). Mr. Edison's theory was that the sooty smoke would gather preferentially on the metal plate rather than the inner walls of his glass bulb.

For the first part of his experiment he made the bulb filament positive (+) and the metal plate negative (−). He then impressed direct current on the filament of the bulb. To his recorded disappointment the bulb inner surface still became sooty and blackened. He then reversed the electrical polarity of the circuit by making the bulb filament negative and the metal plate positive. Again he impressed direct current on the circuit.

He recorded the fact that the bulb glass walls still became sooty and smoky but—low and behold—there was some movement of the galvanometer needle where there had been none before. Since Mr. Edison subscribed to the popular Franklin theory that electric current flowed from positive to negative, and it wasn't part of his experiment anyway, he made note of the galvanometer needle movement but paid no further attention to the incident. It has since been termed the Edison Effect, a notation of which is found in most electrical textbooks.

The electron theory of current flow from negative to positive was again discovered by engineers working with vacuum tubes in the late 1920's. They found that electrical current would not flow through the vacuum tubes from positive to negative but that it would flow through the tubes from negative to positive. Further experiments in electronics proved the electron theory of current flow. The important thing to remember is that electric current will flow in a circuit under proper electrical and mechanical conditions.

Measurement Of Electric Current.

The ampere is the unit of electrical rate measurement. If we determine what the **exact value of one ampere** is we then have the basis for measuring any amperage value in an electrical circuit.

When there is one ampere in an electrical circuit it is the same as saying there is "one coulomb per second" in the circuit. **A coulomb is a unit of electrical quantity.** Since we know that electrons move in a circuit when current (amperage) flows, then there must be a certain number of electrons in one coulomb. Mathematical calculations show **the number of electrons in one coulomb** to be 6,300,000,000,000,000,000. The number is read as 6.3 quintillion or **6.3 billion billion electrons.** One ampere, therefore, equals 6.3 billion billion electrons moving past a given point in an electrical conductor **every second.**

When electric current is flowing there could be a few electrons moving great distances in the electrical conductor or great quantities of electrons moving relatively short distances at moderate speeds. It has been calculated that there are great quantities of electrons moving short distances in electrical conductors.

When electrical power is used for productive purposes it is not "used up". It is actually the **energy of the electrons moving in the conductor** that we use. The electrical energy is derived from the process of creating the difference in the positive and negative electrical charges at specific points in the electrical circuit.

Atomic Movement In Matter.

All atoms of matter are made of sub-atomic particles such as electrons, protons and neutrons. Atoms are the basis for all matter regardless if it is in solid, liquid or gaseous form. **Atoms are in constant motion at all times.**

In solid materials atomic movement is limited to a rather confined space. Liquid materials provide more freedom for movement of the atoms than does solid material. Gases, being a free form of matter, permit very easy atomic movement since the gases are limited in space only by the confining walls of their container.

When any atom, or atomic particle, is in motion it has some amount of energy. **Energy is the ability to do work.**

The **energy** of an atomic particle is dependent on the **mass** of the particle and the **speed** the particle is moving through space. The relative motion, compared to other atomic particles, determines the energy available in a specific atomic particle. It is important to know that, whenever atomic particles are in motion, **work is being done and energy is being expended.** The result is apparent as **heat** in the particle in motion.

The basic theory of heat is called the **Kinetic Theory**. Kinetic energy is available because of motion of atomic particles. Heat energy may properly be called "kinetic energy". The greater the average energy of atomic particles the higher the temperature of the material of which they are a part.

A good example of kinetic energy is when cold metal is being hammered. In this situation, the metal will get hot. The energy of the hammer blow is transferred to the metal and, therefore, the atoms become excited and move more rapidly. As we have discovered, when there is movement of atomic particles there is work being done and energy being expended. The faster the movement of the atomic particles, within the metal structure, the more release of heat energy there is in the metal. In this circumstance the temperature of the metal will increase.

Electrical Conductor Concepts.

Materials that are classified as electrical conductors are those materials that have minimum electrical resistance to current flow in an electrical circuit. Copper and aluminum are two of the electrical conductor materials in common use. When electric current (amperage) is caused to flow in an electrical circuit, the passage of the tremendous numbers of electrons causes the atoms of the conductor material to be disturbed. In effect, the atoms of the conductor are speeded up in their random motion. They are said to be **excited**; that is, **the atoms have increased motion and increased energy.**

The greater energy of the atoms causes the temperature of the conductor material to rise. As has been stated, atoms that are excited move faster, and therefore have greater energy, than when they are in the normal state. It is a fact that the better the electrical conductivity of the material, the lower the electrical resistance of that material to electric current flow. Of course, if the conductor is large in cross sectional area, measured in circular mils (thousandths), there is less possibility of causing excessive movement of the atoms in the conductor material and so less heat is generated.

We have mentioned electrical resistance in conductors and the fact that every conductor has some measure of resistance. What we have not determined is how electrical resistance in a conductor is measured.

The unit of measure for electrical resistance is the ohm.

The electrical resistance of conductor materials is a constant factor based on the conductor material type, its temperature while working remaining fairly constant, and its cross sectional area as measured in circular mils.

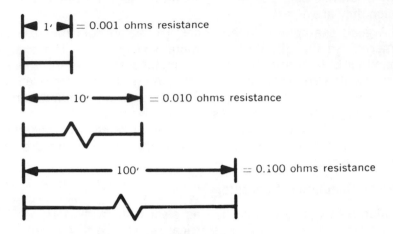

1' = 0.001 ohms resistance

10' = 0.010 ohms resistance

100' = 0.100 ohms resistance

Figure 68. ELECTRICAL RESISTANCE FACTORS.

In the illustration, Figure 68, you can see that if a specific size electrical conductor has 0.001 ohms resistance per linear foot, a ten foot length of the same electrical conductor would have 0.010 ohms resistance (10 times as much). A hundred foot length of the same conductor material would have 0.100 ohms resistance (100 times as much). In all cases the electrical resistance of an electrical conductor will remain the same for a given cross sectional area, given length and specific temperature of the conductor.

Volts, Amperes And Ohms.

In working with electrical power it is necessary to know some basic electrical laws. One of the most important to know and understand is Ohm's Law.

Ohm's Law may be defined as follows:

Ohm's Law = "In any electrical circuit the current flow, in amperes, is directly proportional to the circuit voltage applied, and inversely proportional to the circuit resistance".

Using electrical symbols, Ohm's Law may be stated:

$$E = IR \text{ or, } I = \frac{E}{R} \text{ or, } R = \frac{E}{I}$$ where E = Volts

I = Amperes

R = Resistance (Ohms)

Some basic applications of Ohm's Law will be apparent as this discussion progresses.

Electrical charges are termed positive and negative. Keeping in mind once again that like charges repel and unlike charges attract, it is logical that work must be done and energy expended to create a concentration of negative electrons in one place. The potential energy expended is called **voltage**. It is an interesting fact that voltage is always measured between two conductors. The measurement taken is the potential difference in value between the two conductors. The value that is measured is called voltage. Other names used to describe voltage are "electrical pressure", "electro-motive force" and "electrical potential".

It is important to know that **voltage is an electrical force which does not flow in a conductor** but which causes current flow through electron movement in the conductor. As a matter of fact, voltage is the electrical pressure that forces the electrons in the conductor to move thus creating electric current flow.

The ampere is the unit of electrical rate measurement.

The ampere value indicates the number of electrons flowing past a given point in an electrical conductor at any given time in seconds. **Another name for the ampere is current.** Since electrons move in an electrical conductor, and electrons are electrically charged negative, they will attract, or be attracted to, a positive electrical charge.

Electrical current may be defined as, "**the time rate of charge flow**". To explain this concept, **time** is what it says: the time the current is actually flowing in the electrical conductor in seconds, minutes or hours. The **rate** is how many, and how fast, electrons are moving in the conductor in a given period of time. The **electrical charge** is negative because it is the electrons that are actually moving in the electrical conductor.

The movement of electrons; that is, **the current flow**; is actually **directly proportional to the voltage,** or electrical pressure, in the circuit. For a specific diameter of electrical conductor the higher the voltage applied, the higher the amperage value. This statement, of course, is the first part of Ohm's Law.

When current flows in an electrical conductor there is some electrical resistance to the current flow. **The electrical resistance is mostly due to the reluctance of the electrical conductor atoms to give up their electron particles.** The electrical resistance of a specific type and size of conductor material will remain a constant value if the conductor remains at a constant temperature. The electrical resistance of a conductor is measured in ohms, the practical unit of electrical resistance.

To define the meaning of Ohm:

"The electrical resistance in an electrical conductor that allows the passage of one ampere to flow in a circuit when the impressed voltage is one volt".

In an electrical circuit which has steady, constant voltage applied, the amperage or **current flow is inversely proportional to the electrical resistance in the circuit.** Inversely proportional means that if the circuit resistance is doubled, for example, the current flow in the circuit would be reduced to one-half its original value. This statement is another part of Ohm's Law.

Perhaps an experiment will show the relationship between amperes, volts and resistance. For example, metals are normally in an electrically neutral state because, even in solid form, their electrons tend to be easily removed from the individual atoms. The electrons move in such a manner that a balance of positive and negative electrical charges exists in the metal. The result is an electrically neutral material.

If two metal objects of similar size and material were placed on a wooden table and connected by some type of electrical conductor, there would be little, if any, electron movement between them. What electron movement there might be would be approximately equal in both directions so the result would be **no net movement of electrons** between the two metal objects. For convenience, let us label the two metal objects as "A" and "B".

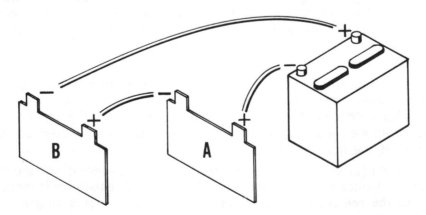

Figure 69. **ELECTRON MOVEMENT CONCEPTS.**

If we connect a battery, which always supplies direct current (dc), to point "A" and complete the electrical circuit by bringing a conductor back to the battery from point "B", we have in effect created an electrical power generator.

There is now the ability to create, and maintain, an excess of electrons at point "A" while having a deficiency of electrons at point "B". In this situation there will be measurable electron movement between the two metal objects; that is, some value of current flow can be recorded. The electron flow would be from point "A" to point "B" since it has been established that electron, and current, flow is from negative to positive.

Remember: **Current, or amperage, is the time rate of charge flow. One ampere equals one coulomb per second.** There is current and time involved in this statement. The **charge flow** is the number of electrons flowing in a conductor at any given second, the electrons being the electrical charge (negative).

In the example, work is being done to maintain the excess of electrons at point "A". When work is being done energy is being expended. The energy is apparent in the circuit as voltage which causes the current to flow in the circuit from points "A" to "B".

The voltage force causes the electrons in the conductor material to move faster; that is, the electrons become excited and therefore have more energy. This energy is partially used to overcome the inherent resistance of the conductor material. The energy expended in this manner appears as heat in the conductor. The temperature of the conductor would increase as the current flows in the circuit.

It would be possible to decrease the electrical resistance of the circuit by using electrical conductors with larger cross sectional area. Conductor size is limited only by the economics of manufacturing the unit.

Some Electrical Energy Concepts.

There are many forms of energy in use today. Some of the **forms of energy** are **mechanical,** as exemplified by power boats, trucks and automobiles; **thermal,** as exemplified by fire; and **electrical,** as exemplified by the welding arc.

Some of the terms used in discussing electrical energy may not be familiar so the following brief definitions are provided. The definitions may, or may not, be complete in all respects.

> **Work** = The transfer of energy from one body to another; the product of force times distance, measured in foot-pounds.

Foot-pound = Normally considered **a unit of work or energy**; the energy required to move one pound of weight through one foot of distance.

Energy = **The ability to do work.**

Watt = The fundamental unit of power; the amount of power required to maintain a current flow of one ampere at an electrical pressure of one volt.
W (watts) = E (volts) × I (amperes). **W = EI**

Power = **The time rate of doing work**; work done in a unit of time; the energy or force available for work; for example, one horsepower equals 33,000 foot-pounds per minute.

Joule = The joule is based on the metric system of measurement; A joule is equal to 10^7 ergs as a **unit of work or energy.**

Erg = The actual work done by one dyne moving through a distance of one centimeter; the unit of work in the cgs system (centimeter-gram-second).

Dyne = The unit of measure of force in the cgs system; the force of one dyne acting on a mass of one gram of substance for one second gives the gram of mass a velocity of one centimeter per second.

These definitions should be studied so that the terms may be familiar to you when you hear them used in welding discussions. In particular, the terms "watt" and "joule" will be used extensively.

What's Watt?

Electrical energy is purchased from the utility company to operate electrical devices in both the home and factory. The consideration is, what are you purchasing?

The watt is the fundamental unit of power. It is the product of **volts times amperes** (dc circuits) as shown in the electrical symbols here:

W = Power in watts.

E = Volts or electrical pressure.

I = Amperes or current.   W = EI (Power equation).

In electrical and mathematical equations, **one watt equals one joule per second. One volt equals one joule per coulomb.** The coulomb is a unit of electrical quantity measuring 6.3 billion billion electrons per second. While the joule is a unit of work or energy based on the metric system, it is similar but not equal to the foot-pound.

You will recall that a horsepower equals 33,000 foot-pounds. In electrical measurement, one horsepower equals 746 watts. The only difference is that one measurement is mechanical while the other is electrical.

One of the metric terms which is in common use in the electrical industry is "kilo". A **kilo** is 1,000 units of something. For example, a kilowatt-hour would be 1,000 watt-hours of electrical power.

If you examine the electric bill you receive from the utility company you will see they have charged you for so many kilowatt-hours. It is logical that a kilowatt-hour would equal 1,000 watt-hours. But what is a watt-hour?

To define: Watt-hour = A commonly used unit of electrical energy; the product of **the average power in watts $\times$ the time, in hours, during which the power is maintained.**

It should be noted that circuit voltage is considered to be a constant factor. For example, electrical circuits in homes are usually 115 volts for operating most appliances, lamps, televisions and so on. There may be 230 volt circuits for operating kitchen ranges, clothes dryers and air conditioners.

The amount of amperage drawn from the utility company pri-mary electrical system depends on the watts the appliance may re-quire to operate. To determine the amperage drawn by the appliance, all you have to do is **divide the total watts** shown on the electrical appliance nameplate by the **primary voltage** for which it is designed to operate. The answer will tell you **how many amperes** are required for proper operation.

For example, an electric clock is rated to operate on 115 volts at 2.5 watts. To obtain the amperage drawn and used by the clock, divide 2.5 watts by 115 volts (Remember: volts times amperes equals watts).

115 (volts) $\overline{)\,2.5\ \text{(watts)}}$ = 0.0217391304 amperes.
In this case, we can settle for 0.02 amperes!

Based on some of the data previously presented we will now break down the watt-hour into energy units. Here is how it is done:

a) One hour = 3,600 seconds or (60 minutes × 60 seconds).

b) One watt = $\dfrac{\text{joule}}{\text{second}}$ or (one joule per second).

c) **One watt-hour =**

= one watt = $\dfrac{3,600 \text{ seconds}}{1}$ or (change hour to seconds).

= One × $\dfrac{\text{joule}}{\text{second}}$ × $\dfrac{3,600 \text{ seconds}}{1}$ (change watt to joules per second).

= One × $\dfrac{\text{joule}}{\cancel{\text{second}}}$ × $\dfrac{3,600 \cancel{\text{seconds}}}{1}$ (cancel the seconds).

= One × joule × 3,600 (multiply)

= 3,600 joules (result)

The joule is a unit of work or energy based on the metric system as previously defined. This is the energy the electric utility company supplies to home and factories. This is the energy they charge for.

The kilowatt-hour is simply a convenient manner in which to write the energy amount (1,000 watt-hours) without all the zeros. In the example discussed above, the answer (3,600 joules) would be multiplied by 1,000 to obtain the energy in a kilowatt-hour. The answer would be 3,600,000 joules.

In welding it is common to discuss the "joules per inch" of weld deposit. The calculation concerns the energy input to the weld joint. It is possible to associate joint penetration, heat-affected zone width, and even weld deposit microstructures, to the energy input to the weld; the "joules per inch".

Electric Energy For Welding.

In electric arc welding, electrical energy is converted into heat and light. This is accomplished by creating an electric arc between the welding electrode tip and the base metal being welded. The electrode may be consumable, as in Shielded Metal Arc welding and Gas Metal Arc welding, or it may be non-consumable as in the Gas Tungsten Arc welding process. The various types of electrodes used in arc welding will be considered later in the chapters about specific welding processes.

There are just two basic values in a welding circuit. They are **voltage** (electrical pressure) and **amperage** (current). Voltage is the force, the electrical potential, that causes current to flow in a circuit. **Voltage is a force which does not flow in an electrical circuit.**

In the welding arc it is voltage that literally "pushes" the welding amperage across the air space between the electrode and the base metal to be welded. Air is a good electrical resistor or, conversely, a poor electrical conductor of current. Let us examine the effect of an air space between the welding electrode tip end and the base metal to be welded.

Welding arc length is the physical distance from the electrode tip end to the base metal. Arc voltage is normally correlated to arc length. (We are concerned here **only** with the electrical fundamentals of the welding arc, not the welding process or procedure. These subjects will be discussed under their specific chapter headings later in this text).

In Figure 70, three different arc lengths are shown. Consider that a welding arc may be maintained in all three instances. For this example there are just two factors involved. They are:

1. The electrical resistance of the air space between the electrode tip and the base metal workpiece.

2. The arc voltage, or electrical pressure, required to overcome the electrical resistance of the air space so that welding current can flow across the intervening space.

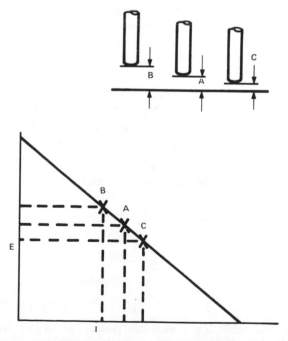

Figure 70. CORRELATING ARC LENGTH AND ARC VOLTAGE.

At point "A" in the illustration the arc length requires some value of voltage to overcome the electrical resistance of the air space thus permitting welding current to flow. The specific voltage is of no particular interest in this example.

At point "B" the arc length is greater. There is more physical air space, and therefore more electrical resistance, between the electrode tip and the base metal workpiece. It will require more voltage to push the welding current across the additional air space to create and maintain the welding arc.

At point "C" of the illustration the arc length is less than at either "A" or "B". There is less physical space across the arc—less air space and, therefore, less electrical resistance. The arc voltage required to maintain the welding arc would be decreased since only the voltage necessary to overcome the electrical resistance of the air space, and maintain the flow of welding current, is required.

It becomes apparent that a welding operator has the opportunity to slightly change the welding amperage and voltage in the welding arc by simply adjusting the arc length. This is shown in the volt-ampere curve in Figure 70.

Resistance Heating In Welding Circuits.

When using a welding power source of any kind there is some value of amperage produced. The amperage is carried to the welding arc through the metal electrical conductors. The amount of heat energy actually produced in the welding arc depends on the current value in the circuit and the time the current flows in the welding arc.

All electrical conductors have some measure of electrical resistance. The amount of electrical resistance, and therefore the power loss resulting from circuit resistance, can easily be calculated.

Remember: The watt is the fundamental unit of electrical power.

When heat is generated in an electrical circuit, including a welding arc, the rate of heat production is measured in watts. Let us examine the equation for calculation of electrical circuit power losses.

Ohm's Law states that volts (E) equals amperes (I) times the circuit resistance (R). In electrical symbols it reads:

$$E = IR \qquad \text{where} \qquad E = \text{volts}$$
$$I = \text{amperes}$$
$$R = \text{circuit resistance in ohms.}$$

If we now consider the electrical **power** equation we find that watts (W) equals volts (E) times amperes (I). In electrical symbols the equation would read:

$$W = EI \qquad \text{where} \qquad W = \text{power in watts}$$
$$I = \text{amperes}$$
$$E = \text{volts}$$

If we keep in mind **both** the equation of Ohm's Law and the power equation for watts it is possible to create a formula for finding the power losses in an electrical circuit. Here is how it is done:

1. $W = EI$ (Watts = Volts × Amperes).
2. From Ohm's Law ($E = IR$) substitute IR for E. We now have:
3. $W = IRI$ (Watts = amperes × resistance × amperes).
4. $W = I^2R$ (Watts = Amperes² × circuit resistance).

Anytime there is current flowing in an electrical circuit there is some measure of electrical resistance. When current flows through an electrical resistance there is energy released in the form of heat. The rate of heat production is measured in watts.

The amount of current used in the welding circuit is determined by the specific welding procedure. It is important to keep the circuit electrical resistance as low as possible. This will prevent excessive heat losses with possible damage to the conductor insulation. Of course, the larger the conductor cross sectional area for a given current value, the lower the electrical resistance of the conductor and the less the amount of heat generated in the conductor.

The power loss formula ($W = I^2R$) shows amperage as a mathematical squared term. It is a simple matter to plug in the known values in a given situation and determine the power losses in the circuit.

$$I^2 = \text{circuit amperage} \times \text{circuit amperage.}$$
$$R = \text{circuit resistance in ohms.}$$
$$W = \text{circuit power loss in watts.}$$

If the circuit amperage is doubled ($I \times 2$) in a specific size of electrical conductor the heat generated will be four times greater ($I^2 \times 2$).

Electric Power Generation.

The fact that electrons flow in an electrical conductor has been well established. The logical question now is, "Where does this movement of electrons start?" In this section of the text we will consider the generation of electrical power for all uses. To accomplish this goal it is necessary to begin at the power generation plant.

Hydro-Electric Systems.

The concept of the hydro-electric power generating plant has been known for many years. Great dams have been built in many parts of the world to harness the water power of major rivers for electric power generation. One of the first and most famous is Grand Coulee Dam in Washington State. Hoover Dam in Nevada, the Aswan Dam in Egypt, and others too numerous to mention provide the same type of capability and perform the same basic function. All hydro-electric power plants operate on the same fundamental principles.

The river is dammed and a huge lake is created. The stored water has **latent energy** simply due to its mass. The energy is released under controlled conditions as the water is permitted to flow through the dam penstocks. The force of the moving water causes the generator turbines to turn and electrical power is generated.

Smaller dams are sometimes used where there are no major rivers for the job. The water is usually channeled from a lake behind the small dam, through aquaducts to large tapered conduits called penstocks. The penstocks are of large diameter at the top, or input, end. They taper to something near half their original diameter at the bottom, or output, end. This type of penstock is usually inclined from top to bottom down the side of a hill or mountain. The pressure created by the column of water in the penstock is very high.

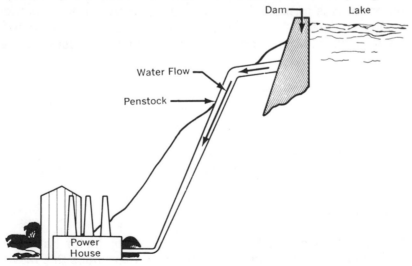

Figure 71. HYDRO-ELECTRIC PLANT.

From the bottom of the penstock the water is directed into the generating plant where it is piped to a water wheel. The water wheel is designed with fixed position "stirrup cups" to catch the water flow. The tremendous pressure of the water column in the penstock causes the water wheel to turn. The water wheel is normally directly coupled to the turbine generators where the electrical power is generated.

Fossil Fuel Power Plants.

Fossil fueled steam generating plants have been used in many parts of the world where water power is not available. Some of the fossil fuels used include natural gas, oil, coal, manufactured gas and other energy-releasing fuels. The energy of the fossil fuels is released through burning, or combustion. The heat of the combustion brings water in the steam boilers to the temperature of super-heated steam. The steam pressure is then applied to turbines in a manner similar to the water used in hydro-electric plants. The actual electric power is generated by the turbines and generators.

Nuclear Energy Power Plants.

The use of nuclear energy for generating electrical power is much the same as in fossil fueled power plants. The main difference is in the type of fuel energy used.

Nuclear energy power plants have been built in many parts of the world. At first the cost of electrical power from nuclear energy plants was considered prohibitive but in recent years the costs have dropped sharply. Nuclear generated electrical power is now able to compete with most other types of electrical power on a reasonable cost of production basis.

There has been some concern about the effect of nuclear energy power plants on the environment of the world. At this writing there does not appear to be any danger from nuclear power plant operations. In fact, nuclear energy power plants offer less waste into the earth's atmosphere than most fossil fueled power plant operations.

Total Electrical Power (KVA).

Electrical power is generated at a pre-determined **specific voltage and amperage.** The generated amperage is usually too high in numerical value for economical power transfer over any great distances. The generated voltage is, therefore, **stepped up** by means of transformers to very high numerical values for transmission purposes.

Stepping up the voltage causes the amperage to decrease proportionately to a numerical value that can be transmitted over great distances without excessive power losses in the system. Remember, it is amperage that flows in an electrical circuit and conductor. It is voltage that is the force which causes amperage to flow.

Multiplying the generated voltage by the amperage results in a value called **volt-amperes.** Dividing the number of volt-amperes by 1,000 provides an answer in kilovolt-amperes (KVA). **KVA is considered to mean total power.**

For a specific KVA value the numerical product of volts times amperes must always equal the same calculated result. This can be illustrated in the following example.

Consider the amount of 10 KVA. This value may be arrived at in several ways:

10 KVA = 10,000 volt-amperes, or
10 KVA = 10,000 volts × one ampere, or
10 KVA = 10,000 amperes × one volt, or
10 KVA = 100 volts × 100 amperes.

It is evident that any numerical combination of volts and amperes that equal 10,000 volt-amperes (10 KVA) may be used.

This data is useful for considering electrical transmissions when it is understood that **amperage flows** in a conductor while **voltage is the force,** or electrical pressure, that causes it to flow. **Voltage does not flow in an electrical conductor.** Since every electrical conductor has some electrical resistance to amperage flow, the smaller the **amperage value** in the circuit, the lower the **power losses** in the electrical circuit. The lower the **power losses** in the circuit, the lower the **heat energy value, in watts,** dissipated in the conductor. The lower heat losses make the total electrical system more efficient for transferring of electrical power.

For a total amount of generated electrical power (KVA), as the voltage is increased, the amperage must proportionately decrease. By stepping the generated voltage **up** to a relatively high value, the generated amperage is reduced to a proportionately low value. It is logical that lower amperage values would permit the use of smaller diameter conductor wires.

Cables used for high voltage transmission lines are normally a composite material structure. For example, a stranded steel core cable may be overwound with various sizes of copper and aluminum conductor wire. This composition of cable provides a high voltage transmission cable having good electrical conductivity with sufficient strength to support its own weight under any reasonable operating conditions. Very often the resulting transmission cable will have an outside diameter of approximately 2 1/8" to 2 1/2".

Electrical Power Transmission.

Electrical power is created by generation at the power plant, transformed to high voltages and carried by transmission cables to its destination. Previous data has shown that it is the movement of electrons in the conductor that constitutes current flow. The electrons actually move relatively short distances from atom to atom in the conductor. It is important to know that **the electron that moves at one end of the conductor is not the same electron that moves at the other end of the same conductor.** Electrical current moves at approximately the speed of light (186,000 miles per second). In some conductors there is a delay in current movement compared to its speed in free space. For all practical purposes it may be assumed that electrical power applied to a conductor will be apparent instantaneously all through the conductor.

At the receiving end of the transmission line the electrical power is transformed again. This time the voltage is decreased through transformers until a reasonable distribution voltage is reached. This is usually 4,160 volts. As the voltage is decreased the amperage capability is raised proportionately.

The electrical power brought into the shop or plant comes from a line transformer normally located near the shop. In some major manufacturing plants, where large amounts of electrical power are used, an electrical sub-station may be required to supply sufficient electrical power for the plant operations.

Special recording meters are used to measure the amount of electrical power used by a customer. As we know, electricity is not used up when current flows and does work. What is paid for is an amount of electrical energy made available to us for use. The only charge the consumer pays is for the **actual electrical energy used.** This is shown as kilowatt-hours on the electric bill. As long as the total electrical system from power generation to power use is maintained and functioning electrical power will be available.

In a sense, a practical consideration of electrical power indicates it is basically nothing more than the excitation of atoms in an electrical conductor which causes the electrons to move faster in the conductor. The greater the input force or **voltage,** the more electrons that may be excited to movement. It is apparent that voltage in the primary supply system must be maintained at a constant level if there is to be good strong supply and regulation of primary amperage.

To provide some additional information which will be used and discussed in later chapters, the following electrical data is offered. In this instance, **the data relates specifically to welding power sources.**

KVA, Primary KW And Secondary KW.

The **total power** demanded from the utility company is termed **kilovolt-amperes**. It is calculated by multiplying volts times amperes and dividing the result by 1,000. KVA includes the power losses dissipated in the plant electrical system as well as the power used in the plant. KVA is sometimes referred to as "apparent power".

Remember: The watt is the fundamental unit of electrical power.

From previous discussion we know that volts times amperes equals watts. Kilowatts are calculated by multiplying volts times amperes and dividing by 1,000. This applies to dc circuits. (In ac circuits the RMS value must also be considered in the calculation). This appears to be the same mathematical formula used for determining KVA. It is. Why then have two different names for what is apparently the same thing?

KVA has been defined in some of the preceeding paragraphs. To properly discuss KW we must specify if it is **primary KW** or **secondary KW** we are talking about. Then this whole thing starts to make a little practical sense!

For welding power sources, **primary kilowatts are the actual power used by the unit when it is producing, and putting out, its rated load.** (Rated load, often called rated output, is the amperage and load voltage the welding power source is designed to produce for a given specific duty cycle period. For example, a typical constant current welding power source may be designed to furnish 300 amperes at 32 load volts, 60% duty cycle).

Secondary kilowatts are the actual power output of the welding power source. In the example cited above, the calculation for KW would be 300 amperes times 32 load volts, divided by 1,000. The result is 9.6 KW secondary.

Summary.

The reasons for using electrical power for welding metals are relatively few in number and basic in concept. Electric power is a reasonably inexpensive source of great energy. Most important, electric power can create the arc with sufficient temperature for melting metals. Electric power is easily controlled and directed for welding applications.

Last, but not least, electrical power is readily available in most areas of the world. When it isn't available from power generating plants, it can be generated with the portable welding power sources and generators now being manufactured.

In summarizing this chapter, it is well to remember that there must be some form of energy to start with before electrical power can be generated. Water power uses the physical mass and weight of the water to gain the pressure necessary for electrical power generation. Other power generating plants use the latent energy of fossil fuels or nuclear energy as the source of power to generate what we know as electricity. If you think about it, all we are doing is exchanging one type of energy for another type. But look at the results!

Chapter 5

WELDING POWER SOURCE TRANSFORMERS

There are several types of transformers in general use in industry today. Some are called "power transformers" and are used for primary power circuits. Others, such as those used in welding power sources, are smaller in physical size and have specific characteristics that especially suit them for an individual application.

The function of any transformer is to change electrical power from one voltage to another, with an inverse ratio of amperage change, **without changing frequency**. The voltage at which electrical power is used depends on the power requirements of the specific application. For example, most household requirements demand 115 volt and 230 volt service. Industrial users, on the other hand, may require 208 volt, 230 volt, 380 volt, 460 volt or 575 volt power. Certainly there would be no need for **all** the voltages mentioned at any one plant. Normally a plant will require 115 volt power with one other higher voltage for operating industrial machines.

Any appreciable amount of power transmitted even a few miles at such relatively low voltages, with attendant high amperages, would take an enormous amount of conductor material, either copper or aluminum, in the transmission lines. The cost of providing electrical service would be prohibitive.

Power in alternating current circuits depends on the **voltage** between the conductors, the **amperage** flowing in the conducting circuit and the **power factor** of the overall system.

Voltage is always measured between two conductors. It is the potential difference in value between the two conductors. The value measured is called voltage, or potential. The two terms are synonymous in their meaning when discussing electrical power. Voltage is the electrical force, or pressure, that causes amperage (current) to flow in electrical conductors.

As we have discovered, current flow is simply the movement of electrons in, and through, an electrical conductor. Actually, the electrons are sub-atomic particles of the atoms of the conductor material. When voltage is applied to an electrical conductor, force (electrical pressure) causes some of the electrons to leave their atoms. Literally billions of billions of electrons are moving in an electrical conductor when current is flowing in a circuit.

The use of large power transformers makes it possible to generate power at any convenient voltage, step it up to an economical transmission voltage, and then step it back down to the desired level of voltage at the use end of the transmission system.

For instance, at Hoover Dam in Nevada, power is generated at 13,800 volts. This is stepped up by means of transformers to 287,000 volts for transmission approximately 275 miles to the city of Los Angeles. At the receiving end of the line the voltage is stepped down through transformers to 132,000 volts for delivery to five main receiving stations. At this point, the voltage is again stepped down, this time to 34,000 volts, for feeding distribution sub-stations located around the city. At the sub-stations the voltage is further reduced to 4,160 volts for local city distribution.

The 4,160 volt lines feed pole, or line, transformers in various neighborhoods. These transformers make the final reduction in voltage to 115 volts, 230 volts or 460 volts, as required by the customer.

The power passes through five sets of transformers after it leaves the generators at Hoover Dam and before it reaches the ultimate customer. The total losses in each transformer bank are approximately five percent at rated load. The low loss percentage would indicate the line transformer is a very efficient piece of electrical equipment.

Welding Transformer Fundamentals.

Welding transformers apply the concept that current flowing in a coil wrapped around an iron core of some type and configuration will convert the iron into a magnet and that, if the magnet iron is inserted into another coil of conductor wire, an ammeter connected to the output terminals of the second coil will be actuated when the circuit is completed in that coil.

There is the electrical phenomena of two copper or aluminum circuits carrying electrical power connected by an iron magnetic circuit. The iron circuit, called the transformer core, is made of **thin gauge sheet steel laminations, insulated on both sides** with a film type insulation to prevent the formation of eddy currents in the iron. The steel core material is usually either a high silicon steel or a specially rolled electrical steel that has preferred grain size and grain orientation.

The two copper or aluminum electrical circuits may be entirely independent of each other or they may have some of their turns in common as in an auto-transformer. Illustrations of both types of circuits are shown in the circuit diagrams in Figure 72.

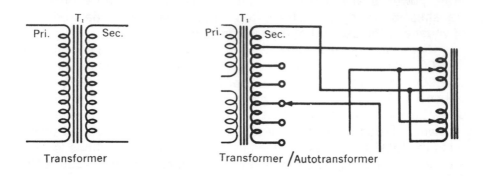

Transformer   Transformer /Autotransformer

Figure 72. TWO WINDING TRANSFORMER AND AUTO-TRANSFORMER.

When an alternating current **voltage** is applied to the terminals of one of the conductor circuits (the "primary"), and the circuit is completed, alternating current will flow. The magnetic field set up in the iron core of the transformer depends on the primary current for **direction and strength.** As the primary alternating current changes direction each half-cycle so must the magnetic lines of force of the magnetic field alternate their lines of direction each half-cycle. Since energy is used to create the magnetic field, but not dissipated or "used up", it must be momentarily stored in the magnetic field. Each half-cycle, as the alternating current decreases in electrical value, so the magnetic field decreases in strength. The energy in the magnetic field must go somewhere—and it does. The energy is returned to the welding power source circuit but not to the primary coil of the transformer. The energy that is stored in the magnetic field is in-duced into the secondary coil, and the total secondary circuit, as voltage. When the secondary circuit is completed, as in striking a welding arc, electric current will flow. If the secondary circuit is not completed, the induced energy will be apparent at the output termi-nals of the power source as **open circuit voltage**.

In some of the discussions the term "winding" has been used. A "winding" is simply an electrical coil. In fact, the terms "coil" and "winding" are synonymous in electrical usage.

In the simple two-winding transformer illustrated in Figure 73 there is shown a useful **theoretical** electrical rule. It is stated as follows, "Theoretically, the ratio of the primary volts to secondary volts in a transformer will be the same as the ratio of primary electrical turns to secondary electrical turns". Remember: an **electrical turn** is one wrap of the conductor wire around the outside of the total coil.

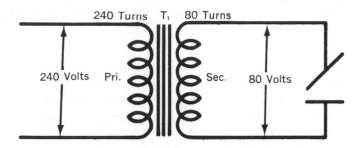

Figure 73. SIMPLE TWO-WINDING TRANSFORMER DIAGRAM.

There is a second part of the rule just proposed. It is stated as, "The ratio of primary amperes to secondary amperes is **inversely** proportional to the ratio of primary to secondary volts (or turns)". This is illustrated in Figure 74.

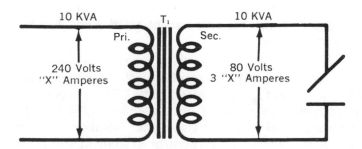

Figure 74. AMPERES AND VOLTS RATIOS, SIMPLE TRANSFORMER.

When amperage flows in an electrical conductor, energy is being expended and heat is being generated, both in the coils and the iron core of the transformer. Too much heat will cause deterioration and physical breakdown of the insulation on the coil wire and the iron core material. This actually presents a major problem in the engineering design of the welding power source. To decrease heat buildup, it is necessary to provide the largest diameter conductor possible for the circuit service. This always poses a cost problem from the standpoint of what you must obtain as a fair market price and its relationship to how much money you have to spend to produce the product item.

The insulation used in welding transformer cores and coils must have a sufficiently high thermal capacity to withstand the operating temperatures of the unit. The insulation must also be able to withstand the heat which may be caused by high transient voltages. High transient, or momentary, voltages are considered to last no more than a few milli-seconds.

The steels used in welding power source transformer cores have high **magnetic permeability** characteristics. "Magnetic permeability" means the ability of a material to accept large quantities of magnetic lines of force. Iron which has high magnetic permeability provides better magnetic and inductive coupling of the primary and secondary coils of the main transformer.

Welding transformer type power sources usually have "step-down" transformers. In this instance, the primary voltage is a rather high value; certainly much too high for safe welding operations. It is stepped down through the power source transformer to a usable welding voltage.

Of course, there are some special applications and process requirements where the power source may be a "step-up" transformer unit. Such applications are usually not for normal commercial welding operations with the possible exception of some types of plasma welding power sources.

Another type of transformer used in some welding power source circuits is called the "isolation transformer". This type of transformer usually has a 1:1 voltage, and turns, ratio. The purpose of an isolation transformer is to isolate a circuit where some voltage value is required for a special purpose. Usually an isolation transformer provides voltage for some type of control system in the power source circuitry.

To properly consider the step-down welding power source transformer we must realize that relatively high voltage, and relatively low amperage, is brought into the primary circuit of the power source. The primary volts times the primary amperes used by the power source result in watts, or more likely, kilowatts. **Primary kilowatts are the actual power used by the welding power source transformer to put out its rated amperage at rated load voltage.**

As illustrated in Figure 74, transforming the primary power through the transformer lowers the secondary voltage (which is called the **open circuit voltage**). If the relatively small power losses are discounted (as in a theoretically perfect transformer) the volts times amperes calculation on the secondary side of the main transformer should equal the same kilowatt value as the primary side of the circuit. Since we know that voltage is lower on the secondary side of the main transformer it is logical that the secondary amperage capability of the transformer **must** be proportionally higher than the primary amperage. Remember: **volts times amperes equals watts.**

The electrical rule of thumb is, **"For a given amount of KVA (kilovolt-amperes), as the voltage is increased the amperage must decrease proportionately. Conversely, as the amperage value increases, the voltage value must decrease proportionately".** It must be remembered, of course, that the **rule only applies when a specific KVA value is used. It does not apply,** for example, to the output volt-ampere curve of welding power sources.

In summary of this section on welding power source transformer characteristics it is evident that a transformer is a relatively simple electrical device. Transformers are designed to perform specific functions for designated purposes. It should be noted that a welding power source transformer is always an inductive electrical device. This simply means that welding power is induced into the secondary circuit of a power source through magnetic action rather than by direct electrical connection of the coil conductors.

Transformer Coil Characteristics.

Transformer coils are made of conductor materials, normally called "magnet wire", which are wound to the necessary coil configuration or shape. Primary alternating current voltage is impressed on the primary circuit of the power source. Since voltage is the electro-motive force, or electrical pressure, that causes current to flow, there is current moving in the circuit. The value of the current that any conductor can carry is limited by the cross-sectional area of the conductor as measured in circular mils.

We learned in the chapter Electrical Fundamentals that every electrical conductor has a certain amount of electrical resistance; that is, resistance to the flow of current in the circuit. When current flows in a circuit, work is being done to overcome the electrical resistance with the result that energy is being expended. The expended energy is apparent as heat in the electrical conductor. This indicates some power loss, in watts, in the electrical conductor. The power losses may easily be calculated by using the power loss formula discussed previously. It is:

$I^2R = P$ where I = circuit amperage

R = circuit resistance

P = circuit power losses in watts.

In welding power sources the losses of power are apparent as heat in the circuit conductors. It is important to any electrical circuit that adequate cooling of the circuit conductors be achieved. The reasons for keeping the electrical conductors as cool as possible relates to both electrical and metallurgical factors. All metals used as conductors in electrical circuits will have an increase in electrical resistance when there is an increase in their operating temperatures. Conversely, there will be a decrease in electrical resistance when there is a decrease in the temperature of the conductor.

It is necessary to maintain the circuit conductors at a temperature which will permit the most effective, and practical, electrical efficiency to be achieved. The electrical efficiency of any device decreases as power losses, and conductor temperatures, increase. By using larger diameter conductors to carry a specific amount of maximum current the electrical "friction" (resistance) of the conductor is reduced. The power losses, dissipated as heat, will also be reduced. It is a fact that the electrical efficiency of an electrical circuit will be improved when power losses are limited to a low value.

Eddy Currents In Conductors.

By definition, an eddy current is a contrary, or circulating, current running against the general flow of current. It could be called a counter-current. In welding power sources, eddy currents are usually induced currents. They would normally appear in the iron cores of power source main transformers or reactors but they can appear in electrical conductors.

In an alternating current magnetic field, eddy currents may be set up in the body of a conductor, increasing the **effective resistance** of the conductor. Eddy currents move on an axis ninety degrees to the longitudinal axis (direction of force) of the magnetic field.

The presence of eddy currents, and their movement, will cause electrical energy to be expended as heat. As the electrical conductor increases in temperature the electrical resistance also increases. The **amount** of electrical resistance increase will depend on the conductor cross-sectional area in circular mils, the frequency of the alternating current and the number of ampere-turns effective in the coil. Conductors used for carrying large amounts of amperage have larger cross-sectional areas than those conductors designed for carrying lower current values.

Eddy currents are minimized in welding cables by using a number of small insulated conductor wires in parallel. Usually the small diameter conductors are twisted at regular intervals so that each conductor has a changing physical position with relation to the other conductors in the group. The twisting of the conductors reduces the possibility of developing circulating currents otherwise caused by different induced voltages, and resulting in different electrical resistance, in the conductor wires.

Eddy current flow would be within the conductor and is not directly reduced by the transposition of the conductor wires. **It is reduced by eliminating the circulating currents.** The reduction of eddy currents will decrease the circuit losses expended as heat and will, of course, improve the electrical efficiency of the welding power source circuitry.

Transformer Iron Core Data.

Any iron core that is magnetized with alternating current must have continuous primary power supplied to make up the core losses and to maintain the magnetic field strength. The input kilovolt-amperes (KVA) required for a specific iron core size, or mass, depends on the alternating current cycling frequency, the permeability of the iron core material, and the magnetic density at which the iron core material functions.

Iron cores used in welding power source transformers have many different shapes or configurations. Some of the common core types are shown in Figure 75.

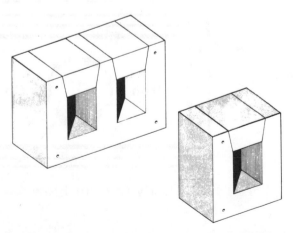

Figure 75. SOME TYPICAL IRON CORE SHAPES.

Iron Core Losses.

Energy losses in welding transformer iron cores usually are caused by two detrimental conditions: (1) eddy current losses and, (2) hysteresis losses.

Eddy currents are caused by induced currents in iron cores. Losses from eddy currents will vary as the mathematical square of the iron core lamination thickness varies. For example, if a given thickness of iron core laminations produced an eddy current loss of 100 watts, reducing the iron core lamination thickness to **one-half** its original value will reduce the eddy current loss to **one-fourth** its original value. The equation for calculating eddy current loss is:

$$\frac{E^2}{R} = \text{Eddy current losses; where } E = \text{circuit voltage}$$

$$R = \text{circuit resistance.}$$

Eddy current losses are almost eliminated in welding transformer power sources by using thin gauge laminations of sheet steel which is insulated on both sides with a film insulation. The possibility of forming eddy currents in material cross-sections of this thickness are almost non-existent.

The formation of eddy currents is shown in Figure 76. Note that the drawing shows a solid iron core rather than a laminated core. Such a solid iron core would not be usable for a welding power source transformer core.

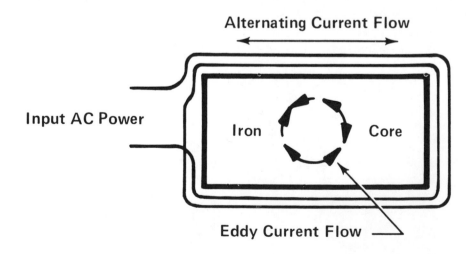

Figure 76. EDDY CURRENT FORMATION IN AN IRON CORE.

The direction of flow of the primary alternating current in the primary coil is on the same geometric plane as the flow of the induced eddy current. **The magnetic field lines of force are always at right angles to the current flow.** The results of the arrangement shown would be an induction heater rather than a welding power source. There would be no welding output from a welding power source designed as this illustration shows. Within a matter of minutes, as a matter of fact, there would be catastrophic deterioration of the circuit insulation due to the induced current and subsequent heating of the circuit components. It would be noticeable by the smoke coming from the power source as it burns.

Hysteresis Losses.

In welding power source transformer iron cores, hysteresis means the loss of energy due to the cycling of alternating current power. As the polarity (current flow direction) changes each 1/120th of a second, for 60 hertz power, so must the polarity of the molecules of the iron core material change with the alternation of direction of the magnetic lines of force. Again, work is being done and energy is being expended. The expended energy is dissipated as heat in the transformer iron core.

Hysteresis losses may be reduced by the use of transformer core steel having a relatively high silicon content. The element silicon helps to increase the magnetic permeability of the steel core material. The increase in magnetic permeability reduces the energy losses due to hysteresis. In a sense, hysteresis may be defined as "magnetic friction" or "magnetic flux inertia".

New technology in metals manufacturing has succeeded in producing materials classed as electrical iron. Modern techniques of rolling steel are used to control the metal grain size and grain orientation. Steels with controlled grain orientation will have low hysteresis loss characteristics without requiring the high silicon content of other types of iron core materials.

Air Gaps In Iron Cores.

It is well to keep in mind that any air gap between the transformer core laminations will make it necessary to apply much more magnetizing current in order for the transformer to function correctly. Air acts as a very effective insulator for both electrical current flow and magnetic lines of force in a magnetic field. **Air has much lower magnetic permeability than iron core material used in welding power sources.** This helps to explain why the sheet steel laminations of welding power source transformer cores are fitted extremely close togther. This would not necessarily apply to other iron electrical devices in welding power sources such as control reactors, current stabilizers, inductors and so on. These component parts may be designed to have a specific air gap so they will help produce a certain arc welding characteristic.

Welding Power Source Efficiency.

The calculation of welding power source efficiency is based on the **primary kilowatts used** with relation to the **secondary kilowatts produced** as output to the welding arc. Constant current, or conventional, transformer type power sources are purposely designed to be relatively inefficient from the electrical standpoint. The conventional power source could be made much more electrically efficient but it would probably result in extremely poor welding arc characteristics.

Constant current, or conventional, type welding power sources have "loose magnetic coupling" in the primary-secondary relationship of the main transformer. The loose magnetic coupling is accomplished by physically separating the coils with an air space. It is this arrangement of the primary-secondary coils of the main transformer that produces the so-called "drooping" volt-ampere characteristic required for Shielded Metal Arc welding. Figure 77 shows a single range, constant current welding power source volt-ampere curve.

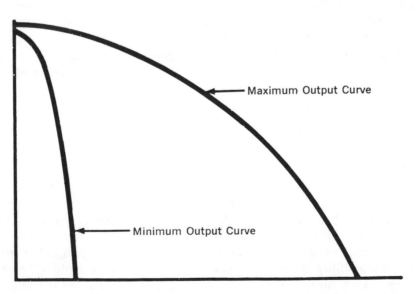

Maximum Output Curve

Minimum Output Curve

Figure 77. CONVENTIONAL POWER SOURCE VOLT-AMPERE CURVE.

The shape of the curve is correctly called a "negative curve" although the term "drooper" seems to be more popular with welding people. The important thing to recognize is that the **maximum short circuit current output of the power source is limited.** This protects the power source in case a dead short-circuit welding condition should occur, such as a welder sticking a large, coated electrode to the base metal workpiece.

There is an example of how welding power source efficiency is calculated. Let us consider a NEMA Class 1 power source having an output rating of 300 amperes, 32 load volts, 60% duty cycle. Keep in mind that this rating is only for one specific power source. It could just as easily be a 400 or 500 ampere rated unit. For the 300 ampere rated power source we multiply the 300 amperes times the 32 load volts and have a result of 9.6 secondary KW.

If the input, or primary, KW is 15.2 kilowatts it is apparent that this is the power required for this power source to put out its rated 300 amperes, 32 load volts.

To determine the electrical efficiency of the welding power source transformer the secondary KW (9.6) is divided by the primary KW (15.2) as shown in the following illustration:

$$\frac{9.6 \ \text{KW (Secondary)}}{15.2 \ \text{KW (Primary)}} = 63\% \ \text{electrical efficiency.}$$

If the electrical efficiency is **too good** in a constant current welding power source the transformer will probabiy have poor arc welding characteristics. It would be a much more efficient piece of electrical equipment, however. It just wouldn't work for welding.

In summary, a welding power source should be designed to provide the best arc welding characteristic possible regardless of the electrical efficiency of the unit.

Chapter 6

POWER FACTOR

Any discussion of electrical distribution invariably leads to the subject of power factor and power factor correction. Discussions of power factor are usually conducted in mathematical and electrical engineering terms that are almost incomprehensible to the average layman in welding. In an effort to make the subject of power factor more easily understood by more people various analogies have been used with some success.

It is the intent of this short chapter to explain power factor and power factor correction in relatively simple electrical terms. To do this properly we will first provide some definitions of terms used in discussing power factor. The terms and definitions are related here to **power factor.** In other usage the same terms may have a different meaning.

Kilo = 1,000 units. (One kilowatt = 1,000 watts).

Primary Kilowatts = The actual primary power used by the welding power source to produce and deliver its rated amperage at rated load voltage (rated load). Usually measured by a wattmeter, primary kilowatts are the portion of total demand power registered on the kilowatt-hour meter. Primary kilowatts are the power that can be metered, and charged for, by the electric utility company.

Primary Kilovolt-Amperes (KVA).

The total demand power that is taken from the primary power system is termed kilovolt-amperes. For single phase power, KVA is determined by multiplying primary volts times primary amperes times the effective (RMS) value. The product is divided by 1,000 to obtain kilovolt-amperes (KVA). The following illustration shows the mathematics involved.

a) Primary Line Voltage
    $\times$ **Primary Amperage**
    = **V-A Product**

c) 1,000) net V-A Product

b) V-A Product
    $\times$ **Effective Value (0.707)**
    = **net V-A Product**
    = **KVA**

For three phase primary power, KVA is determined by multiplying primary volts times primary amperes times the effective (RMS) value times 1.73. The product is divided by 1,000 to obtain kilovolt-amperes (KVA). ($\sqrt{3} = 1.73$).

Vector = Direction and magnitude values based on a stable reference point.

Vector Sum = The sum obtained in vector addition. It may be derived by either mechanical or mathematical means.

Apparent Power = The vector sum of the real power and the kilovars is the apparent power. (Apparent power = KVA).

Kilovars = Kilovars may be either inductive or capacitive. The abbreviated term for kilovars is KVAR. **Inductive KVAR** always has current lagging voltage by up to 90 electrical degrees. This is shown in Figure 78.

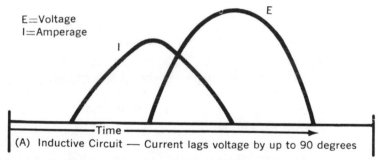

(A) Inductive Circuit — Current lags voltage by up to 90 degrees

Figure 78. INDUCTIVE CIRCUIT KVAR.

Capacitive KVAR is illustrated in Figure 79. Note that current **leads** voltage by 90 electrical degrees.

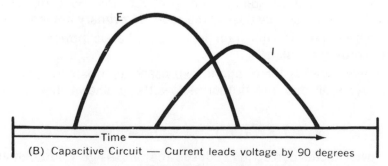

(B) Capacitive Circuit — Current leads voltage by 90 degrees

Figure 79. CAPACITIVE CIRCUIT KVAR.

Kilovars are the reactive, or "wattless", component in electrical systems. The magnetizing current required to produce the magnetic field necessary for the operation of any inductive equipment, such as welding power source transformers, multiplied by the line voltage is termed **inductive KVAR.** Without a small percentage of inductive KVAR, energy could not transform through the iron core of the main transformer. Specific KVAR values are determined by a wattmeter which has a special phasing transformer. This type of calculation is normally done in welding power source manufacturer's development laboratories.

It can be stated that the unit of measure of **real,** or **used, power** is the primary kilowatt. The unit of measure for **total, or apparent, power** is the kilovolt-ampere.

This discussion of power factor relates to its determination and use in the primary circuitry of welding power sources. Although some of the observations made will refer to the affect of power factor on electrical distribution systems they will also apply to shop power loading.

Probably the most asked question concerning power factor and power factor correction is, "Why do we need it?"

There are several good and correct answers to the question, some of which are presented here. Some of the benefits are for the customer-user of electrical power and some are for the benefit of the electric utility company. Some are actually beneficial to **both** the customer and the utility company.

Some of the benefits are as follows:
REDUCED ELECTRICAL SYSTEM LOSSES (Utility and customer).
BETTER PRIMARY VOLTAGE REGULATION (Utility company).
INCREASED ELECTRICAL SYSTEM CAPACITY (Customer).
SMALLER LINE TRANSFORMER (Utility company and customer).
REDUCED PRIMARY AMPERAGE DRAW (Customer).

Of the various reasons stated, probably the most important to the welding power source user is the **reduced primary amperage draw.**

Another popular question is, "Just what is power factor and power factor correction?"

Power factor is the quotient, expressed as a percentage, of kilo-volt-amperes divided into primary kilowatts as shown in the calculation:

$$\text{KVA} \overline{)\ \text{KW (Pri.)}} = \% \text{ Power Factor.}$$

Power factor correction reduces the amperage value demanded from the primary power line. Primary shop wiring carries the total power drawn from the primary distribution system of the utility company. The shop wiring would naturally have to carry **less primary current** when power factor corrected welding power sources are used.

As a general rule, power factor correction capacitors and allied circuitry are found only in single phase based ac and ac/dc welding power sources. Three phase power sources have inherently good power factor and do not require power factor correction.

In the following step-by-step analysis of power factor and power factor correction we will use data taken from published literature for a NEMA Class 1 ac power source. Calculations are based on the published data.

The primary voltage is the voltage supplied by the electric utility company at the shop distribution point. Normally this is a main central control panel and disconnect switch to which the primary line voltage is brought into the shop and from which it is distributed throughout the shop. For the purposes of this discussion the primary voltage is considered to be a constant factor in value. The primary amperage, or current, draw is the current required to operate the welding power source at its rated output. The primary amperage value will vary according to the ampere rating of the welding power source.

The examples shown in the following steps are for a typical heavy duty NEMA Class 1 AC welding power source of the transformer type. The data given is for a non-power factor corrected power source unless otherwise noted in the text. As indicated in the power source specifications shown in Figure 80 the unit operates from single phase primary electrical power.

Primary Voltage	= 230
Primary Amperage	= 106
Primary KVA	= 24.4
Primary KW	= 15.2
Single Phase	= 50/60 Hertz

Figure 80. AC POWER SOURCE SPECIFICATIQNS.

STEP 1.

Primary KVA is determined by multiplying primary voltage times primary amperage and dividing by 1,000 for single phase primary power. This is illustrated below:

Primary Voltage   = 230 volts
Primary Amperage = 106 amperes
Primary volt-amps = 24,400 volt-amperes

Calculate KVA   = 1,000 $\overline{)\,24,000}$ = 24.4
Primary KVA   = 24.4 KVA

STEP 2.

Primary kilowatts (KW) are measured by a wattmeter. The measurement is made while the welding power source is under a simulated load (producing rated amperes at rated load voltage). Primary KW for this power source is 15.2 KW. The primary KW is indicated in the welding power source specifications noted in Figure 80. It is the primary KW, rather than the secondary KW, that is used when calculating the power factor of a welding power source.

STEP 3.

The vector diagram shown in Figure 81 makes use of the data previously developed. Normal industry practice is to correct welding power sources for a nominal 75% power factor at rated load output. Small variations from this value, either plus or minus, will be noted in actual practice due to the increment value of power factor correction capacitors.

The vector diagram, Figure 81, illustrates a method for determining the necessary power factor correction for this NEMA Class 1 AC power source. In the illustration we will use the same numerical measurement value to equal one KVA, one KW or one KVAR. The following data are noted in the vector diagram:

	w/o PFC	w/PFC
Primary Amperage (I)	106	88
Primary Voltage (E)	230	230
Primary KVA	24.4	20.2
Primary KW	15.2	15.2

To summarize this discussion, power factor correction is a method of making an electrical distribution system more efficient. Power factor and power factor correction concerns only the primary power drawn from the primary supply by a welding power source. Secondary output characteristics of a welding power source are not affected by power factor or power factor correction.

Power factor correction of any electrical device, such as a welding power source transformer, reduces the primary amperage draw, and the primary KVA demand, from the utility company supply lines. Since the total current demanded from the primary supply system must be carried through the shop wiring circuits, the conductors used in an electrical system with good power factor correction can usually be smaller in cross sectional area than conductors in a similar system without power factor correction.

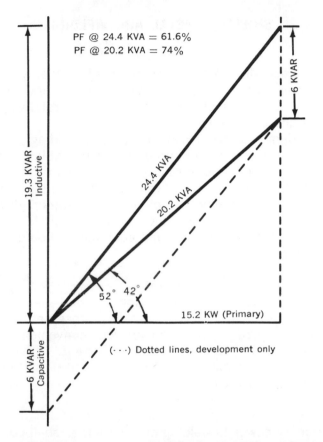

Figure 81. VECTOR DIAGRAM FOR POWER FACTOR DETERMINATION, NEMA CLASS 1 AC POWER SOURCE.

It is significant that, in some areas, the unit cost of electrical power in kilowatt-hours is less where a high average power factor is maintained.

The main advantages of power factor correction are in reduced electrical system losses, increased electrical system capacity, smaller line transformer requirements and reduced primary current demand from the primary power system.

Remember, power factor correction is **not** a line voltage compensator for fluctuating line voltage!

Chapter 7.

SHIELDED METAL ARC WELDING

The **Shielded Metal Arc Welding** process is probably better known as the "stick electrode" process. The equipment and materials necessary for welding with shielded metal arc are relatively simple in construction and easy to use.

The fundamental equipment includes a welding power source (either ac or dc), electrode and ground welding cables, a substantial ground clamp and a properly rated electrode holder. The personal working and protective equipment of the welder will include a welding helmet with properly numbered colored lens, leather gloves, a wire brush, a chipping hammer and such protective clothing as the job calls for. For safety, all clothing and equipment should comply with the OSHA regulations.

Safety.

Much of the safety data noted for shielded metal arc welding will be applicable for other electric welding processes. Some of the fundamental safety rules are noted for your convenience. Although these rules are rather general, they are important. In no case should they be construed as superseding any local plant or statuatory safety rules in effect in your area.

Rule 1.

All primary power to welding power sources shall come through a fused line disconnect switch. **There are no exceptions to this rule!**

Rule 2.

Welding power sources shall be placed in a well ventilated area with fan discharge output areas not closer than 18 inches from the wall. This permits the cooling air to circulate over the power source components and cool them properly.

Rule 3.

All welding cables shall be checked for electrical continuity on a regular basis to insure that no conductor breakage has occured. This may be done easily with an ohm meter and a resistance check.

Rule 4.

All electrical connections shall be clean and tight. In particular, ground clamps shall be of adequate amperage rating for the power source with which they are used. Electrode holders shall be reasonably clean and in good operating order.

Rule 5.

Protective clothing, such as leather gloves, clothing, welding helmets and safety shoes, shall be worn as necessary for the work being performed. Proper colored lens shade shall be used and the clear cover lens shall be kept clear of spatter and other dirt for better visibility.

Rule 6.

All metal in a welding shop should be considered "hot". When you have proven otherwise, then it is all right to handle it. Many men have been severely burned for picking up a piece of "cold" metal in the shop.

Rule 7.

High voltage is used to power welding power sources and other shop electrical equipment. **Always be sure the fused line disconnect switch is in the "off" position before working on any part of a welding power source's circuitry.** If it is not, it is entirely possible that you could wind up with a handful of volts and no place to put them!

Rule 8.

Most welding power sources will have either a polarity switch or a multiple range switch. Under no conditions should either of these switches be moved when the power source is being used for welding; that is, when it is under output load. The switches carry FULL welding power and would be seriously damaged if switched under load.

Rule 9.

ALL electrically operated equipment, including welding power sources, should be connected to a primary power ground. This will prevent transient voltages and currents from energizing the power source case and possibly causing injury to a person.

Rule 10.

The welding arc supplies both heat and light to the work area. The heat is for welding, of course, and the light is only incidental.

The light can damage the eyes if they are not protected. Never look directly at a welding arc without proper colored lens and a welding helmet.

Rule 11.

There are compressed gas cylinders in most welding shops. Since the various gases are either highly volatile, such as acetylene, or under high pressure, such as oxygen, **a welding arc should never be struck on a gas cylinder.** Severe fires or explosions could result if this rule is not observed.

Rule 12.

Common sense will tell you that you have only one pair of eyes. If one or both are lost for any reason there is no replacement. **Always wear hardened safety glasses in the shop area.** If grinding or chipping, use either protective goggles or a protective face shield in addition to safety glasses.

There are many other rules of safety that could be discussed here but I believe the important thing to remember is, "Think before you act", use plain common sense and don't, under any circumstances, take part in any horseplay in the shop. Always vent closed areas so that adequate ventilation is present for the workmen in the area. Above all else, don't weld on containers that have held combustible materials, such as gas tanks, without first cleaning them out according to the American Welding Society Standards on "Cutting and Welding of Containers Which Have Held Combustible Materials". **Be safe** in all working operations. If in doubt, do not weld but obtain clearance from a supervisor. If you still are not sure, don't strike the arc!

Electrodes.

Although the first metal electrodes used for arc welding were without any type of flux coating it was soon found that better arc stability could be achieved by having some type of coating (even rust) on the electrodes. This led to the development of very thin coatings on the electrodes and, even today, some manufacturers of electrodes still produce what is called a "sull-coated" electrode product line.

Eventually electrodes were developed which had extruded flux coatings. Although the idea of coating electrodes was patented as far back as 1907 (in Sweden), it was not until about the late 1920's that extruded coated welding electrodes were commercially available. For some metals, notably aluminum, dipped coatings were still in popular use until the late 1950's.

With the shielded metal arc welding process heat is produced between the tip of an electrode and the base metal workpiece. Shielding of the molten weld puddle and the electrode molten tip end is accomplished by the decomposition of the electrode flux under the heat of the welding arc. It was the analysis of the atmosphere around the welding arc that called attention to the fact that there was a substantial amount of carbon dioxide (CO_2) in the arc area. From this has come the use of CO_2 welding grade shielding gas for use with mild steel and the Gas Metal Arc Welding process.

With the shielded metal arc welding process, the arc is initiated by touching the electrode tip to the base metal and then withdrawing the electrode about $1/4''$. The arc length is then adjusted to the proper distance for the specific electrode type and diameter. The "proper arc length" is usually considered to be the same as the diameter of the core wire of the electrode. Arc length is the physical distance from the end of the electrode tip to the surface of the base metal workpiece.

Once the arc is initiated the consumable electrode melts. The molten metal from the electrode is moved across the arc column by the voltage force in the arc. The molten electrode filler metal fuses with the molten base metal in the weld joint to form the weld puddle. Upon cooling to the solid state a properly made weld will be one homogenous mass which joins two, or more, separate pieces into a single weldment.

The illustration shown in Figure 82 is an example of what is occuring in the shielded metal arc welding process. Note that with this process the depth of penetration is relatively shallow. As a matter of fact, the depth of penetration with shielded metal arc welding electrodes of the E-6010 classification is only about $1/8''$ maximum into the base metal. Since the E-6010 electrode operates with direct current, reverse polarity, and is considered the deepest penetration electrode type by most welding people, it makes sense that other electrode types and alloys would have even less penetration into the base metal.

The electrode flux coating performs several functions when it is melted and thermally decomposed under the heat of the welding arc. It promotes electrical conductivity across the arc column by ionization of the developed gases; it produces a shielding gas (basically CO_2) that excludes the atmosphere from the weld puddle; it adds slag-forming materials to the molten weld puddle for grain refinement and, in some cases, for alloy addition to the weld; it provides materials for controlling bead shape and width. The slag residue that is formed over the molten weld metal deposit helps prevent rapid oxidation of the weld metal while it is cooling through the liquid, plastic and solid states. The slag is usually easy to remove with a slag chipping tool when it has cooled and solidified. A good application of a wire brush will normally remove any last residue of slag.

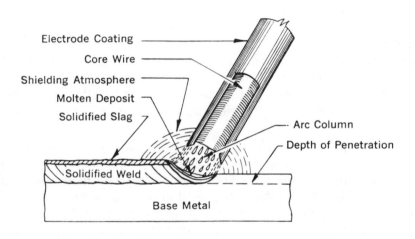

Figure 82. SHIELDED METAL ARC WELDING.

The electrode core wire material may, or may not, be similar in element content to the base metal being welded. In the case of some low alloy steels, for example, the electrode core wire may be plain carbon steel. The alloying elements, and the deoxidizing elements, are contained in the flux coating of the electrode. The resulting mixture of deposited weld metal and base metal will normally have the same physical and metallurgical characteristics as the original base metal.

The melting, or "burnoff", rate of the electrode is directly related to the amount of heat energy (electrical energy) in the welding arc. The arc energy is divided with a portion used to heat the base metal and a portion used to heat, and melt, the electrode. The electrical polarity (if using dc welding power), and the flux coating constituents, determine where the heat of the welding arc is concentrated. Figure 83 shows the approximate thermal (heat) energy distribution in the welding arc.

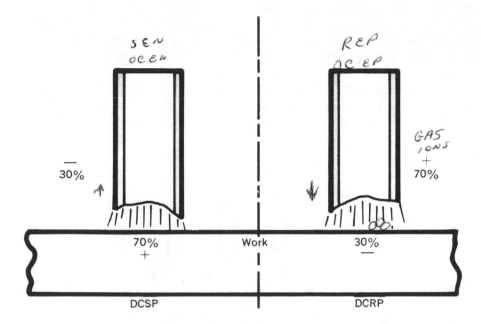

SEN
OC EN

REP
AC EP

GAS
IONS
+

30%
—

70%
+

70%
+

Work

30%
—

DCSP

DCRP

Figure 83. HEAT DISTRIBUTION IN ARC WELDING.

Usually the deeper penetrating arc of direct current, reverse polarity (dcrp) is used where penetration and out-of-position capability are required. Typical of this type of electrode is the AWS E-6010 classification.

Some arc welding applications require fast welding speeds with relatively high deposition rates. Direct current, straight polarity (dcsp) is usually specified in these circumstances. Much of this type of welding is done in the downhand (flat) or horizontal position because of the high fluidity of the welding puddle.

Alternating current has gained widespread use in many shielded metal arc welding applications. One of the outstanding advantages of ac arc welding is the almost total elimination of magnetic arc blow. (Magnetic arc blow is caused by magnetic fields which are set up in the workpiece due to the welding current flowing in the circuit. Magnetic arc blow causes the welding arc to literally blow wildly about and, in some severe cases, has actually caused molten weld metal to be expelled from the weld puddle). Most ac welding electrodes are considered all position electrodes. Some classifications, however, are designed only for flat and horizontal welding.

Welding current is the factor that controls the electrode melt-rate, and flow, into the weld joint. Arc voltage is set **only** at the welding arc and is closely correlated to arc length. (Arc length, you recall, is the physical distance from the electrode tip end to the surface of the base metal being welded). Voltage is an electrical force which does not flow itself but which is the motivating energy causing current to flow in a completed welding circuit.

A brief discussion of arc length, arc voltage and arc amperage is made in Figure 84. We ask you to consider that in all three of the shielded metal arc welding electrode positions a welding arc can be maintained.

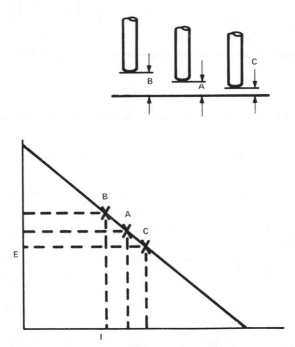

Figure 84. ARC VOLTAGE AND ARC LENGTH CORRELATION.

At point "A" in the illustration we see there is a normal arc length indicated. The current and voltage are considered to provide the best welding conditions for the electrode type and diameter. There is some measure of arc length shown. The important fact is that there is an **air space** between the electrode tip and the base metal surface. You may remember that air is a good resistor to electrical current flow. It will, therefore, take some measure of voltage to overcome the electrical resistance of the air space. Only the voltage required is applied through the welding arc.

Moving now to point "B" in the illustration, it is apparent that there has been an increase in the arc length. This simply means that there is more air space between the electrode tip end and the base metal surface. We can logically expect there to be more electrical resistance in the arc. More voltage will be required to overcome the increased electrical resistance of the welding arc. As shown in the volt-ampere curve point "B" has higher arc voltage but, in this case, less amperage. This is typical of a constant current, or conventional, welding power source.

At point "C" the welding arc is illustrated as a tight arc. The air space between the electrode tip end and the base metal surface is less and, therefore, the welding arc voltage is lower. Remember, less air space means less arc voltage is necessary. Please note, however, that arc amperage has increased due to the lower voltage and the shape of the volt-ampere curve.

It is the air space between the electrode tip end and the surface of the base metal being welded that actually determines arc voltage. It must be noted, however, that the electrical resistance of the electrode body will have some effect on the total circuit resistance and arc voltage.

As the current value is increased when using a given diameter of electrode, the current density also is increased. **Current density is calculated by dividing the electrode cross sectional area (in square inches) into the welding amperage used.** The result is amperes per square inch of electrode area. A chart in the appendix of this book (Data Chart 3) provides the formula and other information for determining current density for electrodes from 0.020" diameter to 3/8" diameter.

It is current density that determines the melt-rate of an electrode. It should be noted that, if current density is too high for a given diameter of electrode, there will be a rapid build-up of heat in the body of the electrode. This can cause the electrode to overheat thus destroying the flux coating and making the electrode unusable for welding.

There are actually three areas of metals that are affected by the arc welding action. They are: (1) the weld deposit, usually classed as a cast structure metallurgically, which is a combination of the deposited filler metal and the diluted base metal; (2) the heat-affected zone (HAZ) where the base metal has not been melted but where the physical properties and characteristics have been altered by the heat the welding arc introduced to the metal area; (3) the base metal un-affected by the welding heat.

Metallurgically, the deposited weld metal and the heat-affected zone have a sharp demarcation line between them. Between the heat-affected zone and the un-affected base metal, however, the line is not so finely drawn. It is usually difficult to tell where the heat-affected zone ends and the base metal not affected by the welding heat begins. This is usually brought out, however, under what is called a **macro examination.** (Macro examinations may be magnified up to ten diameters under the microscope). A typical macro examina-tion of a steel weld is shown in Figure 85.

Figure 85. A TYPICAL MACRO EXAMINATION.

Welding techniques used with the shielded metal arc welding process will vary widely. The type of material being welded, the joint design used, the classification of the electrode and the position in which the weld is to be made will all have a determining effect on the welding technique to be used. Other factors certainly enter into the development of a welding procedure but these are primary con-siderations.

Types Of Welding Power Sources Used.

Welding power sources used with the shielded metal arc welding process may be either the rotating type equipment, such as motor-generators and engine driven equipment, or static, non-moving equip-ment such as transformers and transformer-rectifiers. Rotating type equipment includes electric driven motor generators and engine driven units.

All welding power sources designed for the shielded metal arc welding process are the conventional, constant current type power sources having a "drooping" volt-ampere output characteristic curve. Figure 86 illustrates typical "drooping" volt-ampere output curves.

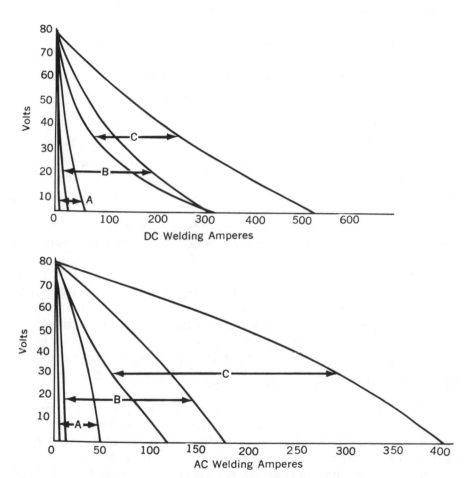

Figure 86. AC AND DC VOLT-AMPERE CURVES.

Note that the ac volt-ampere curve is slightly convex while the dc volt-ampere curve is slightly concave. The dc curve takes into account the small voltage loss which occurs when the electrical power goes through the main power rectifier of the power source. The loss is insignificant for this process.

Alternating Current Power Sources.

Alternating current welding power sources have found many applications in the welding industry for the shielded metal arc welding process. Both industrial and utility type ac power sources have increased in popularity with the welding public. In particular, smaller 180 and 225 ampere rated units are used by home hobbyists, metal sculptors and others who just want the capability of arc welding in their home workshops.

There are several different classifications of welding power sources which are listed in the National Electrical Manufacturers Association (NEMA) Standard EW-1. The following data includes three general classifications of welding power sources.

NEMA Class 3 Welding Power Sources.

The NEMA Class 3 power sources were previously classed as "limited input" units. Known as "utility" or "farm" welders, they have a limit on the maximum primary amperage they may draw from a given primary voltage. Under the NEMA Class 3 rating, the duty cycle for this type of power source is 20%. This means they can operate at rated amperes for two minutes out of every ten minutes. The other eight minutes of the time period they must be permitted to idle and cool. Most power sources of this type are designed so they may be de-rated for current with higher duty cycles. This is explained later in this book under the chapter title "Welding Power Sources For Gas Tungsten Arc Welding". The one design of power source that **cannot be de-rated for amperage** is the tapped secondary coil design. This type of power source is 20% duty cycle in all ranges which makes it difficult to work with in many applications.

NEMA Class 2 Welding Power Sources.

NEMA Class 2 welding power sources were previously classed as "limited duty" power sources. Under the new classifications a NEMA Class 2 unit may have either 30%, 40% or 50% duty cycle rating. Amperage on this type of power source is normally limited to ratings of less than 300 amperes. While there is no limitation on the amount of primary amperage they may draw while welding, there is an automatic limitation because of their lower amperage output rating and load voltage rating while welding.

NEMA Class 1 Welding Power Sources.

The NEMA Class 1 welding power sources were formerly called "industrial" type power sources. They are presently rated at 60%,

80% and 100% duty cycle at a specific amperage output. They are designed for heavy duty industrial type operations including submerged arc and some air carbon-arc applications. Class 1 welding power sources are normally used for those industrial applications where production welding is done.

Advantages Of AC Welding Power Sources.

Some manufacturers prefer ac welding power sources because of their initial low cost and negligible maintenance requirements. Very little maintenance is required for this transformer type welding equipment since there are no moving parts to wear out. Many companies specifically order ac welding power supplies because there is an almost total lack of magnetic arc blow with this type of welding equipment. The constantly changing current, which is alternating current, precludes the possibility of setting up magnetic fields that would disrupt the welding arc. The fact that **all ac welding power sources operate from single phase primary power** is an asset in many welding areas.

Direct Current Welding Power.

The use of direct current (current which flows in one direction only) is as old as electric welding itself. The first method of obtaining dc welding power was to draw it from rather large battery banks which were, in turn, charged by dynamos. An early welding shop is illustrated in Figure 87.

The electric motor was then an industrial tool. From the cumbersome battery was developed the idea of an electric motor-driven generator designed specifically for welding operations. In early models of motor-generator welding power sources, the rotor consisted of a through shaft, the armature iron core material, the armature copper coils for current generation, and the commutator (a set of copper bars, connected to the armature coils, which are used as current collectors). It is through a set of carbon brushes and the commutator that the ac welding current generated in the armature is changed to dc welding power. Although many years have passed the principle is still the same for all motor-generator welding power sources manufactured today.

The development and use of fuel-powered engine driven welding power generators provided mobility of welding power sources and opened up new areas for the arc welding processes. Even though no primary power was available from generating power plants electric welding could be done in the field.

— 153 —

This is the first commercial welding shop of record. It was located in LaChappelle, France, July 9, 1887.
The welding current is taken from wet cell storage batteries. The batteries are charged with a generator operated from a line shaft. The welding current is adjusted with a series of switches and by using a series of resistance grids.

Figure 87. BATTERY POWERED DC WELDING.

Through the years both the engines and the generators have been improved in performance and reduced in physical size. In some models, the design concept has changed radically from a revolving armature design to a revolving field coil design. Far more efficient in its concept and design, the revolving field coil type unit has proven itself to be the welding power source for field applications. An ac/dc welding power source, with engine drive for mechanical energy, is illustrated in Figure 88.

In recent years the use of static electrical devices for changing ac to dc have come into prominence in the welding industry. This type of welding power source is called a transformer-rectifier unit. In this instance, alternating current is supplied to the rectifier from the main transformer of the welding power source. The primary power may be either single phase or three phase although three phase is considered to provide the highest electrical efficiency and the best

Figure 88. ENGINE DRIVEN AC/DC POWER SOURCE.

welding conditions. Three phase primary power also provides the highest average power in the welding arc. A typical three phase powered transformer-rectifier welding power source, having dc output only, is illustrated in Figure 89.

Output volt-amperage control, commonly called amperage control, may be accomplished by several methods. They may be separated into two broad categories of **mechanical** and **electrical** controls.

Mechanical controls are used primarily because of lower manufacturing costs and their ability to reproduce previously set welding conditions. Electric control of welding current provides the important factor of **remote control** over the entire range of the welding power source. Electric controls are more expensive to manufacture than most types of mechanical controls. Either method of amperage control is satisfactory with selection depending on the application requirements of the welding power source.

Figure 89. TRANSFORMER-RECTIFIER POWER SOURCE, DC.

There are two specific materials used for rectifying welding current from ac to dc. **They are selenium and silicon.** The relative merits and uses of the two rectifier elements are discussed in the chapter in this text entitled "Rectifiers". At this point it is only necessary that we know what a rectifier does with relation to the rest of the welding power source circuitry. A rectifier, regardless whether it is selenium or silicon, does only one thing and that is to change ac to dc. The alternating current may be either single phase or three phase. There is no phase with direct current.

The use of rectifiers is not limited to static transformer type welding power sources. Rectifiers are used on rotating type engine driven portable equipment, particularly those that feature the revolving field coil design. With this design concept, the armature coils are located in the stator, or stationary part, of the generator. The magnetic field coils are located on the rotor assembly of the unit. The generated ac welding power is changed to dc by a rectifier on those welding power sources having dc welding output. There is no requirement for the commutator-brush arrangement with this type of current generation and rectification.

Many of the engine driven welding power sources available today provide auxiliary 60 hertz ac power for emergency lighting, operation of tools, etc. In addition, up to one KW of 115 volt dc power is available while welding on some models.

Electrical Safety.

Wherever welding people gather there is bound to be some discussion of safety in welding. Sooner or later the question of which is safer, alternating current or direct current, is brought up. The clear cut answer is that **neither is safer!** Electrical power, either ac or dc, can be deadly if handled carelessly. The greatest asset to anyone working with, or around, electrical power is **common sense.** Following normal precautions when working with electric welding power will assure you that any probability of electrical shock will be minimized. It is possible, however, that a shock may be received from **any exposed electrode** that is electrically energized.

It is certainly true that ac will give more of an electrical shock than will dc. The reason for this is that alternating current changes direction of flow each half-cycle (each 1/120th of a second for 60 hertz power). This change of electrical polarity each half-cycle, coupled with the constantly changing current value in the circuit, is the cause of the electrical shock with ac. Conventional protective clothing normally worn when welding is usually sufficient protection against electrical shocks from welding equipment. This is based on the assumption that the welding equipment is properly grounded and installed.

It has often been said that electrical power will "reach out and grab" an individual. This is not true. A brief discussion of what actually happens will be useful in clearing up this misconception.

As you may recall from high school or college biology, the nerves and muscles of animals may be **stimulated** by an electrical current and caused to move and quiver. The same type of reaction is typical of the human body when it is stimulated by electrical current and voltage.

It has been well documented that the human brain sends out electrical impulses to various parts of the body which cause us to do things as our thoughts direct. In essence, this is a form of mind over matter, or, a form of levitation of the heavy body parts. The act of moving the legs when walking or running; the arms for any reason such as reaching or for balance when running, etc., are dictated by this incredible electrical system in the human body. It may be said that the human body is stimulated to do these things. When you consider it, the human body is a marvel of engineering in more ways than one!

Basing our thoughts on this premise of electrical stimulation, it is logical that a **greater electrical charge** will override the smaller electrical impulses sent out from the brain. This is what happens when a person comes in contact with an electrical power line. The stronger ac charge causes the muscles and tendons to tense and become rigid. In such a circumstance the person affected **cannot let go** of the electrical conductor. The result can be very serious if the electrical power to the conductor is not shut off immediately. As a matter of fact, serious injury or death can result.

NOTE: NEVER UNDER ANY CIRCUMSTANCES TOUCH A PERSON WHO IS IN CONTACT WITH AN ELECTRICAL POWER LINE. THE POWER WILL BE TRANSMITTED TO YOU AND YOU MAY BE SEVERELY INJURED OR KILLED!

This portion of the discussion can best be summed up by saying that, if you are working with electrical circuitry carrying over six volts, **keep one hand in your pocket!**

At the beginning of this chapter we talked a bit about safety and safety equipment. The use of leather jackets, leather sleeves and aprons, gloves, safety shoes, spats, and welding helmets is necessary when arc welding. The protection is for the welding operator and anyone who may be working near him. Severe radiation burns, as well as injuries from hot metal, can be avoided if the proper safety precautions are taken.

Welding helmets are manufactured in a variety of sizes and shapes. Try several and use the one that is comfortable for you. Be sure the colored lens (dark glass) is the proper shade for the welding process being used and the arc welding current employed.

REMEMBER: ONE PAIR OF EYES IS ALL YOU GET!

The higher the amperage for welding, the darker the colored lens should be. Take the time to make a lens change if you must go to higher amperages for a given application. You will see better and work safer. Figure 90 shows some of the safety equipment worn by welders.

Figure 90. WELDING SAFETY EQUIPMENT.

Chapter 8

AC TRANSFORMER WELDING POWER SOURCES

Alternating current welding power sources are basically simple electrical devices. The design concepts discussed in this chapter are general in nature and apply to a great number of models in use today. The ac welding power sources described in this text are designed for use with the shielded metal arc welding process although some of them may have application for other welding processes.

The actual function of a transformer type welding power source must be considered before deciding how that function will be accomplished in the design of the power source. The intent of any welding transformer is to change **amperage** and **voltage,** without changing frequency value, or else to isolate a circuit. An isolated electrical circuit in welding power sources is normally a control circuit of some type. The main transformer of a welding power source is designed to take the relatively **high voltage, low amperage** of the primary power and convert, or **transform,** it to **usable welding voltage and amperage.**

The primary voltage provided at the shop electrical distribution point is much too high for use in welding operations. By calculating the necessary open circuit voltage (OCV) required in the welding power source, it is possible to design a step-down welding transformer for shop use. The design goal, of course, is to have a specific open circuit voltage, lower than the primary voltage, that is relatively safe for welding. In addition, the welding amperage required for most applications is greater than is supplied in the primary electrical system. To obtain the necessary secondary circuit values of voltage and amperage it is necessary to transform the primary power to usable welding power.

To gain a proper perspective of an ac transformer type welding power source each electrical component will be illustrated and its function explained. The results will show conclusively what each part does, why they are there, when they are in operation and how they perform their function.

A basic ac transformer design, the "closed U" core, is shown in Figure 91. The sheet steel core material is very thin gauge and is insulated on both sides of each lamination. The insulation between each core lamination prevents the formation of eddy currents. Eddy currents would tend to create heat energy in the iron core thereby decreasing the electrical efficiency of the transformer.

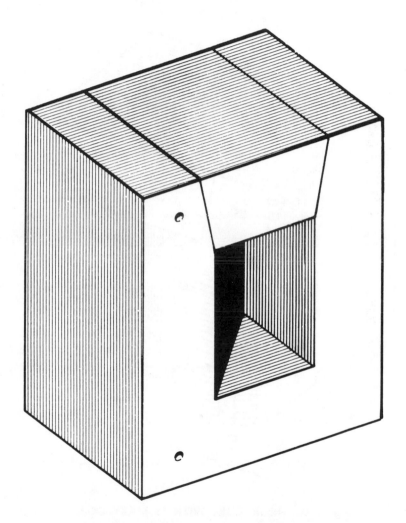

Figure 91. CLOSED "U" IRON CORE DESIGN.

The iron core acts as an intermediate magnetic connecting link between the primary and secondary coils of the main transformer.

The primary and secondary coils carry the actual current in the electrical circuits.

The addition of a primary coil is shown in Figure 92.

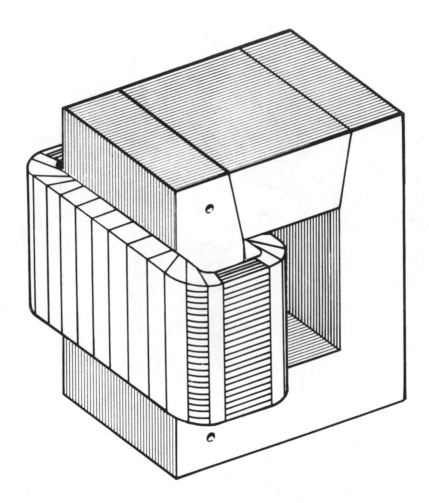

Figure 92. IRON CORE WITH PRIMARY COIL.

The conductor wire of the primary coil is normally relatively small in diameter, or gauge, since it is designed to carry only the rather low primary amperage value. It is the primary coil that carries the primary shop current into the welding power source so it can operate.

At this point there are two of the three necessary components which make up a total transformer. Figure 93 illustrates the third basic part of a welding power source transformer, the secondary coil.

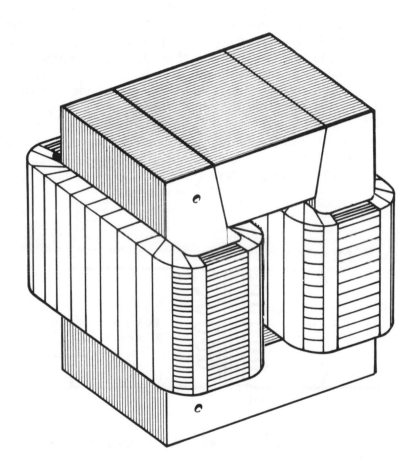

Figure 93. SIMPLE TWO WINDING TRANSFORMER.

The secondary coil is one of the two electric "fields", or circuits, that are joined by the iron core magnetic field circuit. Thus we are discussing the **primary and secondary coils** as being the **electric field circuits** and the **iron core** of the transformer as being the **magnetic field circuit.**

The secondary coil has relatively fewer electrical turns, and is made of substantially heavier conductor wire, than the primary coil. The secondary coil, of course, carries the welding amperage which is considerably higher in numerical value than the primary amperage.

In the illustration, Figure 93, there is no method of current output control shown. The subject of current output control in welding power sources will be discussed later in this chapter.

In the chapter entitled "Welding Power Source Transformers" we developed the theoretical rule that the **ratio of primary to secondary volts will be the same as the ratio of primary to secondary turns. The ratio of primary to secondary amperes is inversely proportional to the voltage, and turns, ratio.** (Inversely proportional means the ratio values are the same but the figures are turned exactly around. For example, if there is a ratio of 3:1, the inverse ratio is 1:3).

To provide a better understanding of ac welding transformers, and how they work, there is a simple circuit diagram shown in Figure 94. A circuit diagram is nothing more than a group of welding power source components shown as electrical symbols.

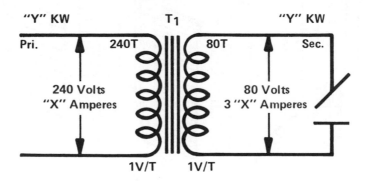

Figure 94. CIRCUIT DIAGRAM, SIMPLE TRANSFORMER.

It is well to remember that a **circuit diagram** will show the **electrical relationship** of welding power source parts without necessarily showing their physical relationship. A **wiring diagram** will show the **physical relationship** of the welding power source parts including placement and location of terminal connections, relays, etc. The electrical symbols used by most welding power source manufacturers are the **American National Standards Institute (ANSI)** approved symbols.

The circuit diagram shown in Figure 94 could be for the semi-pictorial drawing of the simple transformer shown in Figure 93. It is for a simple two coil, or **two winding,** transformer. Comparing the two illustrations will make it easier to visualize the relationship between a circuit diagram and the actual physical part of the welding power source.

An Example Of Welding Transformer Design Concepts.

The circuit diagram shown in Figure 94 indicates the primary voltage as 240 volts. The primary current is shown as "X" since this value will change with the output amperage rating of the welding power source being considered. The product of volts times amperes is either volt-amperes or watts. This value, divided by 1,000, gives a quotient of either kilovolt-amperes (KVA) or kilowatts (KW). **Kilovolt-amperes (KVA)** are **the total power demanded** from the primary power supply system furnished by the utility company. **Primary kilowatts (KW)** are the actual **primary power used** by the welding power source to produce its rated load amperes and load voltage.

For this exercise we will consider that there are no power losses through the theoretical transformer with which we are working. The primary and secondary KW will, therefore, have the same value. (A perfect electrical transformer is impossible to construct because of the electrical circuit conductors involved and their resistance to the passage of electrical current. Also, the inductive and magnetic coupling can never be 100% in a transformer).

Welding power source design engineers begin the development of a welding transformer by calculating the magnetic density of the iron transformer core. (Magnetic density is the number of magnetic lines of force the iron core will take per square inch of core lamination surface area). Once that is established, the engineers develop the voltage ratio by knowing the primary voltage to be used. They then determine the secondary, or open circuit voltage, the power source will require. The mathematical calculation to achieve the voltage ratio is then relatively simple. From this data it is no problem to develop the electrical turns ratio and, subsequently, the amperage ratio.

In the circuit diagram example, Figure 94, we will develop the power source design a little differently since this is not an engineering design textbook. Let's look at the facts:

If we consider that the primary coil has 240 electrical turns then it is evident that there is **one volt per turn (1V/T)** in the primary coil.

This is calculated as follows:

Primary voltage = 240 volts; primary turns = 240 turns.

240 (turns) $\overline{)\,240(\text{volts})}$ = 1 V/T

It is a theoretical fact that the volts per turn on the secondary coil of the transformer will be the same as the volts per turn on the primary coil. In this case, it results in one volt per turn on the secondary coil. If the **open circuit voltage** of the power source is to be 80 volts open circuit, then there must be 80 turns on the secondary coil of the main transformer. A little thought will show that the secondary **open circuit voltage value is controlled by the number of effective electrical turns on the secondary coil.** (Effective turns are those electrical turns of a coil through which electrical current is actually flowing).

The maximum open circuit voltage allowable by the NEMA (National Electrical Manufacturers Association) EW-1 Standard is 80 volts for alternating current welding power sources. The circuit diagram shows 80 volts potential between the two conductors from the secondary coil of the transformer. At one volt per turn it is apparent that there must be 80 electrical turns on the secondary coil.

NOTE: It is important to remember that this is an example only. The numbers used as ratios, turns, etc., are indicative of no particular type or model of welding power source.

Some checking of the figures developed so far will show that the turns ratio and the voltage ratio are exactly the same (3:1). This brings up an interesting fact. We stated that we had a theoretically "perfect" transformer and that the KW would be the same on the secondary circuit as on the primary circuit. The circuit diagram shows "Y" KW on the primary circuit and "Y" KW on the secondary circuit. The "Y" value, whatever its numerical equivalent, is the same for both the primary and secondary circuits.

Lets look at the volt-ampere relationship in both the primary and secondary circuits. The primary voltage is 240 volts with "X" primary amperes. The secondary circuit voltage is only 80 volts or one-third the primary voltage. To have "Y" KW in the secondary circuit there must be an increase in the amperage value of the circuit. To arrive at the same numerical value of KW in the secondary circuit the volts times the amperes, divided by 1,000, must equal "Y" KW. It is apparent that there must be 3 "X" amperes times 80 volts in the secondary circuit of the power source. This would satisfy the numerical part of the problem and also prove the rule stated previously about amperage. That is, **"The amperage ratio of a welding transformer is inversely proportional to the voltage, and turns, ratio".**

We have proved the fundamental rules of design for welding power source transformers. Carrying the thought a bit further, let us examine the electrical circumstances that occur when the welding transformer is in actual operation. The reader is cautioned to assimilate and understand the information presented here for it will be the basis for considering methods of output amperage control in electrically controlled welding power sources. These same concepts will also be used to describe "slope control" in constant potential type welding power sources.

A circuit diagram similar to the one previously used is illustrated in Figure 95. Again there is the primary coil shown to the left of the drawing; the transformer iron core (represented by the three straight lines) at the center, and the secondary coil shown at the right of the iron core.

Primary current flows in the primary coil of the power source transformer when it is energized. The primary voltage is the electrical pressure, or force, that causes the current to flow when the circuit is complete. Remember: **voltage does not flow.** It is only a motivating force that causes current to flow.

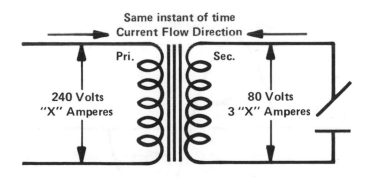

Figure 95. TRANSFORMER ELECTRICAL CONCEPTS.

The primary coil is wrapped around one leg of the transformer iron core. The primary amperage and voltage is considered to be an electric "field", or circuit. Any time there is current flowing in a coil wrapped around an iron core, as there is here, there will be a magnetic field created. The strength of the magnetic field will depend on three factors:

1) The mass and type of iron in the transformer core.
2) The number of effective electrical turns in the coil.
3) The numerical value of the amperage flowing in the coil.

In the welding power source shown in the drawing, Figure 95, we can consider that the mass of iron will remain a constant factor once the core is set in the power source. The number of effective turns in the primary coil will also be a constant factor once it is placed around the iron core. It is apparent that the only changing variable will be the **amount of current flowing in the primary coil.** It is evident that the strength of the magnetic field is dependent on the constantly changing amperage value in the primary coil. As the ac sine wave form increases to maximum current, so the magnetic field increases to maximum strength. When the amperage decreases, as it does each half-cycle, the magnetic field must decrease in strength also. Keep in mind that **all transformers** operate with alternating current.

Energy is used to create the magnetic field. The energy is not consumed; rather, **it is stored in the magnetic field** momentarily. It is logical that, since the magnetic field strength is based on amperage flow and value, and since amperage flow is at zero each 1/120th of a second for 60 hertz ac power, the magnetic field must be at zero each half-cycle also.

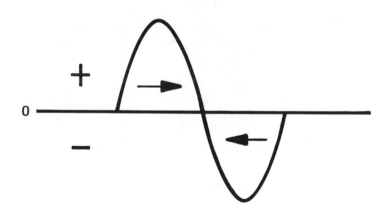

Figure 96. AC SINE WAVE AND MAGNETIC FIELD TRACE.

Referring to the ac sine wave trace in Figure 96 it is apparent that the amperage starts at zero, reaches maximum strength, or amplitude, at 90 electrical degrees and decreases in strength until it reaches zero at the 180 electrical degree point. The magnetic field stores energy through the first quarter-cycle (from "0" to 90 degrees) until maximum current flow is obtained. As the amperage value decreases in the second quarter-cycle (90-180 degrees), the **magnetic field collapses.** The energy stored in the magnetic field is released and is induced into the secondary coil and circuit of the transformer power source. The energy is apparent as open circuit voltage at the output terminals of the power source. The second, or negative, half-cycle follows the same pattern except that current flow has alternated in direction as it does each half-cycle with alternating current.

Figure 95 shows the electrical relationship of the primary coil, the iron core and the secondary coil of the main transformer. Note that the primary amperage is brought into the primary coil; the iron core creates the magnetic field and the result is an induced voltage in the secondary coil and circuitry.

It is logical to conclude that the primary coil is part of the electric field (the electrical circuit), the iron core is part of the magnetic field circuit and the secondary coil and circuit is another electric field.

The preceeding discussion has shown the fundamental relationships of the various transformer components to one another. It requires all three pieces for the transformer to function. Remember, however, that without eletcrical power they are just three pieces of oddly shaped metal, some insulation and a base carriage. It is the electrical power that causes the inanimate parts to perform their functions and provide welding power.

Duty Cycle And Service Use.

The National Electrical Manufacturers Association (NEMA) provides a set of standards within which welding power sources may be rated at a specific amperage and load voltage. While not all welding power sources are NEMA rated, most manufacturers use the NEMA standards as guidelines for design and manufacturing purposes.

Until 1962, industrial welding power sources were arbitrarily rated at a specific amperage and 40 load volts. The NEMA Standard EW-1, revised and published in 1962 and re-affirmed in 1972, proclaimed a graduated load voltage chart based on the power source amperage ratings. Figure 97 lists the voltage and amperage ratings of the various NEMA rated welding power sources. The chart applies to all NEMA rated constant current welding power sources, either ac or dc output.

AMPERE RATING	VOLTAGE RATING
200	28
300	32
400	36
500	40
600 and up	44

Figure 97. NEMA CLASS 1 VOLTAGE RATINGS.

Duty Cycle.

Duty cycle is the actual operating time that a welding power source may be used at rated load without exceeding the temperature limits of the insulation of the component parts. Most NEMA Class 1 welding power sources are rated at 60% duty cycle. Some welding power sources, notably those used for automatic and semi-automatic processes, are rated at 100% duty cycle.

To explain the ten minute duty cycle period we will show the following example. If a welding power source is designed to operate at 60% duty cycle, 300 amperes, 32 load volts, it has been built to provide the rated amperage, at rated load voltage, for six minutes out of every ten minutes. The other four minutes the unit must idle and cool the power source components. **The full ten minute duty cycle period must be completed for proper operation.**

Duty cycle is not accumulative. To operate a welding power source for 36 minutes and then let it idle and cool for 24 minutes is **not** 60% duty cycle! DUTY CYCLE IS BASED ON A TEN MINUTE PERIOD OF TIME. It is important that the full ten minute period be observed. In the example cited, it is **not** 60% duty cycle to operate the welding power source for six minutes at rated load and then turn the unit off! The residual heat retained in the transformer coils and core will cause heat deterioration of the insulating materials.

Power sources assigned 100% duty cycle may be operated continuously at their rated amperage. If however, the amperage drawn from the power source **exceeds** the rated amperage, the duty cycle will be less than the 100% duty cycle assigned. For example, a power source rated at 300 amperes, 32 load volts, 100% duty cycle, would normally have about 55% duty cycle when operated at 400 amperes.

The reason for the difference in duty cycle is the conductor size and diameter. Certain sizes and types of conductors are used in specific amperage rated welding power sources. If the amperage used is above the normal rated capacity of the electrical conductors, excessive heat will develop due to the electrical resistance of the conductor. The more heat in an electrical conductor, of course, the higher the electrical resistance. The result is overheating, and possibly burning, of the insulation of the conductor.

In the 1972 NEMA EW-1 Standard, welding power sources have been re-classified according to duty cycle rating. This makes it relatively easy for the power source user to determine exactly what he is buying in an electric welding power source.

Types Of Service Use.

There are three general categories of welding power sources from the service standpoint. This statement concerns only the transformer type power source used for shielded metal arc and gas tungsten arc welding processes.

NEMA Class 1.

According to the NEMA EW-1 Standard, those welding power sources having 60%, 80% and 100% duty cycle are classed as NEMA Class I welding power sources. Such power sources may be ac, dc, or ac/dc output. The Class 1 power sources are considered to be the "work horse" of the industry.

NEMA Class 2.

NEMA Class 2 welding power sources carry either a 30%, 40% or 50% duty cycle rating. Such welding power sources normally are rated at less than 300 amperes output. Again, the welding power sources may be either ac, dc, or ac/dc output. Primary amperage draw is limited by the duty cycle and amperage rating of the power source.

NEMA Class 3.

The NEMA Class 3 power sources are limited in the amount of primary amperage they can draw from the primary power system. They are rated at 20% duty cycle by the NEMA Standard EW-1-72. Normally referred to as "farm" or utility welding power sources, this type of ac power source is excellent and inexpensive. In those areas where the primary power distribution is limited or restricted, as in rural areas, residential areas, etc., the Class 3 power sources perform a real service. They operate from single phase primary power.

Methods Of Output Volt-Ampere Control.

Once the power transformation, from primary to secondary, is achieved control of the output volt-ampere characteristic becomes of prime importance. Commonly called amperage control, the output of a welding power source may be controlled in several ways. The method used will be dictated by the process requirements, the economics of manufacturing, the necessity for remote control capability, etc. Some of the methods used for output control from ac welding power sources are:

Movable Coil Control
Movable Shunt Control
Saturable Reactor Electric Control
Magnetic Amplifier Electric Control
Tapped Reactor Coil Control

One of the most important distinguishing features of a well-built ac welding power source is its facility for infinite amperage output control within the minimum-maximum amperage range. If this feature is provided in two or more output current ranges there should be a generous overlap in the maximum of the lower range and the minimum of the higher range. Most of the methods of output current control may be used for any transformer or transformer-rectifier type power source.

The next portion of the discussion will cover two very popular types of ac welding power sources. They are the movable coil design and the movable shunt design.

Movable Coil AC Welding Power Sources.

The movable coil design is a classic example of an ac welding power source transformer. The primary and secondary coils of the main transformer are **magnetically loosely coupled** to provide the "drooping" (negative) output volt-ampere curve required for shielded metal arc welding applications. A pictorial drawing of a typical moving coil welding transformer is illustrated in Figure 98. Note that the secondary coil is fixed in position in this power source design. The primary coil is on a lead screw which positions it according to the welding amperage output requirements.

There is no mechanical linkage between the primary and secondary coils. Voltage is induced into the secondary coil from the magnetic field created by the iron core and the current flowing in the primary coil which is positioned around the iron core.

The actual current output of the ac power source is regulated by the distance the two coils are apart. The closer together the coils, the higher the current output from the welding power source. Conversely, as the primary coil is moved away from the secondary coil, the output amperage is decreased. This is shown by the volt-ampere output curve in Figure 98.

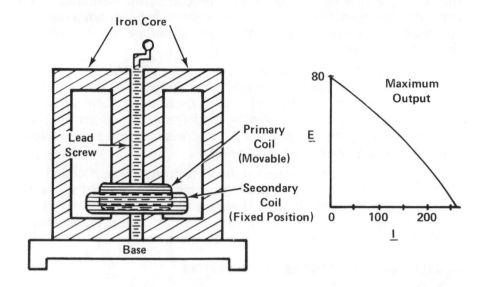

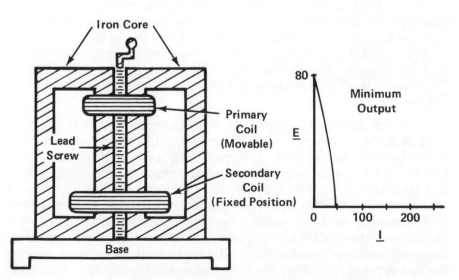

Figure 98. MOVABLE COIL AC POWER SOURCE.

As shown in Figure 98, it is the primary coil that is positioned on a rotatable shaft. The shaft is a lead screw device normally having acme threads and a ball bearing located at the terminal, or bottom, end for easy rotation. Within the primary coil is a positioning bracket, threaded, for moving the coil when the lead screw is rotated. This capability to move the primary coil provides infinite amperage adjustment within the minimum-maximum amperage output range of the power source.

Movable coil welding power sources may be NEMA Class 1, Class 2 or Class 3. There is very little maintenance required for ac welding transformers of the moving coil design because of their simplicity. If the cover is removed periodically, and the accumulated dust and dirt blown away with compressed air, there should be few problems from mechanical breakdown of the unit.

If the handwheel becomes difficult to turn, the application of a small amount of silicone grease to the acme threaded lead screw will provide adequate lubrication. It is always best to remove the dirt and dust from the acme threads before applying the silicone grease. This will help prevent abrasive wear on the threads.

Movable Shunt AC Welding Power Sources.

Movable shunt ac transformer power sources are also loosely magnetically coupled. In the movable shunt design, however, the primary and secondary coils are **both** fixed in position and do not move. Instead, a **laminated iron shunt,** placed within some type of shunt holder assembly, is caused to move between the primary and secondary coils. The iron shunt materials are the same as that used for the main transformer core and are insulated on both sides of each steel lamination. The iron shunt acts as a magnetic "flux" diverter. (The term "flux" is the same as saying "magnetic lines of force" in a magnetic field).

As illustrated in Figure 99, the movement of the magnetic lines of force, or magnetic flux, is unobstructed when the iron shunt is not between the primary and secondary coils of the main transformer. As the iron shunt is caused to move between the primary and secondary coils, the magnetic lines of force are literally diverted to the iron shunt rather than having free access to the secondary coil. In this manner, welding output amperage from the power source can be adjusted from minimum to maximum within the output amperage range of the power source. When the iron shunt is out from between the primary and secondary coils the welding current output is at maximum. As the shunt is moved between the coils the output welding current decreases. Current output adjustment is infinite within the minimum-maximum amperage range of the power sources.

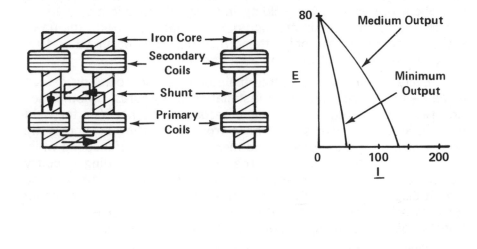

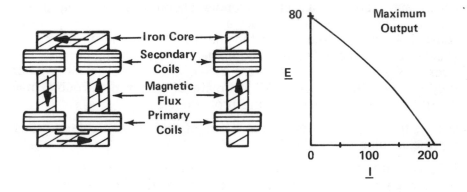

Figure 99. MOVABLE SHUNT AC POWER SOURCE.

Shunt type ac welding power sources may be NEMA Class 1, Class 2, or Class 3. Maintenance is usually confined to blowing out accumulated dirt from time to time with compressed air. If there is any other problem, such as shunt rattle, check the power source manufacturer's operation and maintenance manual for the power source. You will normally find some suggested trouble shooting remedies in the manual.

Trouble Shooting AC Welding Power Sources.

Before any welding power source is connected to primary line power the maintenance and operation manual should be read, **and understood,** by both the installing serviceman and the welding operator. Although this fact is generally understood and acknowledged by just about everyone in the welding industry, it is seldom carried out in actual practice. The instruction manual that comes with the power source is normally read only **after** difficulties have occurred and the power source does not function as it should.

Every welding power source maintenance and operating manual has a section on recommended installation procedures. Another section tells the operator how to make secondary connections for welding. In all cases there should be a trouble shooting section in the book which describes some of the more common malfunctions which could occur with the power source or its related welding circuitry. Suggested remedial repair action is in this section also. An electrical circuit diagram is normally located near the trouble shooting section for easy reference. The circuit diagram will show the electrical relationship of all the welding power source parts to each other.

There are some general questions in the following paragraphs which may serve to assist in trouble shooting ac welding power sources. The discussion topics are applicable to any ac welding power source. As you will see, very often the troubles are not in the power source but in some of the allied equipment in the welding circuit. The questions asked, and the answers given, are general in nature but they will provide a "ball-park" estimate of where the trouble may be and can assist in isolating the problem.

Q.1) Is the welding power source the right type and amperage rating for the welding application?

A.1) The unit may have been the right type and amperage rating for the application it was originally purchased to do. If the application has changed, however, the power source may be too small and be overloaded. This is not at all unusual where E-6011 mild steel electrodes have been replaced by low hydrogen type electrodes. Stainless steel and low hydrogen type electrodes have a globular type transfer of molten metal from the electrode to the base metal. Some of the lime-based flux coatings used on the electrodes make arc initiation and arc re-striking exceedingly difficult with certain ac power sources. In addition, this type of electrode requires higher welding amperage for a given diameter of electrode then does mild steel. The result is a tendency to overload the NEMA Class 2 and Class 3 ac power sources. Overheating of the electrical conductors and the iron core material could cause serious damage to the insulation of those components.

Q.2) Has the welding power source been operating correctly in shop service?

A.2) If the unit has been used right along without trouble chances are that the problem is in the welding circuit **external** to the power source. Many times the problem is simply a poor electrical ground or old and damaged welding cable.

Q.3) Has anything been changed in the welding circuit such as the primary power input, welding electrode size or type, the internal circuitry of the power source, etc.?

A.3) If the primary power has been changed check the primary voltage linkage at the primary connection terminals of the power source. It could be set for the wrong input voltage. If the problem is momentary line voltage fluctuations, it may be due to too small a line transformer supplying power to the shop. It could also be caused by some type of inductive device, such as an inductive electric motor, drawing power at intervals from the same primary line. For example, an air compressor on the same shop wiring circuit would cause a momentary low voltage and amperage condition when its electric motor started.

Welding equipment should always be in good repair. All connections of welding cables to the power source should be clean and tight. Welding cables should be checked periodically for proper conductivity. Ground clamps and electrode holders should be of adequate amperage rating for the power source with which they are used.

In no case should the internal circuitry of the welding power source be changed without authorization from the original manufacturer. The design of the unit provides maximum safety to both the welding operator and the power source itself. Changing the electrical circuitry could cause serious problems.

Q.4) Exactly what is the complaint of the welding operator?

A.4) Many times the welding supervisor does not understand the nature of the welding operators complaint. If the wrong information is supplied to the person charged with troubleshooting the complaint, it cannot be satisfactorily resolved.

Q.5) Does a good secondary circuit exist?

A.5) By secondary circuit we mean that portion of the welding equipment extending from the positive and negative (electrode and ground) output terminals of the power source. Make sure the welding cables are in good repair and are the correct size for the amperage to be carried. Be sure the connections are tight. The ground clamp must make firm and positive contact with the base metal workpiece. A loose, or poorly connected, ground clamp will cause severe resistance heating in the welding circuit with resulting loss of amperage at the welding arc.

When the welding power is ac, **do not have loops or coils** in the electrode or ground cables. Most especially, do not hang loops or coils of welding cable over or around pieces of iron. The resulting magnetic field will cause severe impedance to welding current flow. Although the welding power source may be putting out full amperage for the setting, the result at the welding arc will be very low current.

This is by no means all there is to trouble shooting a welding power source. It may, however, provide you with some useful data from which you can start to correct the problem. Remember: a maintenance and operation manual is provided with each welding power source. **Be smart—be safe—use yours!**

Summary

There has been no attempt to discuss electric control of welding amperage output, tapped reactor control, etc., in this section of the text. Succeeding sections of the book will cover these subjects in detail.

Transformer type ac welding power sources are relatively simple in design, construction and operation. AC welding power sources always operate from single phase primary power. This type of power source is usually low in initial cost and maintenance requirements are minimal.

Although all ac welding power sources are **not** power factor corrected, they may be. Normal ac welding power sources will have 50-55% power factor at rated load as designed. This can be brought up to the welding industry standard of approximately 75% power factor corrected, at rated load, by the addition of capacitors to the primary power circuit. Power factor correction will provide unity power factor (100%) at about half the rated amperage output of the power source. A leading power factor is normal at welding amperages below 50% of the rated amperage output.

Chapter 9

RECTIFIERS FOR WELDING POWER SOURCES

General History.

The function of any rectifier is to change alternating current to direct current. The rectifier may be any one of several materials which will, under the proper circumstances, provide the rectifying action.

Before the rectifier came into prominence in the welding industry other methods were used for providing dc welding power. Perhaps a bit of the historical development of dc welding power sources will provide a setting for the rectifier story.

The first dc welding power came from large battery banks. Power input maintenance for the batteries was provided by high capacity dynamos. The resulting dc welding power output was very stable with no ripple factor at all.

Following this development was the engine driven generator. The first welding power sources of this type, although portable in a sense, were carried on railroad flat cars. The portable generator was a major break-through for the welding industry. For the first time welding power could be generated at the work site where it was needed even though there was no primary electrical power available. This concept is still in use today although the generators and engines have had substantial design changes over the years.

With the engine driven generators came the electric motor-generators, or MG sets as they are better known to industry. This type of welding power source has also undergone many design changes since being introduced to the welding market. The basic revolving armature design, however, is still used.

The next step in the evolution of producing dc welding power was the introduction of rectifiers for welding power sources. This was not the first use of rectifiers by industry as a whole but it was the first use of this type of electrical device to change ac to dc in a welding power source.

While the concept of using transformer type ac welding power sources was not new to the welding industry, the use of rectifier elements to change ac welding power to dc welding power was a novel idea. At that time, several methods of rectification were known but only two of them provided the current carrying capacity required for welding operations. The two types of rectifiers were the **vapor-arc type,** commonly known as ignitrons and thyrotrons and the **metallic types** using such elements as cuprous oxide, selenium, magnesium-cupric-sulphide and germanium.

The vapor-arc rectifier never achieved commercial popularity for use in welding power sources. They function at relatively high voltages and the arc welding processes are essentially a low voltage application. The original cost factor of vapor-arc rectifiers is high and the maintenance required for this type of electrical equipment is expensive. They are used, however, for other types of electrical circuits.

The metallic group of rectifiers have excellent characteristics for welding power source applications. Of the several elements mentioned previously, selenium has been the most successful for this function. Selenium rectifiers have a higher voltage rating per plate than many other metallic materials used. The rectifier plate voltage rating, when properly considered in the design of rectifier stacks, makes it relatively simple to determine the exact rectifier requirements for any welding power source with a known open circuit voltage.

There is another element that has achieved widespread popularity in the manufacture of rectifiers over the past several years. The element silicon is commercially available in silicon diode, or two element, rectifiers. Both silicon and selenium will be discussed in detail in this chapter of the text.

The generation of dc power for welding, whether it be a mechanically driven generator with a brush-commutator arrangement, or by some type of rectifier, is normally based on three phase primary alternating current power. In view of the great design difference in motor-generator and transformer-rectifier welding power sources, though they may use power from the same primary distribution source, the following pages will discuss both types of welding power sources. The reader will then have a better understanding of the methods of operation, the power requirements and other pertinent data involved in selecting a power source for welding.

Motor-Generator Equipment.

In the operation of a motor-generator welding power source

there is primary electrical power, normally three phase, brought into the electric motor portion of the system. **Its sole function is to power the electric motor.** The motor provides the mechanical power to turn the rotor assembly. The electrical power in the electric motor is, therefore, changed to mechanical energy in the rotor assembly as the motor turns the rotor shaft. The rotor is sometimes called the **armature** although this is incorrect terminology. The rotor is actually the total shaft assembly which includes the armature iron core material, the armature coils, the commutator and the shaft end bearings.

The block diagram illustrated in Figure 100 is for a typical motor-generator welding power source. As shown, the primary three phase power is brought to the electric motor. The induction type electric motor causes the rotor shaft to turn. The armature is a mass of iron core material and copper conductors shaped as coils, all located on the rotor shaft. The rotor assembly armature rotates within the **stator,** or stationary part, of the welding power source generator. The magnetic field coils, which are necessary for electrical power generation, are located in the stator portion of the circuit.

The rotation of the armature within the stator generates alternating current. The current is actually generated in the armature coils. It flows from the armature, through electrical conductors, to the commutator.

The commutator is a system of copper collector bars placed concentric to the rotor shaft centerline. (Concentric means the commutator bars are all an equal radius from the centerline of the shaft). The conductor wires from the armature are soft-soldered to the commutator bars. In a sense, the copper collector bars of the commutator collect the generated ac power and present it for pick up by the carbon-type brushes which are part of the circuit.

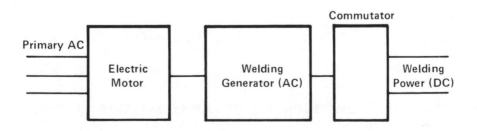

Figure 100. **MOTOR-GENERATOR BLOCK DIAGRAM.**

For a four pole generator system there are four contact brushes positioned at specific points around the commutator. The brushes are so arranged that they pick up specific half-cycles (half-hertz) of the generated alternating current and direct it down a single conductor as direct current. It is in the commutator-brush assembly that the generated ac power is changed to dc welding power. To use an analogy, it could be said that the commutator-brush assembly is a type of mechanical rectifier.

It is at the commutator-brush assembly that mechanical energy is changed back to electrical energy. It is a fact of physics that anytime there is a transfer of energy from one type to another there will be energy losses. The energy losses decrease the electrical efficiency of the power source considerably. This helps to explain why the maximum electrical efficiency of the best set-up and balanced motor-generator set is approximately 55% at rated load. The overall electrical efficiency of the motor-generator set is considerably less. A comparison of motor-generator sets and transformer-rectifier type welding power sources is made in Figure 101.

A three phase transformer-rectifier type welding power source takes the primary power from the input distribution system, transforms it through the transformer to usable welding voltage and amperage, normally has some method of mechanical or electrical output amperage control and then changes the alternating current to direct current through the rectifier. There is no change of energy type as there is with the motor-generator power sources. Electrical efficiency of transformer-rectifier type welding power sources will normally be approximately 73% at rated load.

Power Source	Power Sequence	Approximate Efficiency at Rated Output	Idle (Power Required)	Total Electrical Efficiency
Motor Generator	E-M-E	55%	3,000W	15-18%
Static Rectifier	E	73%	600W	65%

Figure 101. COMPARISON, MG SET AND TRANSFORMER-RECTIFIER.

For the comparison in Figure 101, we selected 300 ampere rated constant current power sources with the old NEMA 40 load volt rating. Both units are rated at 60% duty cycle. This type and amperage rating of power source has been standard in the welding industry and is probably the most popular class of power source manufactured.

Please note that the electrical efficiency of any electrical device is based on the power used (primary) divided into the welding power output (secondary). The percentage of electrical efficiency is one way of determining the power losses in electrical equipment.

The complete story of any welding power source is not told by the electrical efficiency rating **at rated load output.** Analysis of arc welding applications employing the shielded metal arc welding process has shown average welding operator **arc time** as approximately 23% of his working day. The balance of his time is devoted to setting up the work, chipping slag from the welds, changing electrodes, etc.

In an eight hour workday the welding power source is operating at rated load, or less, for about two hours. The rest of the time the unit is idling and cooling. As long as it is energized the power source is still drawing and using current from the primary power system. The amount of primary current drawn but **not used for productive purposes** becomes a vital factor in counting net welding costs. It certainly has an effect on the overall electrical efficiency of the welding power source.

The chart in Figure 101 shows data that has been known for many years. Power sources of similar electrical ratings are compared although they are radically different in design concepts. Some of the data presented has been explored to explain rated load electrical efficiencies of welding power sources. An examination of the idle time efficiencies and costs is necessary since approximately 75% of the work day is spent in the idling condition.

When idling, the welding power sources have a substantially lower power demand from the primary power supply system than when they are operating at rated load.

As shown in Figure 101 the motor-generator power source requires approximately 3,000 watts per hour to maintain the electric motor at its proper speed in rpm. If there is less than the required amount of primary power, due to a 10% or more voltage drop **over a period of time,** the electric motor may stall. A stalled electric motor will not necessarily damage the welding power generator but it could cause serious damage to the electric motor coils because of overheating. Remember: **It takes a certain amount of kilowatts just to keep the electric motor operating.** If there is low primary voltage the primary amperage has to make up the difference in power (volts times amperes equals watts). The excessive amperage, used over a period of time, can cause overheating of the electrical conductors in the electric motor thus destroying their insulation and causing a short circuit in the motor coils.

The primary power required for the 300 ampere rated transformer-rectifier power source is approximately 600 watts per hour. The electrical power is used to operate the fan motor for cooling the power source components and also to provide excitation current for the electrical control circuit. It also furnishes power to compensate for the small electrical losses incurred by resistance heating of the transformer coils and iron core.

Considering the total primary power input to the two types of welding power sources in an eight hour day, it is apparent that the **total electrical efficiencies** of the two power sources will be radically different. The motor-generator set, idling for about six hours each work day, has an electrical efficiency of approximately 15-18%. The transformer-rectifier, working the same hours and idling the same hours, has an electrical efficiency of approximately 65%. The primary current drawn by the transformer-rectifier power source is considerably less than the primary current drawn by the motor-generator unit.

In many areas of the world, utility companies operate on a demand meter basis where the customer pays a rate per KW hour based on the power demanded from the primary distribution system **at any one time.** (It is referred to as "instantaneous demand power"). Any inductive electric motor, such as those used to power motor-generator power sources, will demand **three to ten times their operating amperage for starting purposes.** The demand, since voltage is considered constant at the primary distribution point, is for **more primary current.** For example, if the power source uses 100 amperes at rated output load, **the starting current would be from 300 amperes to 1,000 amperes.** This places an extremely heavy short time load on the primary power system as well as on the shop wiring. In such a situation the user normally has to pay the utility company a penalty factor per kilowatt-hour charged.

The primary current demand is relatively small for the transformer-rectifier type power source. The idling current has a very low value until the power source is actually used for welding. The amount of primary current used is regulated by the amount of welding output current employed. A transformer type welding power source only demands, and uses, the primary current necessary to supply the welding arc with power.

Rectifiers.

The concept of the rectifier has been known for many years. It is only within the past few years, however, that rectifiers have been adapted to welding power source applications.

The function of any rectifier is to change alternating current to direct current. The type of rectifier employed to do the job is of no consequence as long as it will perform the desired objective. Power source design, however, must be different for silicon rectifiers and selenium rectifiers.

In this discussion of rectifiers we will consider only those that are presently used for welding power source applications. As we have said, there are many materials that may be used for rectifiers. For welding applications, however, only silicon and selenium have been found satisfactory for the high amperages and low voltages involved.

Both selenium and silicon have individual advantages that must be considered when applying them to welding power sources. The advantages of each type of rectifier will be discussed under the separate headings of Selenium Rectifiers and Silicon Rectifiers.

Selenium Rectifiers.

Probably the most well known selenium rectifiers built for welding power source applications are the Gold Star main power rectifiers manufactured by Miller Electric Mfg. Co., Appleton, Wisconsin. The following data is basically a modified outline of the manufacturing processes used by this company.

The selenium rectifier single plate **cell** originates with an aluminum base plate which is designated the **back electrode.** An aluminum base plate is usually preferred because of light weight, better heat dissipation and the elimination of the rust problem that would be evident with steel plates.

In preparation for the selenium deposit the surface of the aluminum plates are dry grit blasted to remove surface oxides. The roughened surface provides a greater area for the selenium deposit. Selenium is applied by the vapor deposition method. This operation is performed in vacuum chambers.

In the vacuum chamber the selenium is boiled and the aluminum plate is passed over the molten metal. The selenium vapors condense on the surface of the aluminum plate to a thickness of approximately 0.004″. To achieve this thickness of selenium the plate is passed over the boiling metal 150 times! The selenium layer covers the entire surface of one side of the aluminum base plate.

During and after the vapor deposition the selenium is processed in such a manner as to establish proper electrical conductivity. Figure 102 shows one step in the manufacture of selenium rectifier cells.

Figure 102. **SELENIUM CELL BACK ELECTRODE.**

After removing the selenium-covered plates from the vacuum system, they are placed in a masking device which covers a circular portion of the center of the cell as well as approximately ¼″ around the periphery of the cell. The cell is sprayed with an alloy of low melting metals complimentary to selenium. The application technique is called "thermal spraying" and the deposit is called the **front electrode.** The front electrode materials are usually an alloy of cadmium, bismuth, tin, and other metal elements. The front electrode alloy has a very low melting point. A view of the selenium cell components is shown in Figure 103.

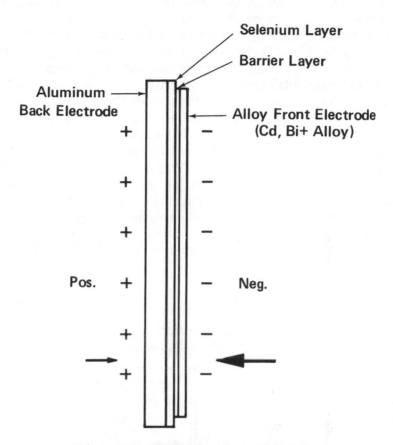

Figure 103. SELENIUM CELL COMPONENTS.

When the masking device is removed, there is a central circular area and the ¼″ outer edge which does not have the front electrode coating. The purpose in leaving this area unsprayed is to provide insulation between the front and back electrode of the plate. (The terms "front electrode" and "counter electrode" are synonymous when discussing rectifier elements). The front, or counter, electrode is the cathode element in selenium rectifier plates. The front electrode alloy materials will have a melting range of approximately 230° F. to 338° F. The element selenium has a melting point of 422° F. This fact is evidence that selenium rectifiers must operate at cooler temperatures than this to maintain rectifier integrity.

Electrical connections made to the back surface of the back electrode are positive while those made to the front surface of the front electrode are negative. The base plate acts as the total assembly support and one electrode. The low melting alloy of the front electrode acts as the other electrode. The combination presents a one piece rectifier requiring no high pressure contact for rectification.

When a dc potential (or one-half of an ac sine wave) is applied to the selenium cell so that the **back electrode is polarized positive** and the **front electrode is polarized negative,** a current of much higher value will flow through the cell than when the polarity of the cell is reversed. The amount of current flow depends directly on the effective rectifying area of the cell and the applied voltage. The relationship of current flow and applied voltage are not linear.

While very few readers will ever have occasion to assemble a selenium rectifier stack, it is interesting to know how they work. In assembling the selenium rectifier cells into "stacks", contact with the base plate (back electrode) is made with a metal washer or terminal member while contact with the front electrode is made with a specially designed brass contact washer. The terminal connection and contact washer are illustrated in Figure 104.

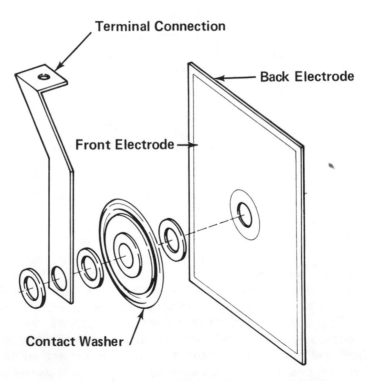

Figure 104. SELENIUM RECTIFIER COMPONENTS.

At this point in the rectifier manufacturing process, the reverse resistance of the rectifier cell falls off too rapidly to permit the cell to be put into practical service. To fully utilize the selenium cell it must be processed further by electro-forming.

A barrier layer of alloy is formed between the selenium layer and the front electrode alloy. The barrier layer is an alloy of the metals in both materials. It is created by applying a dc voltage in the reverse direction while, at the same time, allowing the cell temperature to rise. Additional treatment and "seasoning" completes the formation of, and stabilizes, the barrier layer. The actual barrier layer is only a few molecules thick.

At the beginning of the electro-forming operation, the applied voltage is low to prevent overheating of the cell. As the reverse resistance builds up, the applied voltage is increased until its maximum value is well above the rated voltage of the selenium cell.

The theory is that the electro-forming process increases the effectiveness of the barrier layer and aligns the excess electrons in the front electrode immediately adjacent to the barrier layer. At this point, the front electrode has an excess of electrons and the back electrode has a deficiency of electrons. In this case there is **a high negative charge at the front electrode and a high positive charge at the back electrode.** Since electron flow is from negative to positive current flow will be very good from the front electrode to the back electrode.

Each half-cycle of ac power, however, the current flow is reversed. Now current is trying to flow from the back electrode to the front electrode. This is not a good situation since we are trying to take electrons from an area of electron deficiency and move them to an area having a surplus of electrons. As you might imagine, very little current flows in the reverse direction through a selenium rectifier cell. This is, in fact, what makes a rectifier function properly. The electro-forming process has little effect on the forward characteristics of the cell except to stabilize the forward resistance. A complete selenium rectifier stack is shown in Figure 105.

The principle function of a rectifier is to freely pass electrical current in one direction (forward) while blocking, or greatly limiting, its passage in the opposite direction (reverse). If the rectifier cell easily passes current in the forward direction, it implies that the resistance to current flow in this direction is very low. The resulting forward voltage drop across the cell must also be very low.

If the current flow in the opposite direction is blocked, or held to a minimum value, the reverse electrical resistance must be high.

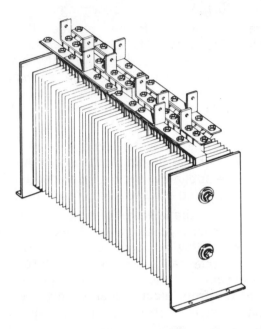

Figure 105. COMPLETE SELENIUM RECTIFIER.

It is logical that the voltage drop across the cell must also be high. The maximum value of the reverse voltage drop is limited by the breakdown potential of the barrier layer of the selenium cell.

Selenium rectifiers have the advantage of being able to accept large voltage value surges, or voltage "spikes", without breaking down. The very large rectifying area tends to spread the voltage over many square inches of rectifier surface and the heat absorbency of the aluminum back electrode dissipates the heat over a wide area. The natural configuration of selenium rectifiers, with their numerous plates or cells, makes cooling of the rectifier relatively easy.

It is simple to add more rectifying area if it is required for a specific application. Just add more selenium cells.

The greater size of the rectifier, of course, requires more physical space in which to fit them into the welding power source. The greater space provision allows more cooling air to pass over the selenium rectifier plates, increasing their work life and decreasing the possibility of rectifier failure.

Silicon Rectifiers.

The silicon diode is a two-element rectifier that has come into prominence in the past few years as a main power rectifier for welding power sources. Silicon rectifiers perform the same basic function as selenium rectifiers which is to change ac to dc. Both selenium and silicon rectifiers are classed as semi-conductors although the basic rectifier materials differ greatly in their natural occurrence.

Silicon is the most common natural element on Earth. It is found on every continent and under every sea. The basic element is, of course, sand.

When sand is melted and refined to a certain state it is known as glass. In this form it is one of the best non-conductors of electrical current known. When used with other elements it can be converted to a semi-conductor. (**A semi-conductor is an element or material that exhibits almost perfect electrical resistance at 0° Kelvin but which increases in electrical conductivity as its temperature rises**).

Before becoming a semi-conductor the basic materials must be refined to an ultra-pure state. There are several methods of refining silicon and often several methods are used, stage by stage, to achieve the purest end item silicon.

One of the ultimate methods of refinement is known as "zone refining". In this process, the ingot of silicon, which is approximately one inch in diameter and twelve inches long, is placed in an inert gas atmosphere. A wave of heat is applied at one end and caused to move the length of the ingot. Any impurities in the ingot are driven out ahead of the applied heat.

Pure silicon must be transformed into a state that will be useful for silicon diode rectifiers. This is done by placing the silicon material in a crucible and heating it to approximately 1450° C. (2462° F.). At this temperature the mass is molten. During the time the material is being brought to the correct temperature, desired impurities are added to convert the silicon from a non-conductor to a semi-conductor of the proper type. A small amount of gallium or arsenic may be added for this purpose. The impurity addition is usually only a few parts per million (ppm).

The crucible is placed in a crystal growing furnace for the purpose of growing a Czochralski crystal. The temperature of the glowing red mass is accurately maintained within one-quarter of a degree centigrade and a "seed" is planted.

The seed is a specially selected crystal of super-silicon which will start the crystal growth phenomenon. The seed is partially immersed in the molten silicon and then **very slowly withdrawn** at a controlled rate. One rectifier engineer compares the crystal seed withdrawal to, "Pulling your foot out of a mudhole over a period of two hours!".

As the silicon seed is withdrawn from the surface of the molten mass, a single crystal forms that eventually reaches approximately one inch in diameter and twelve inches in length. It may take as long as four hours to complete a single crystal growth.

The result is a single crystal of **either** "N" or "P" type silicon. The "N" or "P" describe the type of impurity which was used for the semi-conductor and designates the basic polarity that will be given to the finished silicon device. The silicon ingot will weigh approximately one to two pounds and will provide enough material for several thousand silicon diode rectifiers.

The ingot is sliced into thin wafers with a diamond saw. The wafers are only a few thousandths of an inch thick. The next step is the formation of the "junction" where the actual rectification takes place. Without this junction, there would be equal electrical conductivity in both directions through the silicon wafer.

There are several methods that might be used to create the junction. In this discussion we will consider only the "alloy method". The silicon wafers, or chips, are placed on an inert plate and a solution containing another element, usually arsenic or aluminum, is applied to the upper surface by painting, silk screening or any one of a number of methods. The entire tray of silicon wafers is then sent through a diffusion furnace.

In the diffusion furnace, the temperature of the silicon wafers is brought to a critical point and maintained for a precise period of time. During this time the elements in the applied impurity solution diffuse through the silicon wafers. At a precisely timed moment, when the impurities have diffused approximately half-way through the wafer thickness, the diffusion process is stopped. If the total process has been successful, and it isn't always, the result is a batch of silicon wafers with excess electrons on one side (N) and a deficiency of electrons on the other, or (P), side. These are actually called the "N" and "P" sides of the wafer, respectively. After the silicon wafers have been removed from the diffusion furnace the residue of the upper alloying element is removed in preparation for making the ohmic contact.

The silicon wafer must be mounted on the necessary hardware. The wafer, by nature of its thin cross section and brittleness, is a fragile component and should be treated with the utmost care. Figure 106 shows the various components of the silicon diode in cross section.

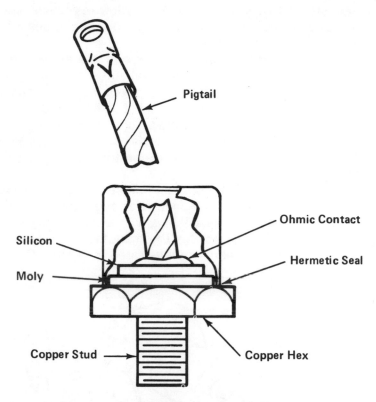

Figure 106. SILICON DIODE COMPONENTS.

When put into service as a rectifier there will be considerable heat generated in the silicon wafer. The heat must be dissipated to prevent fracture or melting. Since copper has excellent thermal conductivity, as well as excellent electrical conductivity, copper stud heat sinks are used. The copper is usually an oxygen-free grade of high purity. The stud and hex-head arrangement is made suitable for mounting to larger aluminum, or copper, plate heat sinks.

On top of the copper hex-head and stud is placed a material which has a coefficient of expansion similar to the silicon wafer. Molybdenum is the material generally used for this purpose. The "moly" is secured to the copper base with some type of hard solder (other than a tin-lead alloy) having a relatively high melting point. The silicon wafer is placed on the molybdenum and hard soldered into place. This is an extremely critical operation in the manufacturing sequence of silicon diodes. Many parts are rejected at this point because of the high standards of quality control imposed by the manufacturers of silicon diode rectifiers.

The next step is to connect the ohmic, or low resistance, contact to the upper side of the silicon wafer. It is to this component that the copper pig-tail lead is attached.

All through the manufacturing operations there are cleaning procedures for the hardware, the copper heat sinks, the silicon wafer, the solders, the ohmic contact area, etc., because it is imperative that no foreign material be present in the finished product. Even the clothing worn by the manufacturing personnel must be lint-free.

Water vapor must also be excluded and, to accomplish this, most of the manufacturing operations are performed in a dry, inert gas atmosphere.

After the connection of the ohmic contact to the silicon wafer, the top cap, or ceramic seal, is applied over the entire device and cold-welded to the copper hex-head. Finally, the copper pig-tail lead is crimped into place. This completes the manufacturing operations. Completed silicon diode rectifiers are illustrated in Figure 107.

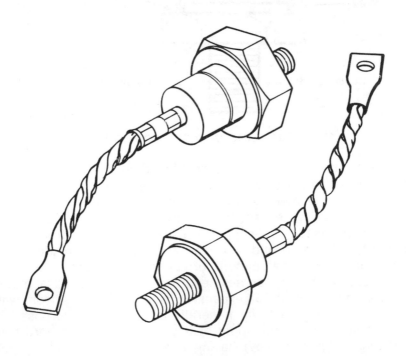

Figure 107. COMPLETED SILICON DIODE RECTIFIERS.

All through the manufacturing process and operations electrical tests are made to eliminate faulty parts as early in the cycle as possible. After the final assembly each silicon diode device is tested for voltage and current rating.

To test the silicon diode for **voltage** capability, a programmed increased direct current (dc) voltage is impressed in the reverse direction and the leakage current observed on an oscilloscope. When the leakage current reaches a certain value the voltage is noted and this becomes the "peak inverse voltage rating" of the silicon diode.

To test the silicon diode for **current** rating, a known forward current value is forced through the rectifier and the voltage drop across the diode cell is observed. The diode is then rated according to its losses. Forward voltage drop should be as low in value as possible.

The result of the total manufacturing effort is an electrical device which, by nature of the rectifying junction, will conduct current in one direction and not in the other. The silicon diodes used for main power rectifiers in NEMA Class 1 welding power sources normally have a peak inverse voltage rating of 250 volts with a non-repetitive transient over-voltage rating of 350 volts. In actual manufacturing practice, the voltage ratings range from 600 to 900 volts.

The two most commonly used silicon rectifiers, or diodes, in welding power source main power rectifiers are the 150 ampere device with a 3/8" or 1/2" diameter stud, and the 275 ampere rated diode with a 3/4" diameter stud.

The silicon diodes are stacked into single phase and three phase bridge rectifiers in much the same manner that selenium cells are used except that fewer diodes are needed for a given rectifier amperage rating.

The aluminum heat sinks used with silicon diodes are large in square inch surface area and thick in depth. These aluminum heat sinks are used to conduct the heat generated in the silicon diode away from the diode and to the air stream created by the welding power source fan. Only one diode is installed on each aluminum heat sink. The mass of the aluminum heat sink will remove any heat generated almost instantaneously. When installing a silicon diode in the aluminum heat sink, a heat sink compound should be used. This material provides excellent thermal (heat) conduction between the interface of the copper hex-head stud and the aluminum heat sink.

Although a torque wrench was required for installing silicon diodes in early silicon rectifier stacks, it is no longer necessary in most cases. Instead, a special Belville washer (a combination washer and compression nut) is used to control the tension on the mounting stud.

The installation of a silicon diode, with Belville washer, is accomplished by putting the diode in place in the aluminum heat sink and bringing the locking nut (Belville washer) up to firm finger tightness. An ordinary wrench of the proper jaw size is applied to the nut and one-half turn (180 degrees), no more, no less, is added. The additional half-turn compresses the calibrated Belville washer the exact amount necessary to insure the proper connection. An installation diagram for silicon diodes is illustrated in Figure 108.

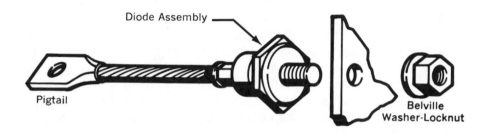

Diode Assembly

Pigtail

Belville
Washer-Locknut

Figure 108. SILICON DIODE INSTALLATION.

Care should be taken when installing a silicon diode, not to strain the copper pig-tail lead. A stress could be introduced within the structure of the diode assembly which could possibly cause the silicon wafer to fracture. Proper installation procedures dictate that the diode should be placed in the heat sink, firmly connected and then the pig-tail should be attached to the proper connection. Only after this has been accomplished can the diode hex-nut be brought up tight and secured. Before the final tightening of the hex-nut the stud should be rotated so no strain is apparent on the pig-tail.

It is true that a rectifier may fail for any one of a number of reasons. Silicon diodes are no exception to this rule. Diodes that are suspected of malfunctioning may be tested in the field using an ohm meter. THE OHM METER SHOULD BE SET ON THE "X1" SCALE OR IT WILL GIVE A FALSE CONTINUITY READING. Simply put one probe on the pig-tail portion of the diode and the other probe on the stud portion of the diode. You should read either **no continuity** or some level of continuity. By reversing the probes on the diode and pig-tail you should obtain an opposite reading to the one you previously had. For example, if you had **no continuity** in the first placement of the ohm meter probes, you should have some level of continuity when reversing the probes.

If a silicon diode is going to fail in service it will normally fail in the **short circuit position.** For this reason it is important to shut off the primary power to the power source as quickly as possible after a suspected diode failure. For example, if a primary fuse in the line disconnect switch failed to blow and interupt the primary current to the power source transformer, there could be serious damage to the power source circuitry. When a diode fails in the short circuit position, a full short circuit load is placed on the secondary ac circuit in that phase, or portion, of the circuit. This would represent only one-half cycle of ac power and the tremendous load placed on the other diode in the same phase would soon cause it to malfunction. We are talking about time in seconds! The total short circuit current would flow through the secondary coils of the main transformer causing it to overheat and possibly damage the insulation. The result could be catastrophic failure of the main transformer.

Silicon diodes should be matched for voltage and amperage rating when replacements are necessary in a welding power source. Supplying the power source model number and serial number when ordering parts will help to assure you of receiving proper diodes for the power source.

Silicon rectifiers are enjoying increased popularity in certain applications for rectifying welding current. This does not mean that selenium rectifiers are a thing of the past in welding power sources. Reputable power source manufacturers will select the rectifier type which will best serve the needs and applications of the customer. Cost, electrical capacity, configuration and space are a few of the considerations to make when selecting the proper rectifier for welding power source applications.

Summary.

The service use of the welding power source will usually determine the type of rectifier that should be used in its manufacture.

All welding power source rectifiers, silicon or selenium, perform the same function which is to change alternating current to direct current.

In view of the OSHA regulations concerning noise pollution the motor-generator type power source may be relatively obsolete as compared to the transformer-rectifier type welding power source.

Chapter 10

AC/DC AND DC TRANSFORMER-RECTIFIER POWER SOURCES

The use of ac/dc welding power sources has gained much favor with the welding industry since World War II. Although this class of power source is a compromise between an ac power source and a dc power source it does have advantages for some applications.

An ac/dc transformer-rectifier welding power source is basically an ac welding power source operating from single phase primary electrical power. The dc welding capability is added by including a rectifier in the secondary portion of the power source circuitry. As you know, rectifiers perform only one function which is to change ac to dc.

Since the ac/dc type power sources operate from single phase primary power, the dc portion of the welding output current will have a relatively high ripple factor percentage. The reason for the high ripple factor percentage is that single phase based dc power theoretically drops to zero current each half-cycle (half-hertz) of ac input (each 1/120th of a second). To provide continuity of current, and stability in the welding arc, a heavy duty stabilizer is normally added to the dc portion of the power source circuitry. The function of the stabilizer is to slow down the rate of response of the power source to changing arc conditions, thereby providing a more even, stable welding current value at the arc.

All ac/dc welding power sources are built in the constant current, or conventional, design and have the "drooping" volt-ampere output characteristics. They are designed for use with the shielded metal arc welding process and the gas tungsten arc welding process. For other dc applications, such as air carbon-arc cutting and gouging, ac/dc power sources do not have the total output power level that three phase based dc power sources have. This is based on similar amperage ratings and load voltage ratings in the power sources.

Typical volt-ampere output curves are shown in Figure 109.

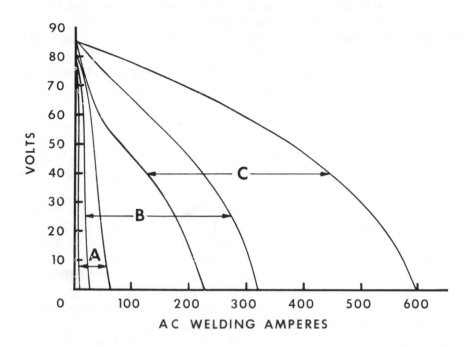

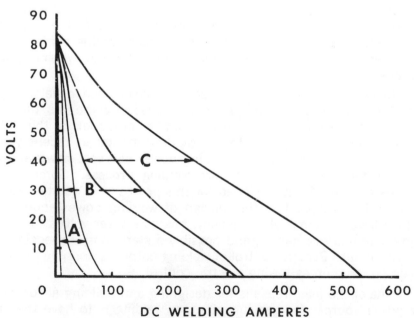

Figure 109. AC AND DC VOLT-AMPERE CURVES.

Note that the ac volt-ampere curve has a definite convex shape while the dc volt-ampere curve is slightly concave. The dc curve will normally have a higher maximum short circuit current, for a given ampere rated power source, than the ac curve. This is characteristic of almost all ac/dc power sources.

The methods of output control of welding power will vary with the model and type of ac/dc welding power source. The control may be either electrical or mechanical.

Mechanical control of welding amperage output is normally provided by a movable coil or a movable shunt system. The control principles are the same as those used for ac welding power sources as previously discussed. Mechanical control may be used for all NEMA Class 1, Class 2 or Class 3 ac/dc power sources.

Electrical control of welding output may be either a saturable reactor, or a magnetic amplifier, control. Electric controls are discussed in detail later in this text. The basic reason for using electric control of welding power output is to obtain remote control capability for the welding operator. This permits him to adjust welding power source output from his work station without having to return to the power source each time a change is required in welding conditions.

The transformer design and function is the same for ac/dc power sources as it is for ac transformer power sources. The addition of a rectifier and the necessary dc circuitry make it possible for the welding operator to have ac, dcsp or dcrp at the flick of a switch. To have this capability, users will tend to sacrifice the better dc arc characteristics obtained with standard three phase dc welding power sources.

Many ac/dc welding power sources have been developed for both the shielded metal arc welding process and the gas tungsten arc welding process. In particular, a number of models of the NEMA Class 2 rating have come into the welding market in recent years. **The first application of ac/dc welding power sources, however, was basically for the gas tungsten arc welding process.** Some power sources are specifically designed for this process. In addition to the ac/dc welding capability, this design of welding power source will usually have built-in high frequency, gas and water valves and solenoids, a primary contactor, and possibly a start control rheostat and control circuit. Remote control of welding output is possible on those ac/dc power sources having electric control.

One of the main reasons for designing and building ac/dc welding power sources for gas tungsten arc welding is to have the capability of ac welding of aluminum and magnesium with the process.

When welding either aluminum or magnesium with gas tungsten arc there will be developed an electrical phenomenon known as "dc component". Some method of dissipating the energy of dc component is usually incorporated in NEMA Class 1 power sources designed for gas tungsten arc welding. This may take the form of ni-chrome resistor bands, capacitors in series with the welding arc, etc. The subject of dc component will be explored thoroughly in the portion of this book relating to gas tungsten arc welding.

In summary, it is evident that the ac/dc transformer-rectifier welding power source has found wide acceptance in the welding industry despite the fact that it is a compromise power source. The fact that it has both ac and dc output welding power appears to offset any disadvantages that may be incurred by single phase rectified dc power. The fact that ac/dc power sources operate from single phase primary power is an advantage in those areas where single phase primary power is all that is available. The increase in the availability and variety of ac shielded metal arc welding electrodes has also done much to make this type of power source popular with smaller shops.

DC Transformer-Rectifier Power Sources.

The use of three phase-based direct current (dc) power for welding has many advantages and characteristics that are desirable or required from the standpoint of the user. Since electrical power distribution is alternating current (ac) in most areas of the world, the ac must be converted to dc in some manner for welding applications.

In the early days of the arc welding processes a welding power generator was driven by either a three phase electric motor or some type of fuel powered engine. Th only real criteria for engine, or electric motor, horsepower and size was its ability to permit the generator to reach full power output. Beyond that the generator motive power had no effect on the welding power output of the welding generator.

Since the inception of the transformer-rectifier type power source in 1950, there has been a definite and predictable trend toward greater and more varied use of this class of welding power source by the welding industry. The substantially reduced noise levels, minimum maintenance costs, stable arc characteristics and ease of remote control (electric control) all make the transformer-rectifier type power source desirable from the user's point of view. Transformer-rectifier power sources may be designed for operation on either single phase, or three phase, primary power systems.

Transformer-rectifier type welding power sources that are designed to supply **dc welding power only** usually operate from three phase primary line power. With few exceptions, ac/dc welding power sources are designed to operate from single phase primary line power. Those ac/dc power sources that claim to operate from three phase primary power function as a three phase transformer-rectifier for the dc welding output power. They function as a single phase transformer for the ac welding power output. Because of the nature of their design, such power sources will have low open circuit voltage when operated as ac power sources. The output current level is also considerably lower than the maximum of a similarly rated ac power source operating from single phase primary power. The reason is that the ac circuit must operate from single phase primary power which, when taken from a three phase line, is less power than if it were taken from a single phase system.

Welding power sources that are designed to supply dc welding power output only, and which operate from three phase primary power, are considered the most electrically efficient while theoretically having the lowest ripple percentage. This is illustrated in Figure 110 which shows the relative characteristics of rectified single phase and rectified three phase power.

Rectified DC

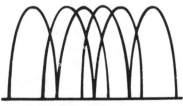

Figure 110. SINGLE, AND THREE, PHASE RECTIFIED DC.

For field welding operations the use of engine driven dc generators having a brush-commutator conversion of the generated alternating current to direct current has been replaced to a great degree by engine driven rectifier type welding power sources. In addition to the greater efficiency of the revolving field-rectifier type power sources, there is no possibility of the welding power source **changing polarity**

while welding current is flowing. This is a relatively common occurrence with motor-generators and engine driven units with the commutator-brush arrangement for changing ac to dc.

Another feature of the transformer-rectifier type power source is that, under parallel conditions with other rectifier type power sources, there can be **no current feedback** into the power source. The rectifier will not permit the welding current to pass in a reverse direction into the power source. This natural protection is not incorporated into the motor-generator type power source. The reverse flow of current into an MG set is possible through the brushes and commutator. From there it goes directly to the armature coils on the rotor assembly where the current is dissipated as heat.

A typical brush-commutator arrangement is shown in Figure 111.

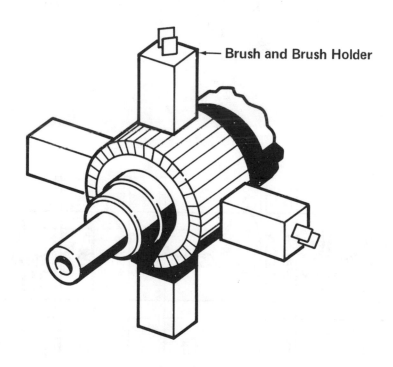

Brush and Brush Holder

Figure 111. TYPICAL BRUSH-COMMUTATOR ARRANGEMENT.

The modern rectifier concept is shown in Figure 112 in which both a selenium and a silicon main power rectifier stack is shown.

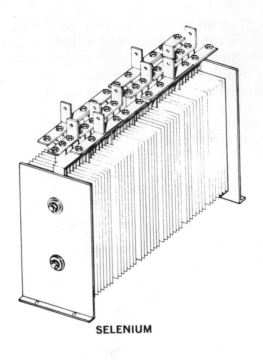

SELENIUM

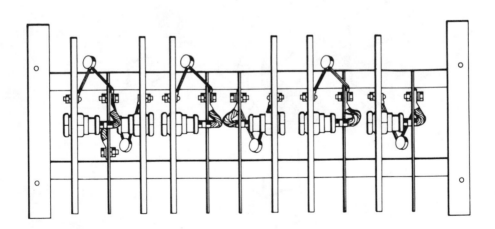

SILICON

Figure 112. MAIN POWER RECTIFIERS.

The use of transformer-rectifier power sources for various welding processes has increased for several reasons. It would not be possible to list them all but included in the list would be the following:

a) More efficient electrical operation.
b) Better power factor.
c) Less demand of primary current for starting.
d) Quiet, smooth operation.
e) Much lower maintenance costs.
f) More versatility in welding applications.
g) Easily re-connectable for different primary voltages.
h) Require less floor space.
i) Exceptionally stable arc characteristics.
j) Less magnetic arc blow.

The comparisons are to the standard motor-generator set.

Chapter 11

KEY CIRCUITS USED IN SOME DC AND AC/DC POWER SOURCES

The term "key circuits" may be defined as "those circuits that are important to the operation of more than one class of welding power source". Key circuits are usually some type of control system for regulating welding current output from a power source. The key circuits that will be described in this text are:
1. SATURABLE REACTOR AMPERAGE CONTROL SYSTEM.
2. CURRENT FEED-BACK CONTROL SYSTEM.
3. STANDARD 30 VOLT CONTROL SYSTEM.
We realize that knowledge of these circuits will not provide the reader with the magic key to understanding every type of power source manufactured for welding applications. It will, however, provide a basis for understanding some of the components of certain NEMA Class 1 industrial welding power sources and why they function as they do.

Saturable Reactor.

The term "saturable reactor" usually causes anyone, not familiar with its concepts, to veer around the problem and continue in ignorance. Although the term has a sense of mystery and majesty, the saturable reactor is a simple electrical device used to regulate the output amperage flow in the secondary circuit of a transformer type welding power source.

The illustrations shown in the next few pages of the text are designed to clarify the functions of the various component parts that make up the saturable reactor. They are intended for illustration only. The drawings do not represent any particular circuitry or welding power source type or model.

We will start with the basic transformer circuitry shown in Figure 113. At the left side of the drawing is shown the **primary coil** symbol. Further to the right are three straight lines which represent the **iron core**. At the right of the iron core symbol is the **secondary coil** symbol. The total group of electrical symbols represents an electrical transformer. The balance of the circuit shows the secondary output leads going to the electrode holder (upper right) and the ground connection (lower right).

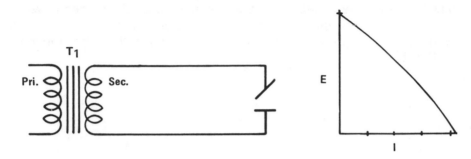

Figure 113. AC WELDING TRANSFORMER WITH ONE OUTPUT.

The drawing indicates the power source has only one output which is both maximum and minimum.

If a coil is added to the secondary circuitry, as shown in Figure 114, the drawing has this added symbol:

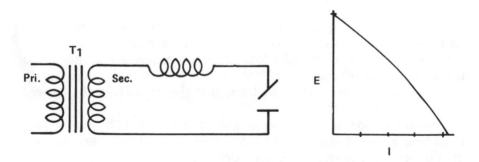

Figure 114. AC TRANSFORMER CIRCUIT WITH COIL.

The addition of the coil to the electrical circuit does create more electrical resistance but only to the value of the added conductor length. The welding current output, therefore, would be the same as before. There might be a slight decrease in the amperage output from the power source.

The next electrical component to be added to the electrical circuit is an iron core. This is indicated by the three straight lines placed under the coil in Figure 115. Note that we have incorporated the letter "Z" near the coil and iron core symbols. The total symbol is the American National Standards Institute (ANSI) symbol for a **reactor.** An important thing to remember is that **the same symbol is used to designate a stabilizer.** Some electrical symbols used in welding power source circuit diagrams, as approved by the American National Standards Institute (ANSI), are illustrated in the appendix of this text.

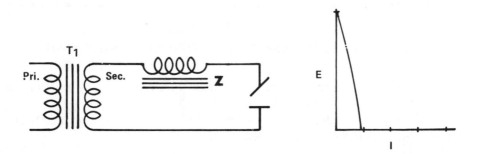

Figure 115. AC TRANSFORMER WITH COIL AND CORE.

The rule of thumb that is used to determine whether the circuit diagram is showing a reactor or a stabilizer is:

a) A **reactor** is normally located in the secondary ac portion of the welding power source circuitry.

b) A **stabilizer** is normally located in the dc portion of the welding power source circuitry.

The creation of a reactor, as shown in Figure 115, in the welding power source circuitry alters the amperage output characteristics of the welding power source considerably.

The letter "Z" is the electrical symbol for impedance. Impedance is a combination of resistance and reactance in the electrical circuit of a power source. While amperage flow could be stopped by impedance this is usually not the design purpose in a welding power source. Instead, impedance is used to limit, or slow down, amperage flow in a welding power source. Impedance does not just happen. It is the result of several separate occurrences that happen almost simultaneously.

The electrical reactor, although the name has very interesting connotations in this nuclear age, is a very simple electrical device. It consists of a conductor, usually copper wire, which is wound as a coil and placed around one leg of the reactor iron core. The two components that make up the reactor are an iron core and a coil of some type electrical conductor. When the conductor wire is not energized and carrying electrical current the two components are just so much copper wire and iron. It is not until amperage is caused to flow in the welding circuit that the reactor comes to life.

Perhaps a bit of review concerning the relationship of current, iron and magnetic fields is in order. You will recall that when current, or amperage, is caused to flow in a coil wrapped around an iron core, a magnetic field will be created. The strength of the magnetic field depends on three factors:

a) The mass and type of iron in the core.

b) The number of effective electrical turns in the coil.

c) The value of alternating current flowing in the coil.

Varying any one of these factors will cause the strength of the magnetic field to vary.

It is well to remember that voltage is the electrical force that causes current to flow in a circuit. Sometimes called "electro-motive force" or "electrical pressure", **voltage does not flow** but is the pressure behind current flow.

There are actually several important steps that take place when current is caused to flow in the welding circuit which has a reactor as part of the system. For clarity, we will number them in the sequence in which they happen. Please keep in mind that **everything listed here takes place within 1/120th of a second, or one half-cycle of ac power.**

1. Voltage is induced into the secondary circuit of the welding power source from the main transformer.

2. An arc is initiated at the electrode-ground area.

3. Voltage causes current to flow in the circuit, including the reactor coil.

4. A magnetic field is created in the reactor.

5. **Energy is used to create the magnetic field.**

6. The energy is not "used up" but is stored in the magnetic field.

7. Since the magnetic field depends on current to increase, and sustain, its strength, and current goes to zero each half-cycle

(each 1/120th of a second), the magnetic field must also collapse and go to zero strength each half-cycle.

8. **The energy stored in the magnetic field is returned to the circuit, and the reactor, where it originated.**

9. The energy is apparent in the circuit as a counter-voltage with force direction exactly opposite that of the impressed ac voltage in the circuit. This is a classic example of Lenz's Law in action.

10. The counter-voltage is the impedance factor in the circuit.

11. **The impedance factor, or counter-voltage, value will determine how much welding current is allowed to pass through the reactor coil.**

The electrical circuit shown in Figure 115 shows the full reactor in the welding power source circuit. The maximum amount of magnetic field strength will be created and, therefore, the maximum amount of counter-voltage, or impedance, in the circuit. The result is that **current flow** in the circuit would be **minimum.**

At this point there has been some significant data assembled. The maximum output current level has been determined. The minimum output current level has been determined. By adding a reactor to the welding power circuit a certain amount of output current control has been achieved.

In this circuitry the reactor seems to be the "key" component. From what we have learned it appears that we may be able to obtain even finer output control if the reactance in the circuit could be varied in some manner.

There are at least three methods in which the output welding current could be controlled in the circuit shown in Figure 115. One way would be to physically **move** the iron core in or out of the confines of the reactor coil. This method of control is cumbersome and not a popular solution to the problem. It would limit the possibility of remote control of welding current output from the power source.

Another method would be to **saturate** the iron core with a magnetic field of force lines (magnetic flux) produced by the current flowing in a dc coil, thereby altering the effect of the iron core. This method would require substantial alteration of the existing circuit shown in Figure 115.

A third method would be to **change the number of effective turns** in the series power coil around the iron core. This may be accomplished by physically tapping the reactor coil at specified electrical turns of the coil. Use of the various taps is provided by either a tapped range switch or plug-in receptacles. A tapped reactor is shown in Figure 116.

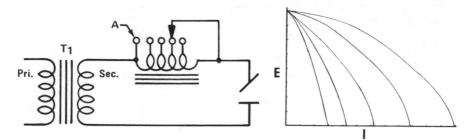

Figure 116. **TAPPED REACTOR CURRENT CONTROL.**

The illustration shows that tap "A" will permit the highest current output from the welding power source. Analysis will show that, under this condition, the entire reactor coil is out of the welding power circuit. There is no impedance to the flow of welding current in the secondary circuit.

The circuit shown in Figure 116 indicates that there are several ranges, or amperage levels, that may be obtained with this method of current control. It is still not possible, however, to achieve the fine current control necessary for most welding applications.

Since it is true that an infinite range of amperage control is highly desirable for welding applications and processes in use today, some method of obtaining this type of current adjustment must be devised for the electrical circuit with which we are working.

Probably the best way to accomplish our purpose is to **change the effective amount of iron in the reactor core.** This may be done either mechanically or electrically. Physical movement of the reactor core would be troublesome from the design standpoint. The alternative is to consider the electrical control concept. The necessary modifications to the circuit are shown in Figure 117.

The circuit diagram is now considerably different than previously illustrated. The basic components are still the same but several more items have been added to the circuit. The equipment additions include a dc control coil (coil "A"), a potentiometer connected as a rheostat (R), and a source of dc power. (In this case the dc power source is indicated as a battery. The battery is used here for illustrative purposes only to denote the electrical circuit is dc). The dc control coil (coil "A") introduces the electric control con-

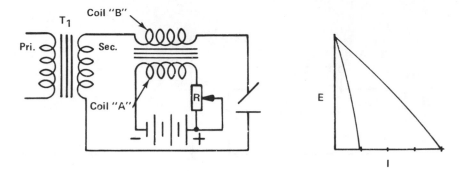

Figure 117. MODIFIED AC TRANSFORMER CIRCUIT.

cept of welding output amperage regulation.

Using "R", a potentiometer connected as a rheostat, dc voltage can be impressed on coil "A", the dc control coil. DC power is used because dc flows in one direction only. A magnetic field established with dc power will maintain its strength until the power is removed from the circuit.

The full use of the saturable reactor can now be explained. We have discussed the reactor and its effect on the welding circuit. Figure 117 shows the reactor with the added dc control circuitry including the dc control coil. The dc control circuit is an isolated circuit since it is common to all three phases in the secondary ac circuit of the power source. There is, therefore, no possibility of the dc impinging on the alternating current welding power circuit. By the same token, there is no possibility of the ac impinging on the direct current control circuit.

The term "saturate" means to soak or impregnate thoroughly and completely. For example, a sponge that has been soaked in water will, by its natural composition, absorb water until it can hold no more. The sponge is then said to be saturated.

The iron reactor core is common to both the ac welding power circuit and the dc control circuit. The iron core has the ability to absorb a certain number of magnetic lines of force (magnetic flux lines). The magnetic lines of force may be provided by either alternating current (ac) or direct current (dc). An important thing to remember is that it is the impedance factor, or counter-voltage, value that determines how much welding current appears at the output terminals of the power source.

To better show the electrical control concept for welding amperage output we will examine Figure 118 which shows a saturation curve for an iron reactor core. The purpose of the saturation curve is to **show the amount of reactor core saturation** by both ac and dc power. The reactor iron core is always **completely saturated** with magnetic lines of force, either ac or dc or a combination of the two, whenever the power source is producing welding current output. Full saturation of the reactor iron core means that the iron has accepted all the magnetic lines of force that it is capable of carrying per square inch of iron lamination surface area. Figure 118 shows the total saturation level of the iron accomplished with dc power.

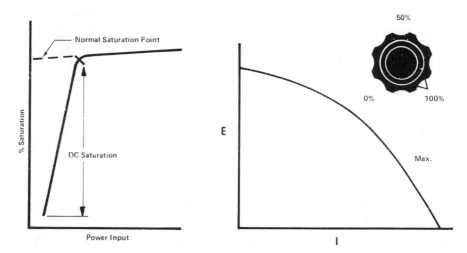

Figure 118. REACTOR CORE SATURATION CURVE.

The rheostat (R) controls the amount of dc power permitted to flow in the dc control circuit, including the coil. When the rheostat is placed in the maximum output position the reactor iron core is **fully saturated with dc power.** As long as the dc power is maintained in the dc control circuit the magnetic field strength created will maintain a steady force. The ac welding power, coming through the welding power circuit at coil "B" at the **same instant of time,** finds the reactor iron core already fully saturated and unable to accept additional magnetic lines of force. In this situation, there can be **no ac magnetic field** created and, therefore, **no counter voltage impedance factor.** The result is maximum welding current output at the terminals of the welding power source.

— 213 —

At this point we can develop a useful rule for the saturable reactor method of welding output current control:

"Maximum dc power in the reactor control circuit provides maximum welding power at the output terminals of the welding power source".

By setting the rheostat (R) at approximately 50% of maximum output the reactor iron core would be approximately half saturated with dc power. This is shown in Figure 119.

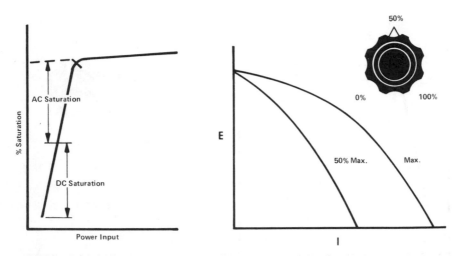

Figure 119. REACTOR SATURATION CURVE USING AC AND DC POWER.

The ac welding current flowing in the series power coil (B) now has the opportunity to create an ac magnetic field. The ac magnetic field strength will be limited to the amount of available iron in the reactor core. The saturation curve, Figure 119, indicates that approximately half the iron core saturation is accomplished with ac. An important point to remember is that **the reactor iron core is always completely saturated when welding** regardless of the welding amperage output value used or the position of the rheostat control.

The energy stored in the ac magnetic field is returned to the originating circuit each half-cycle as counter voltage. The counter voltage is the impedance factor in the welding power circuit. For the example cited, the welding current output would be approximately 50% of the amperage range set on the welding power source.

It becomes apparent that, by varying the amount of dc power in the control circuit, the amount of ac magnetic field strength can also be varied. As the amount of ac magnetic field strength is varied so is the counter voltage value: that is, the impedance value. The affect of the impedance factor on the output amperage value will cause the welding power source to stabilize at any amperage value within the capability of the unit. The control of welding amperage output is infinite within the minimum-maximum current range of the power source.

In summary, the saturable reactor electric control provides full range amperage output control. One other advantage to this method of current control is the fact that it may be operated from a remote rheostat amperage control unit. The welding operator has the capability of changing welding output amperage at his work site by using a remote hand or foot control rheostat instead of having to go back to the welding power source to make current output changes. This is a decided advantage for construction work, shipyard construction and many other welding applications.

Current Feed-Back System.

The current feed-back circuit is used on some types of welding power sources, usually three phase units with dc output. Normally the current feed-back coil is wound on the same basic coil form as the dc control coil and is common to the same reactor iron cores. As was previously stated, the control circuit is actuated by direct current (dc). The current feed-back coil is also dc powered since it is in the dc portion of the welding circuit and reflects the dc amperage output of the power source.

The circuit diagram illustrated in Figure 120 is for a standard three phase transformer-rectifier power source with dc welding output. An analysis of the circuit diagram shows the **current feed-back coil, labelled CF in the drawing**, in the dc portion of the welding power circuit. Of course, dc welding amperage is flowing in the current feed-back coil when welding is in progress.

It is relatively easy to read and comprehend the circuit diagram in Figure 120 if you remember that a welding power source is made up of **component parts** such as coils, iron cores, switches, rectifiers and other electrical devices. A good method of tracing a circuit, and remembering it, is to use various colored pencils for the different parts of the circuit.

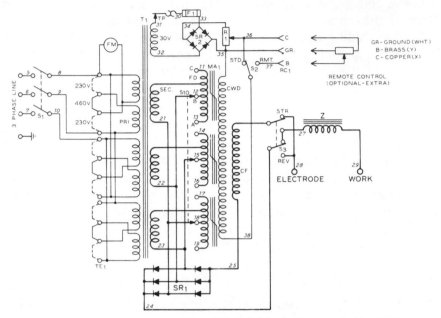

Figure 120. STANDARD THREE PHASE TRANSFORMER-RECTIFIER POWER SOURCE CIRCUIT DIAGRAM.

For example, let us begin at the left side of the circuit diagram which reads "3 phase line". S_1 is the primary switch which turns the welding power source on or off. Proceeding to the right on lines 8, 9, and 10, there are terminals marked "230 V." and "460 V.". This means that the power source primary system is re-connectable for either 230 or 460 volt primary power. The circle extending toward the top of the drawing which has the letters "FM" indicates the power source fan motor. Note that the fan motor is connected across the 230 volt connections regardless of the primary voltage brought to the power source input terminal board.

Continuing to the right as we trace the circuit we find the **primary coils** indicated by the letter "P". There are three sets of coils, each separate from the others, so we can be sure this is a three phase transformer. This completes a total description of the primary ac circuit components.

Just to the right of the primary coils are three straight lines. This is the electrical symbol for an iron core. In this case it is the

main transformer iron core. Immediately adjacent to the right of the main transformer iron core are three coils labeled with the letter "S". They are the **secondary coils** of the main transformer. This is the beginning of the secondary ac portion of the welding power source circuitry. (You should use a different color pencil to identify this portion of the power source circuitry). The coil and related circuitry at the top of the drawing is another, isolated, circuit and will be discussed in detail later. It is a separate and isolated circuit called the **control circuit.**

The secondary coils have conductors numbered 21, 22, and 23 leading to another set of electrical coils labeled "FD". The FD means **flux diverter.** In the circuit diagram they are shown as the **ac reactor control coils.** The FD coils carry welding power since they are a part of the secondary ac welding circuit.

The ac reactor control coils are wrapped around one leg of the **reactor iron cores.** The reactor iron cores are shown as three separate units. The reason for three reactor cores in the circuit is that each of the three electrical phases of secondary ac power has a separate reactor.

The ac reactor coils have three taps indicated by the letters "A", "B" and "C". The arrows pointing to tap "B" are connected by a dotted line which indicates that, when one moves, they all move. This is known as a "gang" switch. The switch is labeled S_{10} and is the **amperage range control switch.**

Following the circuit from the ac reactor control coils the conductors may be traced to SR_1 which is the **main power rectifier.** The rectifier performs only one function which is to change alternating current to direct current. The main power rectifier may be either selenium or silicon. The element used for the rectifier will depend on the design parameters of the welding power source. The secondary ac circuit terminates at this point. Everything in the welding power circuit from this point on is **direct current.**

Using another color of pencil, we will now trace the dc portion of the welding power circuit. In the dc circuit there are two conductors leading from the main power rectifier, SR_1. They are numbered 24 and 25. The number 24 conductor leads to S_3 which is the **polarity switch.** This switch changes the dc welding power from one polarity to another without having to change the welding cables from one output terminal to another. At this point the electrical conductor becomes number 28 and leads to the electrode output terminal of the power source.

Beginning at the work terminal and following line number 29, we come to an electrical symbol showing **a coil, an iron core and the letter "Z".** You may recall the rule of thumb that explained this symbol as being either a reactor or a stabilizer. If it is in the **ac** portion of the circuit it is a **reactor.** If it is shown in the **dc** portion of the circuit it is called a **stabilizer.** The purpose of the stabilizer is to smooth the ripple factor that is always present in rectified dc welding power output. This assists to provide a more stable welding arc.

Line 27 leads to the S_3 polarity switch where it changes to line number 26. Following line 26 takes us to a coil labelled **"CF".** This is the **current feed-back coil.** Its position in the circuit diagram indicates that it is common to all three reactor iron cores. The purpose of the current feed-back coil is discussed in subsequent paragraphs of this section of the text.

Following line number 25 leads us back to the main power rectifier, SR_1. This completes the tracing of the welding power circuit.

The control circuit is the heart of this type of power source as far as amperage output regulation is concerned. The next step is to trace this circuit and relate it to the entire circuit diagram.

Looking at the coil located near the top of the drawing and next to the main transformer iron core we see lines 31 and 32. The coil is a **standard 30 volt ac coil** designed to provide power for the total control circuit. The symbol indicated as "TP" is a **bi-metal thermostat.** In this power source, the thermostat is physically located in the selenium rectifier as a protective device for the welding power source components.

F_1 is a **ten ampere plug type fuse** used for protection of the **control rectifier** shown as SR_2. It is at SR_2 that the ac control voltage and amperage is changed to dc. Lines 34 and 35 extending from the SR_2 control rectifier carry direct current. Line 34 leads to R_1 which is the **main welding rheostat** for controlling welding current output. Lines 35, 36 and 37 lead to a symbol shown as RC_1. This is the **remote amperage control receptacle** into which a remote control device, such as a foot control or a hand controlled rheostat, may be plugged.

The switch indicated as S_2 is the **remote-standard switch** for control of the remote amperage control receptacle. When it is in the standard position welding power is controlled by the main welding rheostat. When it is in the remote position, welding power may be controlled at the work site by either a remote hand rheostat or a remote foot control. Proceed along line 38 to coil CW. The CW coil indicates the **dc control winding,** better known as the dc control coil. The dc control coil is common to all three reactor iron cores.

This completes the examination of the electric control circuitry of a typical three phase transformer-rectifier welding power source having dc output only.

The **current feed-back coil** is physically wound around the outside of the dc control coil. The greatest difference in the two coils is that the dc control coil is made up of many electrical turns of fine gauge conductor wire while the current feed-back coil is made of very few turns of heavy gauge conductor wire. Keep in mind that the current feed-back coil is in the welding power circuit and must be made of conductor wire capable of carrying the current rating of the power source.

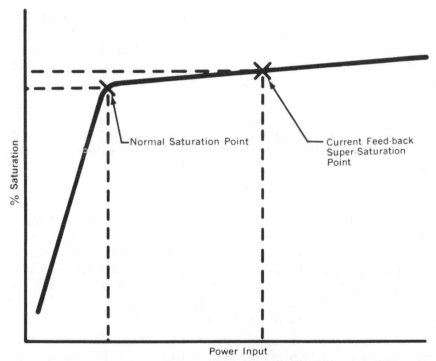

Figure 121. REACTOR CORES SUPER-SATURATION CURVE.

In our discussion of the saturable reactor there was a thorough explanation of the meaning of saturation. The diagram in Figure 121 illustrates the saturation curve of the iron in the reactor cores. The vertical axis indicates the percentage of saturation while the horizontal axis shows the power applied. The normal operating point of saturation is shown with the cross line at the knee of the curve. The saturation point is located here because the iron cores are always fully saturated when the power source is being used for welding. Saturation of the reactor iron cores may be accomplished with direct current (dc) from the **dc control circuit** or by alternating current (ac) from the **ac reactor coils** (the series power coils) in the welding power circuit.

Under normal conditions the current level at the welding arc passes through the current feed-back coil. The magnetic flux lines (magnetic lines of force) produced by the coil being around the reactor iron cores are moving in the same direction as the magnetic lines of force from the dc control coil. The normal saturation point of the reactor iron cores is stable at the knee of the curve. Welding amperage output is controlled by the value of the dc saturation of the reactor iron cores.

If there is a sudden surge of amperage at the electrode, such as would be caused by a short circuit of the electrode to the workpiece, the current feed-back coil would reflect the increased amperage in the dc welding circuit. The greater power input to the reactor iron cores would cause additional magnetic lines of force to supersaturate the reactor cores.

This may be called super-saturation of the reactor cores since a much greater amount of dc power is now applied to the reactor cores. In the diagram, Figure 121, the increased saturation level is shown by the dotted lines. Note that although there is considerable increase in the power applied there is very little increase in the percentage of saturation.

The current feed-back control system provides the small amount of additional saturation which permits a welding power source with this type of control to go to maximum short circuit current almost instantaneously. Immediately there is maximum short circuit current at the welding power source terminals and the electrode tip. This provides the welding operator with electrical assistance in breaking the stuck electrode free of the base metal workpiece.

An output problem could arise if the current feed-back coil is inadvertently wired into the welding power circuit incorrectly. This would be caused by leads 24 and 25 being placed on the wrong terminals of the main power rectifier (SR₁). Such incorrect connections would cause the magnetic lines of force created by the current feed-back coil and the magnetic lines of force created by the dc control coil to cancel each other by moving at 180 degrees relative to each other. The opposing magnetic fields would have the effect of voiding one another. Such an effect would allow the full reactance of the ac control coils to be in the electrical circuit with resulting minimum power output in all amperage ranges of the welding power source.

If this situation should develop in a welding power source with current feed-back control, check to see if the main power rectifier has been recently replaced. If it has, in all probability, the dc connections are on the wrong terminals of the rectifier. Reverse the two terminals (dc) on the rectifier and the power source should provide full power output at the terminals of the power source. If this does not correct the problem, replace the rectifier terminal connections as they were and proceed to check out the welding power source as discussed in the next chapter. Always consult the maintenance and operating manual for the specific power source model when operational characteristics are not normal.

Standard 30 Volt Control Circuit.

The electric control circuit, and its relationship to the rest of the welding power source circuitry, is illustrated in Figure 120. A more simplified drawing of the control circuit only, with conductor

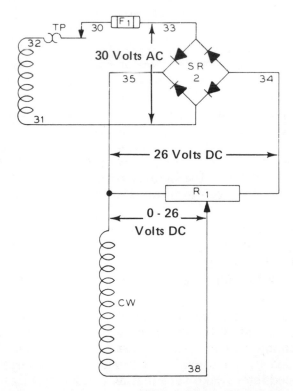

Figure 122. DC CONTROL CIRCUIT DIAGRAM.

wires numbered as in the regular circuit diagram, is illustrated in Figure 122. Although we discussed some of the voltage values when we traced out the power source circuit diagram, you will note that we have inserted the specific voltage values at each point in the standard 30 volt control system. Careful review of the voltage check points in the dc control circuit will assist you in making the actual checks on this type of circuit in a welding power source.

Between lines 31 and 32 there is a coil symbol shown. The coil supplies 30 volts ac to the basic control circuit. Line 31 leads directly to SR_2 which is the control rectifier.

Line 32 shows two safety devices in the control circuit: a thermostat (TP) is indicated as is a fuse (F_1). Note that line 32 becomes line 30 after the thermostat and then changes to line 33 just prior to entering the control rectifier. This completes the ac portion of the control circuit. A good point to remember is that **voltage is always measured between two electrical conductors.**

Lines 34 and 35 are the **dc output conductors** from SR_2, the control rectifier. There are approximately 26 volts dc between these two conductors. The small voltage loss is typical of the energy loss through a rectifier of any type.

Following line 35 vertically we find a black dot which indicates a connecting junction between two or more wires. Below the connection is illustrated a coil, labelled CW, which is the dc control or "winding". Line 38 terminates at R_1 which is the main welding rheostat on the power source. The voltage between lines 35 and 38 should read 0-26 volts dc as the rheostat is rotated through its entire range. The 0-26 volt dc reading is taken by placing the probes of a dc voltmeter as follows: the **positive probe** is placed on the maximum output terminal of the control rheostat (R_1) and the negative probe is placed on the center terminal of the remote-standard switch.

NOTE: Some control circuits have higher control voltages than those indicated in this text. Check the circuit diagram for the power source being tested if the voltage values are substantially different than those shown here. If higher control voltages are used they are normally shown by numerical value in the control circuit portion of the circuit diagram.

If the voltage readings are approximately half the rated voltages in the control circuit, at the given check points, you may assume halfwave rectification is taking place in the control rectifier system. This may be caused by a loose wire, a broken wire, a terminal clip not properly placed, etc. Make sure that all conductor wires are tight on their terminals and that the terminals are properly seated in place.

Check the ac circuit for the proper input voltage level.

In addition to controlling the welding current output of the welding power source, the control circuit may be considered a type of warning system for the welding operator. If something malfunctions, such as a fuse blowing, it will usually indicate trouble somewhere in the welding power source circuitry or the primary input circuit.

In summary, the control circuit is the electrical system that works in conjunction with the reactors to control welding current output. It is also a type of warning device for the welding operator. The control circuit provides remote welding current control capability for electrically controlled welding power sources.

Chapter 12

TROUBLESHOOTING SOME DC POWER SOURCES

The troubleshooting data provided here will be helpful when checking out any transformer-rectifier type welding power source. Although we have discussed the voltage check points of the dc control circuit, there are other areas that you should consider when checking out a power source for possible malfunctions. One thing to keep in mind is that trouble in a welding power source can usually be either **seen** or **smelled!** Visual examination is certainly the first order of business after removing the outer covering from the power source.

In addition to the perceptive senses of the individual, a standard ac/dc volt-ohm meter is useful, particularly to the technical service man or maintenance electrician. One type of meter is shown in Figure 123.

Figure 123. AC/DC VOLT-OHM METER.

The maintenance and operation manual provided with the power source will have troubleshooting information as part of the operational data. The circuit diagram that is part of the manual will assist you in locating the relative check points quite easily.

The following listed check points are given as a guide to help in determining the proper methods for finding trouble in the welding circuitry. This includes the power source internal circuits as well as the external circuit which is comprised of the welding cables, ground clamp and electrode holder. All of the component parts discussed may not be in the particular model power source with which you are working. If not, use those check points that are applicable to **your** power source.

Each of the following listed items will be discussed and explanations given as to what you should look for and what you should **look out** for!

1) Primary Voltage And Phase.
2) Terminal Linkage At The Primary Terminal Panel.
3) Check Open Circuit Voltage (OCV).
4) Welding Power Source Front Panel Components.

It is true that a good craftsman values his tools above almost anything else in his possession. In the welding industry, the least understood tool is the one that is most important to the processes, the **welding power source.**

Primary Voltage And Phase.

This is an area of concern that is often overlooked when troubleshooting welding power sources and other equipment. **Primary voltage and phase** are considered to be **constant values. Primary amperage is a variable value** which is dependent on the welding power source output rating and requirements.

It is well to check the primary distribution system first when checking out a possible problem in a welding power source. The primary voltage may be any one of the common values such as 208, 230, 380, 460 or 575 volts. A voltmeter is used to make the actual check of voltage in the primary system.

For any given primary voltage rating in an electrical system it is not unusual to find the **actual primary voltage** either high or low (usually low). The result would be either high or low amperage output values, for a particular power source setting, from the output terminals of the unit.

The reason for this is quite simple. A welding transformer is a totally electrical device and it can only put out, in welding power, a percentage of the total primary electrical power put into the unit. If the input voltage is low, therefore, the output amperage will be less for a specific setting on the controls. If the primary voltage is high, such as is often found in late night or early morning operations, the output amperage would be high for the same specific setting.

It is important that the primary electrical service be the proper phase required for the welding power source. For example, if the power source is ac or ac/dc output the primary electrical service must be single phase power (or one phase of a three phase system). If the unit produces dc welding output only, it will probably require three phase primary electrical power.

Be sure and check all line fuses in the primary system for continuity. This should be done by removing the fuses and testing them with an ohm meter. PLEASE OPEN THE PRIMARY LINE DISCONNECT SWITCH BEFORE REMOVING THE FUSES! Should one fuse be defective in a three phase system it would be apparent as single phase power input to the welding power source. A slow running fan and low current output from the power source are indications of a defective, or blown, fuse in the primary electrical system.

Terminal Linkage At The Primary Terminal Panel.

In most transformer-rectifier welding power sources there is a primary terminal panel for connection of the primary power leads to the power source. It is necessary that the primary voltage and the terminal linkage be compatible for safe, efficient operation. A typical terminal linkage setup for several different primary voltages is shown in Figure 124. Note the lower right hand terminal strip in each voltage illustration. This would be the typical linkage connection for a **control transformer** if the power source has such a device. It would not appear on all welding power sources.

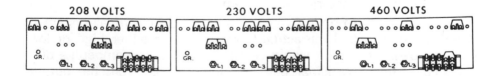

Figure 124. TERMINAL LINKAGE FOR PRIMARY VOLTAGE.

Welding power sources should be shipped from the manufacturers plant set for the highest primary voltage at which they will operate. If the welding power source is a standard unit with 230/460 volt reconnectable linkage it should come to the user set for 460 volt primary power.

A power source that is set for 460 volt primary power and connected to a 230 volt input line will not produce the welding power, for a given setting on the controls, that it should under normal conditions. Conversely, a welding power source connected for 230 volt primary power may burn out the primary coil, due to overheating, if 460 volt primary electrical power is applied.

Check Open Circuit Voltage (OCV).

Most quality built commercial welding power sources have both the primary and secondary electrical ratings listed on the front control panel where the output controls are located. NEMA Class 1 AC power sources normally have 80 volts open circuit maximum. (Certain NEMA Class 1 DC power sources may have up to 100 volts open circuit). NEMA Class 2 and Class 3 power sources, usually either ac/dc or just ac output, may have two open circuit voltages in different output amperage ranges. The low amperage range normally has 80 volts open circuit while the higher amperage range will often have an open circuit voltage of less than 80 volts.

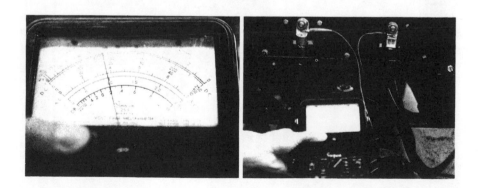

Figure 125. CHECKING OPEN CIRCUIT VOLTAGE.

As shown in Figure 125, open circuit voltage may be checked by placing the voltmeter probes on the positive and negative, or electrode and work, terminals of the power source. The welding power source is, of course, energized for this simple test. It is good and safe practice to remove all welding cables from the terminals of the power source before making the open circuit voltage check.

If the open circuit voltage is low by about one-third the value shown on the control panel, the three phase welding power source may be connected to single phase primary power input. If the unit is connected to three phase primary power the symptoms would indicate that a line fuse was inoperable in the primary electrical supply system. The method of testing primary line fuses has been discussed previously.

Welding Power Source Front Control Panel.

In previous sections of this book we have discussed a standard three phase transformer-rectifier type welding power source which has dc output only. Let us examine the various controls that govern the operation of the power source. Figure 126 illustrates a typical control panel for such a welding power source.

Figure 126. THREE PHASE DC POWER SOURCE CONTROL PANEL.

The various controls, switches and fuses will be described separately as to their function and possible malfunction.

Standard-Remote Switch.

When this switch is in the **standard position** all welding current control output is accomplished at the main welding rheostat on the front panel of the power source. When placed in the **remote position,** amperage output adjustment may be made at the welding operator's work area with either a remote hand control rheostat or a remote foot control rheostat.

Figure 127. STANDARD-REMOTE SWITCH.

CAUTION: DO NOT SWITCH OUT OF THE STANDARD POSITION UNLESS A REMOTE CONTROL DEVICE IS PLUGGED INTO THE REMOTE AMPERAGE CONTROL RECEPTACLE. Such action could cause damage to the switch.

Ten Ampere Fuse.

The ten ampere fuse is a plug type, screw based unit located on the front panel of some welding power sources. It is used to protect the remote control circuit as well as the control rectifier. It may appear to be in good condition and still be defective. If there is any doubt in your mind check the fuse for continuity with an ohm meter. A good fuse will show continuity.

Figure 128. TEN AMPERE PROTECTIVE FUSE.

In some instances people have purposely placed a piece of paper behind the fuse base in the fuse receptacle. The paper acts as an insulator and the fuse does not function in the control circuit. A favorite trick is to place a small piece of plastic tape over the fuse base contact point. Be alert to this type of shop horseplay for it can, and does, occur.

In connection with the ten ampere fuse there is a ceramic fuse block located just inside the front of the welding power source control panel. The ten ampere fuse is threaded into the fuse block. Since it is a ceramic block it can be broken relatively easily. If the fuse has tested out satisfactorily it would be wise to check the fuse block contacts for continuity. This may be done by placing the volt-ohm meter probes at the input and output terminals of the fuse block.

Current Selector Rheostat.

The current selector rheostat may be tested by setting it at maximum output and welding with it. The object is not necessarily to make a weld but to ascertain if the power source is producing full welding current for the setting used. The next step is to set the welding rheostat at minimum output and again weld with the unit. There should be a radical difference in the current output level as well as complete control of the welding current output through the entire range of the rheostat.

If there is little, or no, difference between the minimum and maximum settings for welding current, check the rheostat wiring for breaks or disconnected terminals. Be sure to check the small contact brush (carbon) on the rheostat. It could be worn or chipped and broken. In such a case, the brush would not make firm contact with the rheostat windings and current output would be erratic.

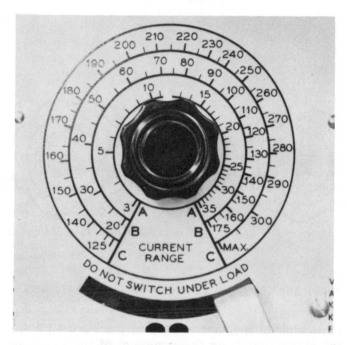

Figure 129. WELDING CURRENT SELECTOR RHEOSTAT.

If the rheostat wiring and brush test satisfactorily check the actual coil windings on the rheostat. In operations where there is a highly corrosive atmosphere, such as salt water vapors or chemical acids, it is possible that the rheostat windings may have developed a layer of oxides on their surface. This condition would cause the rheostat brush to have poor and limited contact with the rheostat windings. The result would be erratic amperage output when the rheostat is rotated through its range.

When installing any welding power source with rheostat control it is good practice to instruct the welding operator to rotate the rheostat control over its full range at least once or twice each day. This will assist in keeping the rheostat windings polished and bright and will promote better welding amperage output control.

IMPORTANT: ON MOST ELECTRICALLY CONTROLLED WELD-ING POWER SOURCES THE REMOTE CURRENT CONTROL IS LIMITED

BY THE MAXIMUM OUTPUT SETTING OF THE MAIN WELDING RHEO-STAT. This is an excellent feature, particularly when the unit is being used with the gas tungsten arc welding process.

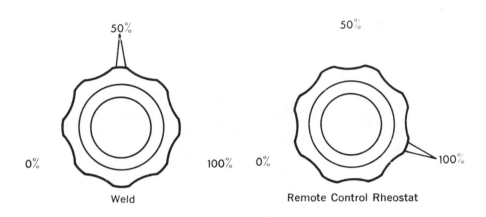

Figure 130. ELECTRIC CONTROL RHEOSTAT SETTINGS.

A remote control rheostat located at the work station will have current control variation **from the minimum of the range set on the welding power source to the maximum set on the main welding current rheostat on the power source panel.** A remote current control device cannot override the main rheostat setting. For example, if the main welding rheostat setting is 50 % of the range, as shown in Figure 130, the maximum available amperage at the remote work station would be the value set at the weld rheostat on the power source. The remote-standard switch would be in the remote position on the control panel.

The benefits of such a control system are substantial. In particular when working with the gas tungsten arc welding process the ability to limit the amperage output of the welding power source can help to maintain the welding current levels within the current carrying capacity of the tungsten electrode. Excessive welding amperage can cause the tungsten electrode to deteriorate rapidly. Such control is helpful to the welding operator since he can make rather substantial physical changes in the remote rheostat settings with relatively small amperage changes at the welding arc.

Start Current Control.

Although not all welding power sources have a start control amperage rheostat and allied circuitry, such a circuit is standard on almost all NEMA Class 1 welding power sources designed for the gas tungsten arc welding process. The start control circuit provides initial amperage at the welding arc that may be either higher or lower than the actual welding current. The specific welding application will determine how the start rheostat will be set or, indeed, if it is used at all.

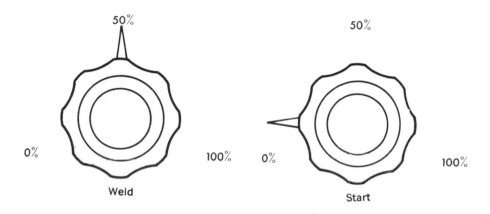

Figure 131. LOW AMPERAGE START RHEOSTAT POSITION.

The low amperage start, sometimes referred to as a "soft start", is usually used when the welding operator wants to begin a weld with a hot tungsten electrode and a stabilized welding arc. A weld began in this manner is excellent for relatively thin gauge sheet metal where arc stability is a prime requirement. In addition, a pre-heated tungsten electrode has better electron emission characteristics and, therefore, provides a more stable arc column when the weld is actually started. Figure 131 illustrates the relative positions of the **start rheostat** and the **main weld current rheostat** for a low temperature weld start.

The high temperature, or high amperage, "hot start" is used primarily for heavier thicknesses of metals or for thinner metal sections having good thermal conductivity. In this manner, a large amount of heat energy is imparted to the metal at the beginning of the weld. The intent, of course, is to pre-heat the base metal arc

start area. Since **heat will travel through hot metal much more slowly than it will through cold metal,** the pre-heat helps to keep the welding heat energy in a smaller area for better weld starts. This technique
will deter the formation of cold laps at the beginning of the weld bead.
Figure 132 shows the relative positions of the **start,** and **main welding,** rheostats for a typical high current weld start.

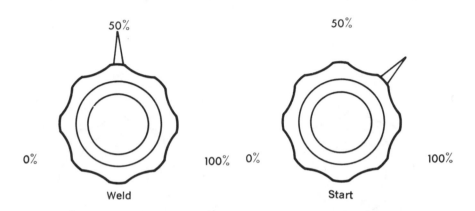

Figure 132. HIGH AMPERAGE START RHEOSTAT POSITION.

This discussion of troubleshooting check points is obviously in-
complete if all types of welding power sources are to be considered.
They are applicable, however, to the welding power sources discussed
in some of the previous sections of this text. The importance of the
power source maintenance and operation manual, for troubleshooting and routine maintenance functions, cannot be stressed too strong-
ly. There should be such a manual for the particular power source
with which you are working. Read it, heed it and USE THE INFORMA-
TION IT CONTAINS!

A parts list showing each component part of the welding power
source is usually incorporated as part of the information furnished
with a welding power source. Use of the parts list when tracing out
a circuit diagram will help you to identify the various component parts
shown as symbols and letters. For example, T_1 shown on the circuit diagram may be found listed in the parts list as the main transformer of the power source.

It is important to realize that information of any kind that is
published by the various manufacturers of welding equipment **costs
money!** The intent in publishing such information is to **assist you**

in using the welding equipment to its fullest potential. If the information and data is not used it has no value to anyone. In such a case the whole welding industry is the loser.

No individual can retain all the information about every welding process and power source in his memory. It is important to the individual to continue to ask questions; continue to be inquisitive about a welding process and what it can do for his application. The man who loses interest in new methods and new welding products is dead in this business.

Continue to learn; continue to ask questions; continue to think, for there is no better way to find out the answers to your questions than to ask those questions!

Chapter 13

GAS TUNGSTEN ARC WELDING

The principles of the gas tungsten arc welding process are relatively simple. The objective is to provide welding heat at the workpiece area without the contaminating influence of the surrounding atmosphere. This is accomplished with a shield of inert gas, either argon or helium or a combination of the two, flowing at a prescribed rate from the torch nozzle.

The equipment used with the gas tungsten arc welding process is somewhat different in appearance than the shielded metal arc welding equipment previously described. Basic gas tungsten arc welding equipment is illustrated in block form in Figure 133.

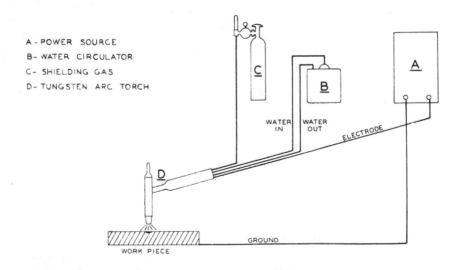

A - POWER SOURCE
B - WATER CIRCULATOR
C - SHIELDING GAS
D - TUNGSTEN ARC TORCH

Figure 133. GAS TUNGSTEN ARC WELDING EQUIPMENT.

The power source used is normally a constant current, or "drooper", class unit. It may be either ac or dc output or a combination of the two (an ac/dc power source). The water circulator is shown for those applications that require a water cooled welding torch. Why spend the money for a water coolant system when you can get water out of the faucets in the shop? It is simple. The water supplied by almost every water utility company **in the world** has some minute chemical particles such as iron in its compound. These particles build up in a very short time and clog the water hoses, pipes or torch passageways. Consider that the reason for using water coolant in the first place is to (1) make the torch lighter and smaller and (2) to make the torch operate cooler. It makes sense that the tubing in the torch which carries the water is very small in internal diameter. If city water is used, therefore, it will eventually build up chemical deposits in the water conduits which will slow down, or stop, the flow of water in the torch body. In this instance there can be no cooling action and the torch body will be destroyed in a very short time. With a water re-circulating system the water used can be distilled water. Distilled water has none of the contaminants of utility-supplied water.

The shielding gas must be either argon (Ar) or helium (He) or a combination of the two gases. Shielding gases and their uses are discussed later in this section of the text.

The tungsten arc welding torch, properly called a "tungsten arc electrode holder", may have any of a number of shapes or configurations. The main thing it does is to channel the shielding gas to the work area and hold the tungsten electrode while passing the welding current through the torch body and collet to the electrode.

The gas tungsten arc electrode holder is often called a "torch". The name torch is actually derived from another method of welding. The oxy-acetylene welding process provides heat only at the welding work area by means of an open flame. The flame is the result of the combustion of the gases oxygen and acetylene which are brought to the work proximity through a torch body and welding tip. As previously discussed, this welding process brings heat only to the base metal workpiece. Any filler metal that might be added to the weld is normally non-energized electrically and, if automatic filler metal feeding systems are used, would be external to the torch.

The gas tungsten arc welding torch performs essentially the same function as the oxy-acetylene torch in that it brings heat only to the base metal workpiece. The difference is that the oxy-acetylene process employs a fuel gas combined with oxygen for generating thermal energy while the gas tungsten arc welding process uses electrical energy for this purpose.

HEAT ONLY	HEAT + MOLTEN METAL
OXY-ACETYLENE	SHIELDED METAL ARC
GAS TUNGSTEN ARC	GAS METAL ARC

Figure 134. **ENERGY INPUT COMPARISON.**

The welding processes that provide heat only at the weld joint and workpiece may be contrasted with the **shielded metal arc,** and **gas metal arc,** welding processes where both heat and molten metal are added to the base metal simultaneously. A comparison of the two types of thermal energy input is shown in Figure 134.

The gas tungsten arc electrode holder (or torch) is made up of several component parts, each of which serves a definite purpose. As illustrated in Figure 135 a typical gas tungsten arc welding electrode holder for manual welding contains the torch body, the collet, the collet cap, the nozzle and hoses for carrying shielding gas, water (if used) and the cable for carrying the welding current. The tungsten electrode diameter will vary with the application of the welding process. The hose and cable assembly is attached to both the electrode holder and the welding power source, the shielding gas source and the source of cooling water (if used).

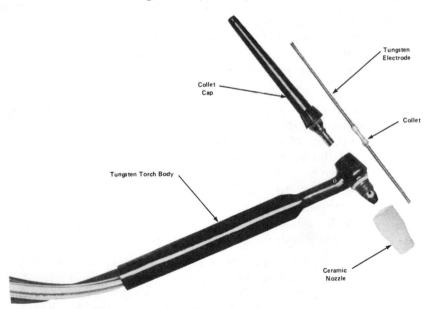

Figure 135. **TYPICAL MANUAL GAS TUNGSTEN ARC TORCH.**

The welding power source used for gas tungsten arc welding may be an ac transformer, an ac/dc transformer-rectifier or a dc transformer-rectifier unit. The type of power source used will depend on how the process is to be applied. The metal to be welded will have a considerable influence on the type of power source selected. For example, alternating current is almost always used for welding aluminum and magnesium because of the cleaning action inherent in the reverse polarity half-cycle. Direct current, straight polarity is normally used for all other metals such as steels, stainless steels, etc.

Shielding Gases.

The shielding gases that may be used with the gas tungsten arc welding process must have certain specific characteristics. They must be inert to the products of the weld zone. They must protect the tungsten electrode from contamination and oxidation. The shielding gases must do the same for the weld puddle and weld deposit. The two gases that are available in commercial quantities to meet these requirements are argon and helium.

Argon (Ar).

Argon is a chemically inert gas that will not combine with the products of the weld zone. As a matter of fact, it will not combine with any other element involved with welding. Even if it is put into the same cylinder with another gas, such as helium, there is only a gas mixture, not a compound gas. This means there are argon gas atoms and helium gas atoms in the same cylinder as a mixed gas. Hopefully, when the cylinder of mixed gases is used, the gases will come out of the cylinder in a reasonably proportionate mixture.

Argon has an ionization potential of 15.7 electron volts. (Ionization potential is the voltage, or force, necessary to remove an electron from the gas atom thereby making it an ion, or an electrically charged gas atom. The ion particles create an electrically charged preferential path for the welding current to follow from the electrode to the workpiece).

Argon has low thermal conductivity. The arc column is somewhat constricted with the result that high arc densities are present. Gas tungsten arc welding with argon gas shielding will result in a weld deposit with a relatively wide top bead. The total deposit will resemble a partial parabola configuration. This is especially true in welding aluminum with this process. The reason for the shape of the weld cross-section lies principally in the fact that the gas tungsten arc welding process brings heat only to the welding area. Any filler metal that might be added would, of course, act as a quench to the weld.

— 239 —

In Figure 136 both the weld cross section (typical) and the effective arc area are diagrammed. The arrows indicate the direction of heat movement through the metal from the rather small area of arc column impingement on the metal surface.

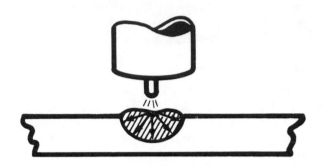

Figure 136. ARGON SHIELDED GAS TUNGSTEN ARC WELD.

Since it is well known that heat will move more readily through cold metal than it will through hot metal, the reason for the shape of the weld deposit cross section becomes apparent. The thermal, or heat, energy is distributed 360 degrees at the work surface to be welded from the effective arc area. It is used to heat, and melt, the upper area of the indicated cross section of the base metal plate. The heat will transfer equally well radially and vertically through the base metal to be welded. The theory propounded applies to the gas tungsten arc welding process where the only function of the arc is to provide a concentrated source of heat energy to the workpiece.

Argon is in abundant supply in the atmosphere since it comprises approximately **8/10 of 1%** of the earth's atmosphere. The gas is obtained as a by-product of the oxygen manufacturing process.

Helium (He).

Helium is another shielding gas that is inert to the products of the weld zone. It is a low density gas that is lighter than air. Helium has an ionization potential of 24.5 electron volts.

Helium has excellent thermal conductivity. The helium arc column will expand under heat, causing thermal ionization of the gas, and reducing the arc density. Remember that arc density is calculated by dividing the arc column cross sectional area, in square inches, into the arc amperage value.

When using helium as the shielding gas there is a simultaneous change in arc voltage where the arc voltage gradient of the **arc length** is increased by the discharge of heat energy from the arc column. This means that considerably more heat energy is lost in the arc area with helium than with argon. Some of this lost arc energy is used to heat the helium shielding gas.

The effective arc area, and a typical cross section of a weld deposit configuration, are shown in Figure 137. Note that the effective arc area is larger than that of argon and the deposit cross section shows deep central penetration.

Figure 137. HELIUM SHIELDED GAS TUNGSTEN ARC WELD.

As was previously stated, helium has excellent thermal conductivity. It may be assumed, correctly, that the helium arc plasma will have a very low temperature gradient from the center of the arc column to the periphery of the arc plasma. Apparently a greater work surface area is being heated by the arc plasma when welding with helium shielding gas. Again basing the premise on the fact that heat moves more readily through cold metal than it does through hot metal, as shown in Figure 137, the heat at the center of the weld deposit can only move to the colder metal at the bottom of the plate being welded. It is this fundamental principle that dictates the deep central penetration achieved with the helium gas shield and gas tungsten arc welding.

Helium is a product of natural gas deposits in the Earth. Until recently helium has been in very limited supply. There is some serious speculation by the Helium Gas Producers Association that there will be no more helium in the earth by the year 2,000.

Gas Flow Rates.

Very often the question is asked, "What is the correct gas flow rate, in cubic feet per hour (CFH), for welding a specific application with gas tungsten arc"? Unfortunately, there is no single correct answer that can be written in a book. The flow rates that are provided by shielding gas manufacturers are generally for a given application and material. They should be used as guides only.

Gas flow rates should be of sufficient volume to provide adequate coverage of the molten weld puddle deposit and the tungsten electrode. Any shielding gas flow in excess of the necessary requirement is wasteful and costly. Excess gas flow can also cause defects in the weld because of the inability of the weld deposit to release the gas internal to the molten metal before the metal solidifies. The result would be random gas porosity in the weld deposit.

Gas flow rates that are insufficient to protect the weld deposit and the tungsten electrode tip will cause oxidation of the weld surface and the deposited metal. The electrode tip will show a dull grey-green color and will undoubtedly erode into the weld deposit as contaminating particles.

Figure 138 illustrates oxidized tungsten electrodes that were ruined by inadequate gas shielding coverage. For comparison, tungsten electrodes that have been properly gas shielded are also shown.

Figure 138. IMPROPERLY, AND PROPERLY, GAS SHIELDED TUNGSTEN ELECTRODES.

Current Density.

The gas tungsten arc welding process has high current density at the electrode tip. This means that the amperage per square inch of electrode cross section is much higher than it would be for the shielded metal arc welding process. The fact becomes apparent when the electrode sizes for the two welding processes are compared at a given amperage.

For example, using amperage as the constant factor, there is this relationship at 120 amperes welding current.

120 amperes = 1/8″ E-6010 mild steel electrode.

120 amperes = 1/16″ tungsten electrode.

Current density is determined by dividing the electrode cross sectional area, in square inches, into the welding amperage value. For the examples cited above the calculations would be as follows:

$$1/8'' = 0.01227 \text{ inch}^2 \overline{)\ 120} = 9{,}780 \text{ amperes per inch}^2.$$

$$1/16'' = 0.00307 \text{ inch}^2 \overline{)\ 120} = 39{,}088 \text{ amperes per inch}^2.$$

The **cross sectional area** of the 1/8″ electrode, in square inches, is **four times greater** than the 1/16″ electrode. The current density of the 1/8″ diameter electrode, however, is only **one-fourth** that of the 1/16″ diameter electrode for the specific amperage used. It becomes evident that the smaller electrode will have greater heat input to the base metal, **in a smaller area,** than the larger electrode.

Current density is certainly one of the deciding factors to consider when selecting a welding process for a specific application. It may also influence the joint design of the base metal. This would be especially true of the metals having high thermal conductivity.

Some of the metals that have high thermal conductivity compared to iron are: aluminum, aluminum alloys, copper, copper alloys, magnesium and magnesium alloys.

Shielding Gas Ionization.

Ionization of the shielding gases occurs when one or more of the atomic electrons is caused to leave the gas atom. The force that causes the electron to leave the gas atom is called **ionization potential.** You will remember from earlier discussions that "potential" is another name for voltage. So it is ionization voltage—a force or pressure—that causes the electrons to leave the gas atom.

Ionization of the shielding gases is necessary to provide a preferred electrical path for the welding current moving from the electrode to the base metal workpiece. The ions are electrically charged positive since a negative electron has been removed from the atom. The ionized shielding gas is, therefore, a good electrical conductor, or preferential path, for the welding current to follow.

Ionization potential is calibrated in electron volts. It is the voltage necessary to remove an electron from the gas atom.

It must be remembered that voltage is a force, or electrical pressure. High frequency voltages, which may reach values in excess of 3,000 volts, are used to promote gas ionization. The open circuit voltage of the welding power source has little to do with shielding gas ionization.

Ionization potentials for some shielding gases are shown in Figure 139. Note that we have included carbon dioxide (CO_2) although this shielding gas is **not used** for gas tungsten arc welding.

Shielding Gas	Ionization Potential (Electron Volts)
Argon	15.7
Helium	24.5
Carbon Dioxide (CO_2)	14.4

Figure 139. SHIELDING GAS IONIZATION POTENTIALS.

The selection of the proper shielding gas to use for a specific welding application requires careful consideration. It is not uncommon for an improper shielding gas to cause poor welds and rejected weldments.

Helium is lighter than air and has low density with a relatively high ionization potential. In this situation it is normally difficult to initiate a gas tungsten arc in a helium atmosphere. Very often argon is used as the gas shield for arc initiation and helium is used for the actual welding operation.

Arc Starting.

One of the prevalent problems that arises with gas tungsten arc welding techniques is arc starting, or arc initiation. There is no simple answer on how to achieve repetitive, reliable arc starts with the gas tungsten arc welding process. Poor arc starts may be caused by improper shielding gas flow rates, erratic high frequency, improper gas nozzle size and diameter, improper or fouled tungsten electrode, or base metal that is improperly cleaned prior to welding. All of the factors listed, plus some others, could cause poor arc initiation. Probably the best advice is to take one variable at a time and check it out carefully.

For instance, the size of the gas nozzle, or cup, may be too large or too small for the application and the gas flow rate. In many cases of poor arc starting the nozzle is too large in diameter, and the gas flow too high in CFH, to permit even partial ionization of the shielding gas. The remedy, of course, is to try a smaller diameter nozzle and a lower shielding gas flow rate. This is especially effective when the tungsten electrode is of relatively small diameter and the amperage is low in value.

Another problem that often occurs, and which causes poor arc initiation, may be traced to the type and diameter of tungsten electrode used with the process. If the tungsten electrode is too large in diameter the electrical resistance factor may be too great and the resulting arc will be erratic. The symptom will be movement of the arc around the periphery of the electrode tip. The result is an unstable arc that wanders from place to place in the weld joint.

High frequency voltage, or the lack of it, may be a deterrent to reliable arc starting with the gas tungsten arc welding process. Nontouch starts with the arc are required when welding aluminum and magnesium to prevent fouling of the tungsten electrode. High frequency is required for this purpose. If, however, the high frequency system and circuit are not functioning properly, the result may be high frequency wander, intermittent high frequency or even complete loss of high frequency at the electrode tip. This subject is discussed in detail in the chapter entitled **High Frequency Systems.**

The best remedy for high frequency loss is a complete check of the total system including all welding lead connections. In particular, **a firm ground, or work, connection is essential** for good high frequency operation at the welding arc. The ground should be connected as close to the weld to be made as is possible.

Other reasons for poor arc initiation include improperly cleaned base metal, low welding current values, low open circuit voltage, faulty connections in the welding circuit, etc.

Thermal Placement In The Arc.

Thermal placement in the arc simply means where the heat of the arc is concentrated for a given polarity. Remember: Each half-cycle of alternating current is actually flowing in one specific direction for 1/120th of a second so it does have polarity for that period of time. Although this subject has been discussed before in this text it is important to relate it to the gas tungsten arc welding process.

In Figure 140 the thermal, or heat, disposition of the arc is shown by percentages. The illustration does not take into account the losses of thermal energy that are due to radiation from the arc to the atmosphere. Note the polarity of the two electrodes in the illustration. Each polarity may be related to one half-cycle of the ac sine wave trace.

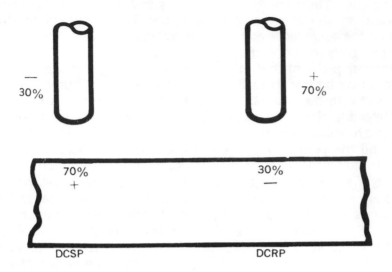

Figure 140. THERMAL DISPOSITION IN THE ARC.

The type of welding current used in the welding process will have a great effect on the penetration pattern of the deposited metal as well as the top and bottom bead configuration. Typical deposit characteristics are shown in Figure 141 for dcsp, dcrp, and ac.

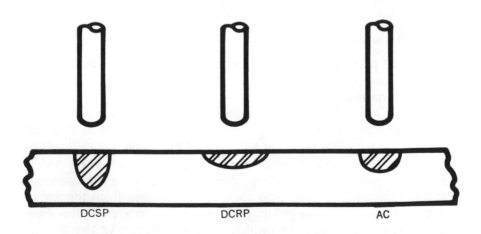

Figure 141. TYPICAL DEPOSITION CHARACTERISTICS.

The normal characteristics of the dcsp gas tungsten arc welding deposit are deep penetration with a relatively narrow weld bead width. Conversely, the dcrp weld deposit will have relatively shallow penetration with wider bead width dimensions. Alternating current (ac) weld deposits will be a composite of the two other types (dcsp and dcrp). The shielding gas employed, of course, will have a modifying effect on the weld deposit shape with any type of welding current. Relative to each other, and with all other factors except the welding current type remaining constant, Figure 141 is correct.

The reason that ac weld deposits have a modified form compared to dc welds lies in the fact that alternating current (ac) is a combination of dc reverse polarity and dc straight polarity. An ac sine wave trace and polarized electrodes are shown in Figure 142.

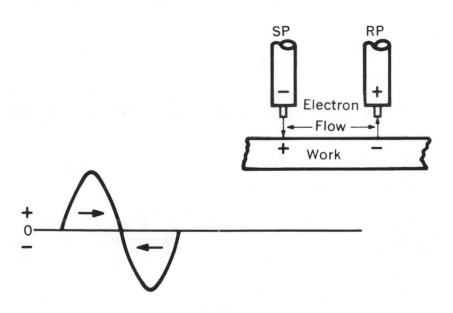

Figure 142. AC SINE WAVE TRACE AND POLARIZED ELECTRODES.

The drawing shows the alternating current wave trace separated into two half-cycles. The zero line remains a time function. Note that the electrode-work portion of the drawing shows the electrical characteristics of both straight and reverse polarity. For example, the straight polarity designation shows the electrode negative and the workpiece positive. The reverse polarity designation is just the opposite with the electrode positive and the workpiece negative.

Tungsten Type And Diameter

Another area of concern when gas tungsten arc welding is the tungsten electrode type and diameter for a given polarity and shielding gas. In Figure 143 the tungsten electrode sizes and applicable current ranges are given. The chart is designed to be used as a guide only. All settings indicated have been used successfully with a shielding gas mixture of 75% argon and 25% helium.

PURE TUNGSTEN		Current Range	
Electrode Dia. (In.)	ACHF-Argon	DCSP-Argon	DCSP-Helium
.010	Up to 15	Up to 15	Up to 20
.020	10 to 30	15 to 50	20 to 60
.040	20 to 70	25 to 70	30 to 90
1/16	50 to 125	50 to 135	60 to 150
3/32	100 to 160	125 to 225	140 to 250
1/8	150 to 210	215 to 360	240 to 400
5/32	190 to 280	350 to 450	390 to 500
3/16	250 to 350	450 to 720	500 to 800
1/4	300 to 500	720 to 990	800 to 1100
1% and 2% THORIATED TUNGSTEN			
.010	Up to 20	Up to 25	Up to 30
.020	15 to 35	15 to 40	20 to 50
.040	20 to 80	25 to 80	30 to 100
1/16	50 to 140	50 to 145	60 to 160
3/32	130 to 250	135 to 235	150 to 260
1/8	225 to 350	225 to 360	250 to 400
5/32	300 to 450	360 to 450	400 to 500
3/16	400 to 550	450 to 720	500 to 800
1/4	500 to 800	720 to 990	800 to 1100

Figure 143. TUNGSTEN DIAMETERS AND CURRENT RANGES.

It should be remembered that electrode diameters for dcrp and dcsp are considerably different at the same welding current value. The thermal distribution and electron flow characteristics in the arc are some of the reasons for the difference.

For example, a 1/16″ diameter tungsten electrode easily has the capacity to carry 125 amperes when welding with dcsp. When welding with dcrp, however, a 1/4″ diameter tungsten electrode is required to carry 125 amperes. At this current level the 1/4″ tungsten electrode is at its maximum current carrying capacity. Any more welding current and it will start to throw tungsten from the electrode into the weld deposit.

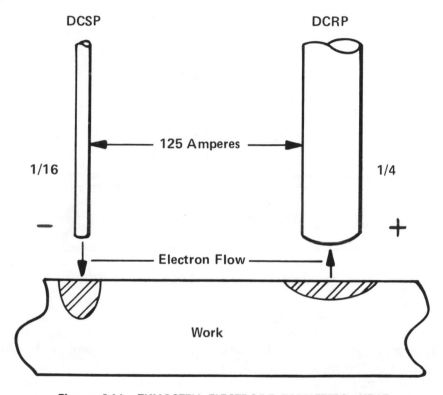

Figure 144. TUNGSTEN ELECTRODE DIAMETERS, HEAT DISTRIBUTION AND PENETRATION PATTERNS.

There is very little dcrp gas tungsten arc welding in the welding industry today because of the limiting factor of electrode heating. Usually dcsp is used for steels, low alloy steels, stainless steels, copper and other relatively dense metals. Alternating current is normally used for aluminum and magnesium and their alloys. Heat distribution and penetration characteristics for direct current, straight polarity and direct current, reverse polarity are shown in Figure 144. Note the difference in sizes of the electrodes used in the illustration.

Cleaning Action.

The term "cleaning action" refers to the removal of surface oxides from materials such as aluminum and magnesium during the actual welding operation. The oxide removal occurs during the reverse polarity half-cycle when welding with the gas tungsten arc welding process and alternating current. Figure 145 shows the action that takes place during the cleaning half-cycle when welding aluminum with gas tungsten arc and ac. In particular, note the direction of electron flow as well as the direction of gas ion movement.

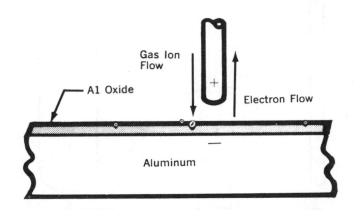

Figure 145. DCRP CLEANING ACTION ON ALUMINUM.

As indicated in Figure 145 electron flow is toward the electrode and gas ion flow is toward the work when welding with dcrp. A gas ion is a gas atom that is deficient in electrons. Remember that almost all of the mass of the atom, and all of the positive electrical charge, is contained in the nucleus of an atom. It stands to reason, therefore, that the gas ion has much greater weight than the electron. The gas ion has some impact value when it strikes the workpiece. The force exerted by the gas ion colliding with the surface oxide of the base metal is thought to promote the physical breakup of the oxide layer. The movement of electrons away from the workpiece and toward the tungsten electrode actually lifts the surface oxides from the work surface.

Cleaning action is important when welding aluminum and magnesium because both materials have dense oxide layers that form very rapidly upon exposure to the atmosphere. The rate of oxide formation is accelerated with a rise in temperature of the material such as would occur when gas tungsten arc welding. Inclusion of the oxide, or oxide residue, in the weld deposit would decrease the strength and integrity of the weldment.

The oxides that form on the surface of aluminum and aluminum alloys are called refractory oxides. One definition of refractory material is, "Any material that has a melting point in excess of 3,600° F.". The melting point of aluminum is considerably lower than that of aluminum oxide. It is apparent, then, that the oxide is still in the solid state after the aluminum base metal has assumed liquid form (melted). The cleaning action that takes place when gas tungsten arc welding with alternating current accomplishes the removal of the oxides.

DC Component.

The subject of dc component, what it is and what it does has been discussed at length since the introduction of the gas tungsten arc welding process. DC component is peculiar to ac welding with the gas tungsten arc welding process and is normally associated with the welding of aluminum and magnesium. Unfortunately, not many shop welding people have taken the time or trouble to find out exactly what dc component is or what causes it to occur when gas tungsten arc welding with alternating current.

Actually, dc component is created during the reverse polarity half-cycle when welding with ac and gas tungsten arc. Referring to Figure 146 the electron flow characteristics are again shown. The workpiece illustrated is aluminum and the oxide layer is shown at the surface of the metal as a darker area.

The zero line is a time function with the area above the line termed positive and the area below the line termed negative. For the purpose of this discussion we will consider the area above the zero line as the reverse polarity half-cycle and the area below the line as the straight polarity half-cycle.

Starting with the dcsp half-cycle (negative side of the line) we will follow the welding current as it goes through two full cycles of ac welding power. The negative charged tungsten electrode has excellent electron emission characteristics so the first half-cycle (dcsp) provides almost perfect wave form shape. This is illustrated in Figure 146.

The second half-cycle is the positive half-cycle. It is shown as dcrp and the aluminum workpiece is the negative pole (cathode) in the welding arc. Aluminum is not nearly as good an emitter of electrons as tungsten and, of course, it has a surface oxide layer that is

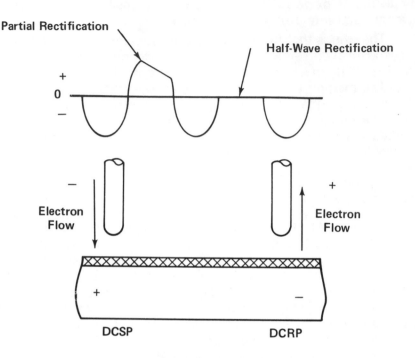

Figure 146. FORMATION OF DC COMPONENT.

very dense. The electron flow is retarded to some degree during the reverse polarity half-cycle by two factors: (1) less electron emission from the cathode, or negative, pole and, (2) dense surface aluminum oxides. The ac sine wave trace is distorted by the lesser amount of amperage that is allowed to pass through the arc in this situation. It is commonly said that the reverse polarity half-cycle sine wave trace is "attenuated". (To define: **Attenuate** = To weaken; to lessen the force or value of something).

If some amperage flows during the reverse polarity half-cycle the effect is termed "partial rectification in the arc". If the oxide layer is heavy enough, and **no amperage** flow is evident, the effect is half-wave rectification in the welding arc.

It is possible to weld when there is partial rectification in the welding arc but it is virtually impossible to weld when complete half-wave rectification is taking place. The symptom of half-wave rectification in the welding arc would be arc flutter and arc instability with possible "pop-outs".

It is at the time of partial, or full half-wave, rectification that dc component is formed. The term "rectification" is the key word because to rectify ac power is to change it from ac to dc power. This is the function of any rectifier element.

The dc component is energy that is not used in the welding arc. Rather, it becomes what is known as a circulating current. Unless it is filtered out of the circuit, or dissipated in some manner, it flows back into the welding power source circuitry and the main transformer coils where it generates heat. It is for this reason that conventional ac welding power sources, not specifically designed for the gas tungsten arc welding process, must be de-rated for current output. If the ac power sources are not de-rated for current the main transformer coils may overheat and be destroyed. The overheating is caused by the dc component "circulating current" dissipating in the secondary coils of the power source as heat. Data is provided in the text book chapter "Power Sources For Gas Tungsten Arc Welding" telling how ac power sources should be de-rated for ac gas tungsten arc welding.

In those power sources specifically designed for industrial applications with gas tungsten arc welding there are usually filtering devices of some sort to dissipate the dc component. Most NEMA Class 1 industrial welding power sources built for the gas tungsten arc welding process have some sort of resistance, such as ni-chrome resistor bands, in which the dc component is evolved as heat—but harmlessly. Another method used, which produces a balanced ac wave form in the amperage output circuit, is capacitor banks in the power source. The capacitors are connected in series with the welding arc. The beauty of the capacitors is that they will pass alternating current but will not pass direct current component. Still another method that has been used is to place wet cell storage batteries in series with the welding arc. While it is effective, it is a cumbersome and awkward system.

Summary.

The gas tungsten arc welding process was first developed at the beginning of World War II. It was invented by Mr. Russell Merideth, an eminent Welding Engineer. The process has found much use in all phases of the welding industry. All weldable metals may be welded with the gas tungsten arc welding process.

The basic objective of the process is to bring welding heat to the base metal workpiece without the contaminating influence of the surrounding atmosphere. This is accomplished by using an inert gas shield of argon or helium to protect the weld puddle and the tungsten electrode.

Both alternating current and direct current are used for gas tungsten arc welding. AC is normally used for aluminum and magnesium and their alloys. Direct current, straight polarity is normally used for all other metals such as steels, stainless steels, titanium, copper, etc. DCRP is not used by the welding industry except in special cases of minimum penetration.

Chapter 14

TUNGSTEN ELECTRODES

The development of the gas tungsten arc welding process was predicated on the use of an electrode material that would not vaporize in the heat of the welding arc. The element tungsten was selected for reasons that will be explored in this chapter.

Tungsten Manufacturing.

Although most welding people never have occasion to be concerned with the manufacturing processes involved with tungsten, or tungsten electrodes for welding, it is important to have an understanding of how the material is put together. From this knowledge may come better methods of preparing and using the tungsten electrodes necessary to the gas tungsten arc welding and cutting processes.

The process of manufacturing tungsten electrodes is a form of **powder metallurgy.** Tungsten particles that are in powder form and purified to a minimum $99.95+\%$ are used for electrode manufacture. To maintain purity of the base metal a continuous program of analysis and inspection is maintained throughout the various manufacturing steps. Quality control is very rigid at all stages of the manufacturing process.

The high purity powdered tungsten is pressed into ingots, or "compacts", under many tons of pressure per square inch. The as-pressed ingots have very little strength and are very fragile.

The next step is "sintering" the ingots in a hydrogen atmosphere at a temperature high enough to provide adequate strength to support the weight of the ingot. To define:

"SINTER" = To heat a mass of fine particles for a prolonged period of time, at a temperature below the melting point of the material, usually to cause agglomeration''.

"SINTERING" = The bonding of adjacent surfaces of particles in a mass of metal powders, or a compact, by heating''.

The material to be sintered is supported and suspended between electrical contacts and a controlled electrical current is passed through the ingot. In this manner the ingot is heated to very near its melting point (6,170°F.).

The "treated ingot", so called because of the electric heating operation, is tested for crystal structure and density. Spectrographic analysis is used to assist in maintaining the quality level of the product.

The tungsten ingots are then swagged into rod form. This is a mechanical metal working process in which the ingot is forged hot, at an initial temperature of 2,750° F. (1,500° C.) approximately, in a rotary hammer. The sintered ingot must be heated before mechanical working. Tungsten ingots are brittle at room temperature and cannot be worked to any extent without fracture.

The swagging operation develops a type of fibrous structure that imparts some measure of ductility and toughness to the tungsten rod. Rods that are reduced in diameter to arc welding electrodes are drawn through sizing dies of hardened steel with a final sizing finish draw through an industrial diamond die. This is the usual procedure followed. As drawn, the electrodes have a dense black oxide coating.

Tungsten Electrodes For Arc Welding Processes.

Tungsten is employed as a non-consumable electrode for the atomic-hydrogen welding process, the gas tungsten arc welding process and various plasma welding and cutting processes. By "non-consumable" we mean that the tungsten electrode is not intended to be part of the filler metal in the weld deposit. This is not to say that tungsten doesn't get into the weld at times! Tungsten that is included in welds is normally considered a defect since it will raise stress points in the weld due to its low ductility and characteristic hardness. Remember: The only time tungsten should be in the weld deposit is when you are actually welding tungsten metal!

Element*	Symbol	Melting Point		Boiling Point	
		Degrees C.	Degrees F.	Degrees C.	Degrees F.
Carbon	C	3,727‡	6,740‡	4,830	8,730
Tungsten	W	3,410	6,170	5,930	10,706

*Metals Handbook, AMS, Eighth Ed., 1961
‡Sublimes

Figure 147. CARBON AND TUNGSTEN DATA.

The element tungsten (W) has the highest melting point of any of the metals. In fact, of all the elements, it is second only to the element carbon which has a melting point of 6,740° F. (3,727° C.). The relative melting and boiling points for carbon and tungsten are illustrated in Figure 147.

Although carbon has a considerably higher melting point than tungsten the boiling point is substantially lower. Since carbon also **sublimes** at its melting point, it is not suitable for most present day welding applications where tungsten is used: **To define:**

SUBLIME = To pass directly from the solid state to the gaseous state without the intermediate liquid state of matter.

SUBLIMATION PRESSURE = The pressure of equilibrium, at a definite temperature, of a vapor in contact with its solid.

The metal tungsten, with a melting point of 6,170° F., has a boiling point of 10,706° F. The temperature gradient (4,536° F.) is such that it is virtually impossible to vaporize the electrode in the heat of the welding arc. This statement is based on the welding amperage being within the current carrying capacity of the specific type and diameter of tungsten electrode.

Tungsten is a good emitter of electrons. This factor assists in initiating and stabilizing welding arcs. In some tungsten electrodes thoria and zirconium have been added in small percentages to facilitate arc starting. Tungsten electrodes are available in either 1% or 2% thoria additions or approximately 1% zirconium addition.

The addition of thoria and zirconium was originally planned to promote better arc starting characteristics in the tungsten electrode. In the process, however, it was found that the addition of these elements provided greater current carrying capacity for the tungsten electrodes. Investigation proved that the addition of up to 0.6% thoria (approximately) to the tungsten matrix would increase the current carrying capability of the electrodes. Additional amounts of thoria will not increase the current carrying capability of the tungsten electrodes although it will improve the arc starting characteristics.

Tungsten electrodes for arc welding come in a variety of diameters ranging from 0.020″ diameter to 1/4″ diameter. They are available as pure tungsten, 1% and 2% thoriated tungsten and 1% zirconium bearing tungsten electrodes. They are supplied in either the clean finish or the centerless ground finish.

Clean finish tungsten electrodes are normally a bright grey color and have had all surface contaminants, oxides, etc., removed. They may or may not be concentric to the electrode centerline. **Centerless ground** tungsten electrodes, as the name implies, are ground to a bright, shiny finish and they are concentric to the centerline of the electrode. Centerless ground electrodes, being perfectly round, make the best possible electrical contact between the torch collet and the tungsten electrode. This means there is the least electrical resistance and, therefore, the least amount of resistance heating in the torch body. It is natural that centerless ground tungsten electrodes are used where minimum electrical resistance losses at the collet-electrode contact point is desired.

Grinding Tungsten Electrodes.

Many welding authorities are of the opinion that tungsten welding electrodes should not be ground on an abrasive wheel. The reasons for this opinion are many and varied. Some of them are examined and discussed in this chapter of the text.

Other equally competent welding authorities are of the opinion that grinding tungsten electrodes is perfectly all right. Probably both sides have some merit in their arguments.

The requirement for grinding a tungsten electrode suggests that the electrode is probably too large in diameter for the welding application. Usually the only time a tungsten electrode should require grinding is when the welding operator is joining extremely thin materials. An approximate range of material thickness, where grinding of tungsten electrodes would be acceptable, would be approximately 0.001″ to 0.050″.

If grinding tungsten electrodes is required it should be done within certain criteria. The proper technique for grinding tungsten electrodes is shown in Figure 148. The incorrect method of grinding tungsten electrodes is also shown.

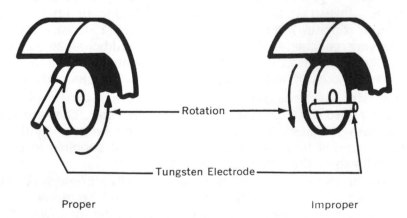

Proper   Improper

Figure 148. TUNGSTEN ELECTRODE GRINDING METHODS.

The illustration is probably familiar to many welding operators who use the gas tungsten arc welding process. Probably most of them grind tungsten electrodes using improper techniques and grinding wheels that are too coarse.

If tungsten electrodes are ground the work should be done on a special fine-grit, extra hard abrasive wheel. **The abrasive wheel should be used for no other material except tungsten.** If other materials are ground on the wheel the electrode could become contaminated with foreign particles or substances.

The question may be asked, "What results may be expected from using improper grinding methods and techniques when pointing a tungsten electrode?" Several things can happen, any one of which could cost money in either loss of the electrode material or repair of the weld.

Many times when a welding operator is grinding a tungsten electrode it will splinter into many pieces and shards. Tungsten is a hard and brittle material with very little ductility. It is certainly harder than any grinding wheel you may purchase. The abrasive wheel is a relatively soft material, compared to the tungsten, and therefore the tungsten particles are not really ground off. They are literally chipped away, particle by particle. You may remember that the tungsten electrode began as very fine powdered particles.

The action that takes place when grinding a tungsten electrode may be compared to an Indian chipping a stone arrow point. The particles of tungsten are removed by a percussive action, or series of blows, made by the grinding wheel. The shock of the impact on the tungsten electrode, which has very little ductility, can cause the splintering effect.

Any grinding operation is bound to leave machining marks on the material being ground and tungsten is no exception. When the electrode is improperly ground the machining marks will be concentric to the longitudinal axis centerline of the electrode. The danger lies in the possibility of the ridges at the periphery of the electrode melting and migrating across the welding arc to the weld puddle. Such tungsten spatter from the electrode would show up in an X-ray film as white spots because of the fact that tungsten is a very dense material. The tungsten is, of course, classed as an unwanted inclusion and is a weld defect. It is entirely possible that small portions of the grinding wheel would become lodged in the ground part of the tungsten electrode between the machining grooves and, under the heat of the arc, would be transferred across the arc column to the weld deposit. This too, would cause defects in the weld.

Welding codes permit certain allowable defects in a weld deposit. If the defects are greater than the specific code permits, they must be removed and repaired. Such work is expensive in time, money and material. Even if the repair is successful it will set up localized stresses in the weldment that could cause problems in service. In some critical applications only one repair weld is permitted on a defect. If inspection shows the part defective after the weld repair the part is scrapped. This is true in some missile skin applications, for example. It is to the advantage of all concerned that weld deposits be as free of defects as possible.

The question often arises as to which tungsten type to use for a specific application. There are a number of variables which may be used as criteria in tungsten electrode selection. Included in the list of variables would be the allowable cost for electrodes, the type of metal to be welded, the type of welding current to be used, the size of weld bead desired, the depth of penetration desired, etc.

Pure tungsten is unalloyed and has the lowest current carrying capacity of any of the tungsten electrodes. It is also lower in purchase price than the alloyed electrodes. Pure tungsten melts at the electrode tip immediately when the arc is initiated. The molten end forms a ball which is hemispherical in shape. When the arc is extinguished the pure electrode end will have a ball that is bright and shiny. Thoriated electrodes do not melt under the heat of the welding arc. This can cause problems because the peripheral edge of the tungsten electrode tip does not have sufficient mass to carry the heat necessary for welding. The result is tungsten "spitting", another name for

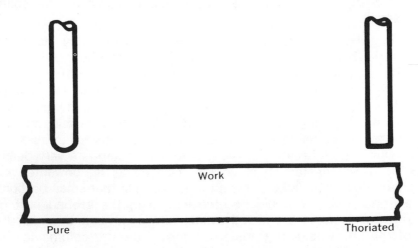

Pure   Work   Thoriated

Figure 149. TUNGSTEN ELECTRODE SHAPES.

tungsten migration across the welding arc column. Tungsten electrodes of both pure and thoriated matrix are shown in Figure 149. Note the rounded end on the pure tungsten electrode.

Pure tungsten electrodes are usually used for welding both aluminum and magnesium with the gas tungsten arc welding process. Some welding authorities believe that thoriated tungsten electrodes will contaminate the light metals. There is no specific evidence that such action will, or does, occur. In fact, the use of alloyed electrodes for welding aluminum and magnesium and their alloys has increased greatly over the past several years.

Another method of pointing thoriated electrodes is to "ball" them electrically. This may be accomplished quite easily by setting the welding power source on direct current, **reverse polarity,** lowest amperage range, minimum amperage setting and using "start" high frequency. A small piece of clean copper plate or bar stock should be used as the base metal when initiating the welding arc. At no time should the electrode touch the workpiece. The high frequency will provide an ionized gas path for the welding current to follow. UNDER NO CIRCUMSTANCES SHOULD A TUNGSTEN ARC BE INITIATED ON CARBON OR GRAPHITE STARTING BLOCKS. The carbon will vaporize and immediately contaminate the tungsten electrode.

Using a remote foot rheostat control, attempt to initiate the welding arc. The welding arc may not start immediately. The electrode diameter may be too large for the low amperage setting on the power source. If the welding arc does not start when the remote control (usually a remote foot rheostat control) is fully opened, stop the operation and increase the rheostat setting on the power source slightly. Again attempt arc initiation.

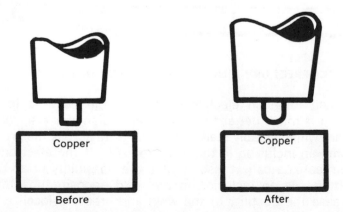

Figure 150. POINTING TUNGSTEN ELECTRODES WITH DCRP.

As soon as the welding arc is established, the electrode end will show a bright orange color. This will change to a brilliant white as more current is added to the arc by depressing the remote foot rheostat control. Increasing the amperage even more will turn the arc column to an incandescent white. This is the time to take the welding power off the arc before the melted tungsten electrode tip becomes too large. The ball formed by this method, or any other, should never be larger than the original diameter of the electrode. Figure 150 illustrates the tungsten electrode before and after pointing as discussed in the preceeding paragraph.

The resulting ball formation at the electrode tip is excellent because the semi-spherical shape allows the welding current to find its own level on the electrode end. It is a fact that electrical current, like water, will seek its own level on an electrode. With a properly "balled" electrode, the arc will be stable and the current level at the electrode tip will remain concentric to the longitudinal axis of the electrode.

Incorrectly pointing a tungsten electrode by grinding, or any other method, may cause defects in the weld deposit when the electrode is used for welding. The correct shape and dimensions for grinding tungsten electrodes is illustrated in Figure 151. Note that the length of the taper is two "d" where "d" is the diameter of the tungsten electrode.

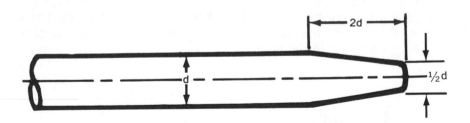

Figure 151. CORRECT GRINDING DIMENSIONS FOR TUNGSTEN ELECTRODES.

Grinding tungsten electrodes to a needle point is to be discouraged. It is much better to use a smaller diameter electrode. The needle point will inevitably melt back or migrate across the arc column and become an inclusion in the weld deposit. If the electrode melts back and creates a ball, it also creates the possibility of an unstable arc condition. As an inclusion in the weld deposit, tungsten particles decrease the ductility of the weld and create a localized stress point. In either case the results are not desirable.

Summary.

The gas tungsten arc welding process may be used with either manual or automatic process equipment. It concentrates thermal energy in a restricted area of the weldment because it has high current density. The shielding gases used with the gas tungsten arc welding process are argon (Ar) and helium (He). Different deposit and welding characteristics may be expected from the use of either one of the gases.

Tungsten electrodes are used because tungsten inherently has good electron emission. Thoria is added to some tungsten electrodes to promote electron emissivity and better arc starting. Thoriated tungsten electrodes have higher current carrying capacity, for a specific diameter, than pure tungsten electrodes.

Grinding tungsten electrodes should be done only when there is no other way to achieve the weld in the joint. High current density is one of the plus features of the gas tungsten arc welding process. This factor permits the welding of metals that have high thermal conductivity with a minimum of heat distribution in the part. Another benefit is the minimizing of weldment distortion.

The gas tungsten arc welding process is used primarily to weld metals of 1/4" thickness or less. It is not uncommon, however, for materials of heavier thicknesses to be welded with the process, especially the root pass. Some relatively new techniques of welding aluminum, using dcsp and helium shielding gas, have produced excellent welds in up to one inch thick aluminum metal using the gas tungsten arc welding process and not more than two passes.

Chapter 15

HIGH FREQUENCY SYSTEMS

The term high frequency, as applied to welding processes and power sources, usually refers to electrical pulses in the frequency range of 50,000 to 3,000,000 hertz (cycles) per second. Actually, high frequency covers a major portion of the total frequency spectrum.

There are a number of reasons why high frequency energy is used for welding applications and with certain welding power sources. Some of them are discussed in this chapter with relative comments. There are also some things to be careful of when using high frequency energy and these, too, are discussed in this chapter. To be very frank, high frequency energy is a necessary evil in some welding operations. If a method could be developed that would perform the functions of high frequency without having to use high frequency energy it would be the greatest thing since sliced bread! In truth, high frequency causes almost as many problems as it solves for welding.

Arc Stabilization.

Arc stabilization is probably the most important function of high frequency power. When welding with alternating current (ac) there is an arc outage each 1/120th of a second (each half-cycle). It occurs each time the ac sine wave trace passes through the zero line. The actual time of arc outage will depend somewhat on the re-initiation characteristics of the welding power source. High frequency, as a total part of the welding power source circuitry, provides the stable arc re-initiation effect necessary to maintain a steady, stable arc condition.

Gas Ionization.

The ionization of shielding gases occurs when an electron is caused to leave the gas atom because of some force exerted on the gas atom. In electric arc welding the force is voltage. That is the reason that ionization potential of a shielding gas is always measured in electron volts. A gas atom that has had one or more electrons removed is called an **ion**. Remember: the nucleus of an atom contains almost all of the mass and weight of the atom and **all of the positive electrical charge.** It is logical, therefore, that the ion would provide a preferential electrical path for welding current to follow.

The open circuit voltage of a welding power source does not have sufficient power to have an effect on the ionization of shielding gases. There are literally billions of billions of gas atoms passing through the intended arc zone every second.

High frequency power provides extremely high voltage at the electrode tip. Since the ionization potentials of the two commonly used shielding gases, argon and helium, are relatively low (24.5 electron volts for helium and 15.7 electron volts for argon) the high frequency voltage does create a minimum ionized gas path for the welding current to follow. High frequency voltage is considered relatively safe for the welding operator to use in making non-touch starts with the tungsten electrode. Non-touch starts are preferable when welding materials that might contaminate the tungsten electrode.

Arc Initiation.

High frequency energy, by ionizing at least a minimum gas path between the electrode and the workpiece, helps bridge the physical distance making non-touch starts possible. This is certainly an effective help in arc starting. High frequency promotes electron emission from the cathode element (tungsten) for more stable arc initiation.

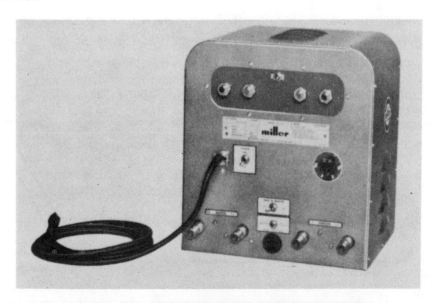

Figure 152. SEPARATE HIGH FREQUENCY UNIT.

A typical spark gap oscillator high frequency unit is shown in Figure 152. Units of this type are the most popular in use by the welding industry today. Although relatively small in physical size, they lend themselves to placement near the actual operation. This helps to minimize the losses of high frequency energy between the high frequency generator and the welding arc.

Most welding power sources that are designed and manufactured for use with the gas tungsten arc welding process have built-in high frequency systems and circuitry. The working parts of the high frequency circuit are usually easily accessible from the front or top of the power source and require no special access door.

Separate high frequency systems, such as that shown in Figure 152, are used with those conventional power sources, either ac or dc, which may be converted from shielded metal arc welding to gas tungsten arc welding. They have excellent mobility and may be used anywhere it is convenient to place them in the circuit.

Spark gap oscillator type high frequency systems are practical because they are relatively inexpensive to manufacture in the frequency ranges necessary for the welding processes. Rugged construction enables the spark gap type unit to perform under normal shop operating conditions without the shock mounting that would probably be required for vacuum tube type oscillators. This does not mean that spark gap type oscillators should be subjected to rough treatment. Common sense handling of any welding equipment will prolong its useful life.

High frequency energy transfers at the surface of a conductor via the so-called "skin effect". High frequency power for welding ap-

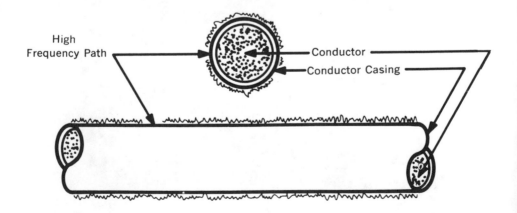

Figure 153. HIGH FREQUENCY POWER TRANSFER FOR WELDING.

plications is usually rated in milli-amperes at several thousand volts. High frequency power transfer is illustrated in Figure 153.

The high frequency power is usually super-imposed on the welding current conductor through air core coupling coils. The high frequency power is induced into the welding power circuit just before the welding current reaches the output terminals of a welding power source. For separate high frequency units, such as is shown in Figure 152, the high frequency is normally super-imposed closer to the work area. This is because the high frequency unit can be placed nearer the welding job.

The circuitry of the high frequency system is simple and yet very unique as compared to a welding power source transformer. The discussion of high frequency circuitry is not necessarily applicable to one manufacturer only. The general circuit diagram shown in Figure 154 is considered a basic high frequency system.

Primary voltage for the high frequency system may be any of the commonly supplied voltages such as 208, 230, 380, 460 or 575 volts. Separate high frequency systems may operate on the primary power available. Usually the high frequency systems built into a welding power source will function from 115 volt primary power although this is not a necessity.

The power flow sequence through the high frequency system is a simple, straightforward operation. The input voltage is impressed at 60 hertz (cycles). (We arbitrarily say "60 hertz" because most commercially generated electrical power is 60 hertz). The primary power energizes the primary coil of a high leakage, **step-up transformer** to bring the secondary voltage to approximately 3,000 volts no load. This voltage is at the secondary coil of the high frequency transformer. Remember that the **frequency stays the same** on both the primary and secondary coils of any transformer.

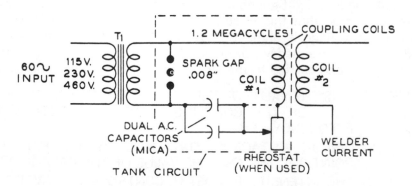

Figure 154. HIGH FREQUENCY CIRCUIT DIAGRAM.

The secondary voltage is applied to two dry-type mica capacitors. The capacitors accept the charge until there is enough voltage, or emf, to overcome the air gap resistance at the spark gap points. When this occurs the capacitors discharge their stored energy. An arc is created at the spark gap points and the stored energy is transmitted to the other side of the same mica capacitors. In other words, the high frequency current flows **from** one side of the mica capacitors, **through** the spark gap points, and **to** the other side of the same capacitors.

When the capacitors have again charged, and the voltage is again sufficient to force the arc at the spark gap points, the stored energy flows in reverse to its previous direction. An arc is established at the spark gap points and the stored energy is returned to its original starting point. To state the proposition plainly, the high voltage and milli-amperes is applied to the capacitors and oscillated back and forth through the spark gap points.

The high frequency generated in this manner is impressed on the air core coupling coils which transfer the high frequency energy to the welding conductor. In this manner relatively safe high voltage is brought to the electrode tip where it is required to perform its functions in the gas tungsten arc welding process.

The drawing of the high frequency system shown in Figure 154 indicates that the oscillatory system, or area, is within the dotted lines. The total area within the dotted lines is called the "tank circuit". This name is derived from the fact that it is within this area that the high frequency current oscillates from one side of the capacitors to the other.

The air core coupling coils, shown in Figure 154, are designated "coil #1" and "coil #2". Coil #1 is the high frequency coil and is normally made of light gauge insulated wire. This makes sense when it is considered that this circuit carries very low current values although the voltage is high. Coil #2 is the welding current carrying coil. The high frequency power is induced into the welding current carrying coil and carried to the electrode through the welding cables.

NOTE: 1) IT IS THE VALUE OF THE CAPACITORS AND THE INDUCTANCE OF THE COUPLING COILS THAT DETERMINES FREQUENCY.

2) CIRCUIT VOLTAGE IS CONTROLLED BY THE SPARK GAP POINT SETTING AND THE VALUE OF THE CAPACITORS.

3) THE INTENSITY RHEOSTAT CONTROLS THE AVAILABLE CURRENT IN THE HIGH FREQUENCY CIRCUIT.

4) THE AVAILABLE CURRENT IN THE HIGH FREQUENCY CIRCUIT IS DETERMINED BY THE SIZE OF THE CAPACITORS AND THE VOLTAGE APPLIED.

A major manufacturer of high frequency equipment sets the spark gap points of all high frequency systems at 0.008". This is done at the factory. The spark gap distance may be increased to approximately 0.012" although this is normally only done in cases of excessive losses of high frequency power in the secondary welding circuit. REMEMBER: As the spark gap point distance is increased the voltage at the capacitors becomes higher in value. Excessive voltage will cause the capacitors to fail in service.

Fundamental problems With High Frequency.

High frequency power is necessary for some gas tungsten arc welding applications and very useful for others. In addition to knowing where it is used to good advantage, such as for ac welding of aluminum and magnesium, it is advantageous to know some of the problems that occur with high frequency. The problems cited are some of the more common ones that occur.

Two of the most common general problems that plague the owners of high frequency equipment for welding are lack of high frequency energy at the welding arc and radio-TV interference from high frequency radiation.

Lack Of High Frequency At The Arc.

The symptom would be complete lack of high frequency power at the electrode tip. The logical place to check first would be the high frequency spark gap points of the high frequency oscillator. If there is no high frequency spark at the points the problem can be pin-pointed to the high frequency circuit.

The first part of the high frequency circuit that should be checked would be the two ac mica capacitors. The proper procedure for testing the capacitors is to remove them from the high frequency circuit one at a time. It is not necessary to physically remove them from the high frequency panel for the test. First, **using insulated tools,** "bleed off" the stored energy by short circuiting across the two capacitor terminals. Then disconnect the circuit conductors from the terminals of one capacitor.

Use an ohmmeter to test each capacitor for continuity. This is done by placing the ohmmeter probes on both terminals of the capacitor. If there is no continuity the capacitor is good. If there is some measure of continuity, the capacitor is short circuited and is faulty. If there is no ohmmeter available the capacitors can still be tested. Remove the circuit conductors from one capacitor as previously directed. Don't forget to bleed off the stored energy before attempting to disconnect the terminals. Turn on the power source and the high frequency circuit. If there is high frequency at the spark gap points, the disconnected capacitor is faulty. If there is still no high frequency spark, shut everything down, discharge the capacitor as before, and re-connect the first capacitor. Disconnect the second capacitor and try the power source again.

If the trouble is in the capacitors, usually only one of them will show continuity on the ohmmeter. The solution, if another capacitor is not readily available, is to put the remaining capacitor back in the circuit, remove the faulty capacitor totally from the circuit and re-space the spark gap points to 0.004″. The high frequency system will operate satisfactorily in these conditions. The second capacitor should be installed as soon as possible and the spark gap points re-set to 0.008″. Figure 155 shows a drawing of typical high frequency spark gap points and their holders.

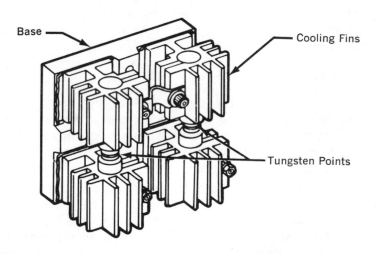

Figure 155. HIGH FREQUENCY SPARK GAP POINTS.

If there is high frequency energy at the spark gap points when you make the first check, the problem is probably that high frequency energy is going to ground somewhere in the secondary welding circuit. A careful check of the circuit from the welding power source terminals to the electrode holder and ground clamp assembly will usually show where the high frequency loss is occuring.

CAUTION: DO NOT ATTACH A VOLTMETER TO THE POWER SOURCE TERMINALS WHEN THE HIGH FREQUENCY SYSTEM IS IN OPERATION. THE EXTREMELY HIGH VOLTAGE WILL RUIN THE METER.

Be sure to de-energize the primary input power circuit when work is to be done on any part of a welding power source, including the high frequency panel. The easiest method is to disconnect the primary power switch at the primary input disconnect switch box.

Radio Interference And Radiation.

Radio and TV interference from high frequency welding installations is always a difficult problem with which to cope. To minimize the interference it must first be established just where the source of radiation is coming from and how it is transferred to the affected area.

Every high frequency stabilized welding power source built by major manufacturers has an FCC certification procedure as part of the operation and maintenance manual shipped with the power source. The certification procedure discusses in detail the proper method of installing the power source and high frequency unit in compliance with FCC regulations.

There are four common ways that high frequency energy may escape from the welding area. Some discussion of these points will perhaps give some ideas of where to look for the possible source of trouble when it strikes.

Direct Radiation From The Power Source.

High frequency energy has no preferred orientation of direction for radiation. It can, and does, go through any opening and in any direction. The radiation level is very pronounced in the immediate area if the power source doors and access panels are left open. The same is true if the power source is not properly grounded. It is necessary, therefore, that all openings in any high frequency stabilized welding power source be kept closed. When proper installation and operating procedures are followed according to the manufacturers directions there should be no problem with this type of radiation.

Direct Radiation From The Welding Leads.

Direct high frequency radiation from the welding and ground power cables, or leads, is very pronounced in the immediate area of the cables. The intensity decreases quite rapidly with distance from the affected area. By keeping the welding cables as short as possible this type of interference can be minimized. It is best to keep the electrode and ground leads as close together as possible.

The use of welding cables having a foil wrapping over the current carrying conductor which, in turn, is covered by the insulating neoprene or Hypolon or other material, has gained wide acceptance in the welding industry. The purpose of the foil shield is to deter grease and oil from getting into the actual welding current conductor. Hypolon coverings are normally recommended by most manufacturers for welding cable that carries high frequency power.

Although placement of the high frequency carrying welding cables close together is recommended to minimize radiation interference, it can cause a certain amount of high frequency power dissipation between the welding cables. This could be evidenced by arc fluttering, arc instability and possibly complete loss of high frequency at the welding arc.

Welding power cables that carry high frequency power should not be suspended overhead since they would have the effect of being radiating antenna. If extensive runs of cable are necessary to the welding operation it is suggested that fastening the welding cables to dry boards (using plastic cable clamps) will lessen the amount of high frequency loss to ground.

A point not generally realized is that high frequency radiation intensity may be altered considerably by changing the relative position of the ground connection and the electrode lead to the work area.

Direct Feed-Back To The Power Lines.

High frequency radiation may leak to the primary power lines by direct coupling inside the welding power source. The power line then serves as a radiating antenna. By proper installation of the primary power to the welding power source (this means through solid conduit) direct coupling can be avoided. Most manufacturers of high frequency stabilized power sources, or separate high frequency systems, specifically state in their instruction manuals that no rubber covered primary cords should be used with this type of welding equipment. **High frequency power will cause rapid deterioration of rubber.** The recommended procedure for bringing primary power to the welding power source is to run the primary power conductors through solid conduit.

Pick-up And Re-Radiation From Power Lines.

While high frequency radiation intensity will decrease rapidly with distance, the field strength in the immediate area of the welding leads will be very high. Unshielded wiring and ungrounded metal objects within this strong magnetic field may pick up the radiation directly from the welding circuit, conduct the high frequency for some distance, and produce a strong interference in a totally unrelated area.

This type of high frequency interference can be troublesome and hard to locate. It can be minimized by carefully following the installation procedures outlined in the instruction manual provided with the welding power source.

Summary.

It is not possible to write a simple formula to follow when dealing with high frequency problems. Each interference case is unique. The solution to the problem may be a simple thing or it may require considerable thought and work. Following the installation procedures as outlined by the manufacturer and the Federal Communications Commission (FCC) will certainly decrease the possibility of high frequency interference above the allowable maximum values.

Good grounding, short welding cables, insulated welding cables, tight electrical connections, properly shielded primary wiring and plain common sense are the best assets to have when dealing with high frequency problems.

Chapter 16

POWER SOURCES FOR GAS TUNGSTEN ARC WELDING

Almost any type of welding power source may be used for the gas tungsten arc welding process. It is true, however, that a specific type may be preferred for particular metals. For example, alternating current power sources are normally used for welding aluminum and magnesium with this process.

Some welding power sources are designed specifically for the gas tungsten arc welding process. They may be either ac or dc output units or even a combination ac/dc power source. The output control may be either a simple mechanical system or it may be electric control with remote output control capabilities. Normally the power sources used for gas tungsten arc welding are the conventional, so-called constant current type. The unit may be a very basic ac power source with high frequency stabilization or a three phase transformer-rectifier dc power source that has rather sophisticated facilities for completely programming the gas tungsten arc welds.

It is important that the difference between ac and dc welding power be recognized. The fact that all ac and most ac/dc welding power sources operate from single phase primary power, while **power sources with dc welding output only** normally operate from three phase primary power, makes a great difference in the arc characteristic produced by the specific power sources.

The welding power sources referred to in this chapter are all single operator units. Much of the data will, however, be applicable to multi-operator units.

As noted all ac welding power sources and most ac/dc welding power sources operate from single phase primary power. It should be noted that there have been ac/dc welding power sources on the market which do operate from three phase primary power. The ac part of the unit actually functions on **single phase primary power only** while the dc portion of the unit functions from the full three phase primary power. In this type of welding power source design, one leg of the main three phase transformer is wound with extremely heavy primary and secondary coil conductor material for the ac welding power output. As might be expected, the open circuit voltage for the ac portion of the power source output is relatively low. This makes it difficult to initiate an arc as well as providing longer arc outage times through the zero zone of the ac sine wave trace. It is not the most efficient power source design for this type of welding equipment.

Both alternating current (ac) and direct current (dc) are used for gas tungsten arc welding. As noted, ac is normally used for aluminum and its alloys and magnesium and its alloys. Direct current, straight polarity (dcsp) is normally used for all other metals including steels, stainless steels, low alloy steels, nickel alloys, etc. DCRP is seldom, if ever, used for production gas tungsten arc welding.

Probably the most common time cycle frequency of electrical power in use today is 60 hertz power although there is quite a bit of 50 hertz power in Britain and Europe. Either frequency is suitable for arc welding although the power source design, particularly the cooling system, is somewhat different. We will examine both single phase ac, and rectified single phase, plus three phase ac and rectified three phase power.

In Figure 156 an ac sine wave trace is shown. The ac power is changed to dc by a rectifier, normally either silicon or selenium. Rectified direct current is shown in the same illustration.

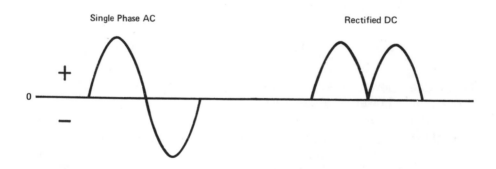

Figure 156. AC SINE WAVE AND RECTIFIED SINGLE PHASE.

As illustrated, ac power has **both** positive and negative values while the dc welding power has **either** a positive or a negative value. Since direct current flows in one direction only, the direction of current flow can only be changed by either switching the welding cables

— 275 —

at the power source output terminals or by positioning the polarity switch if a device of this type is provided.

Three phase ac power is illustrated in Figure 157. Rectified three phase power is also shown in the same illustration. Note that three phase alternating current has three separate sine wave traces but **within the same time span (1/60 second)** as the single phase ac wave trace. The three ac phases are 120 electrical degrees relative to each other. The rectified three phase dc welding power exhibits very smooth arc characteristics and a substantially **higher average power level** than the rectified single phase dc welding power.

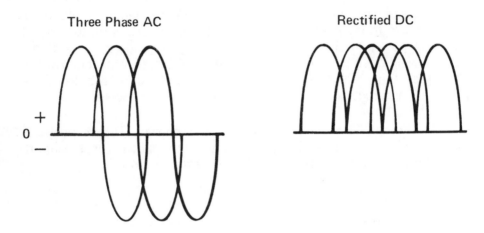

Figure 157. **THREE PHASE AC AND RECTIFIED DC.**

The higher power level is to be expected since one phase of the three phase power is always ascending to maximum strength.

NEMA Class 1 industrial ac welding power sources that are designed for the gas tungsten arc welding process are normally equipped with a high frequency control system, gas and water valves and solenoids, some method of filtering out the dc component, a post-flow timer for gas and water, etc. This type of welding power source does not require de-rating for amperage when welding as would be the case with ac welding power sources without the filtering capability for eliminating, or minimizing, dc component. Most conventional welding power sources designed for gas tungsten arc welding are rated at 60% duty cycle. This means they may be operated at their rated amperage for six full minutes out of ten. The other four minutes of the ten minute cycle they must idle and cool.

Conventional, or "drooping", output characteristic ac power sources that are not designed for gas tungsten arc welding must be de-rated for current in their welding output. By de-rating the welding power source from the rated current output level to a safer amperage level for gas tungsten arc welding the ac power source is protected from the phenomena known as dc component.

DC component is unique since it only occurs when welding with ac power and the gas tungsten arc welding process. It may be caused by partial rectification in the welding arc when welding materials with refractory oxides or by the difference in electron emission characteristics of the tungsten electrode and the base metals (for example, aluminum). Partial rectification is probably a combination of the two factors mentioned.

Tungsten has excellent electron emission characteristics which actually improve when the electrode is heated. Conversely, aluminum has comparatively poor electron emission, part of which is due to the refractory oxide found at the surface of the material. Aluminum oxide forms rapidly at any temperature and even more rapidly when the base metal is heated.

There is no particular problem in de-rating an ac welding power source for amperage when welding with the gas tungsten arc welding process. There are, in fact, two methods which may be used for the amperage de-rating procedure. The ac welding power source may be de-rated for current while retaining the same duty cycle or it may be de-rated to a current value at 100% duty cycle for processes other than gas tungsten arc. The same power source may then be further de-rated to 100% duty cycle with gas tungsten arc welding processes. This is only necessary for ac welding since there is no dc component with dc welding power and the gas tungsten arc welding process. De-rating ac welding power sources for 100% duty cycle and the gas tungsten arc welding process is a little more involved but it does guarantee that the welding power source will not be overloaded when welding. It is the safest method to use.

NOTE: DERATING A CONVENTIONAL CONSTANT CURRENT AC WELDING POWER SOURCE BY 30% FROM ITS RATED AMPERAGE WILL PROVIDE A SAFE WELDING CURRENT VALUE FOR GAS TUNGSTEN ARC WELDING BUT **AT THE SAME DUTY CYCLE** AT WHICH THE POWER SOURCE WAS ORIGINALLY RATED.

The chart shown in Figure 158 lists the factors for derating almost all types of conventional "drooper" output power sources to 100% duty cycle. The chart applies to all ac or dc welding power sources with the following exception: If the ac power source is a **tapped secondary coil** (main transformer) design, it will remain at its original duty cycle, normally 20%, in all taps or plug-in settings. This normally applies to some NEMA Class 3 ac power sources with limited primary power input and the tapped secondary coil design only.

Present Duty Cycle	Rated Amps Times:	Duty Cycle	Derated Duty Cycle For AC-TIG—Derated Amps Times:
60%	75% =	100%	70%
50%	70% =	100%	70%
40%	55% =	100%	70%
30%	50% =	100%	70%
20%	45% =	100%	70%

DC   150 amp. × .45 = 67 Amps.

Figure 158. DUTY CYCLE FACTOR CHART.

At the extreme right of the chart there is provided the data necessary for derating any conventional ac welding power source to be used for gas tungsten arc welding. The additional 30% taken from the current rating provides the safety margin that will keep the main transformer from over-heating due to dc component in the welding circuit.

Figure 159. BURNED COIL CAUSED BY DC COMPONENT.

Transformer heating can occur because of the circulating current called dc component. Remember, this is energy that is generated in the welding arc but which is not used in the arc. It becomes a circulating current in the welding circuit and is dissipated, as heat, in the main transformer secondary coils.

Heating in the main transformer may cause at least two serious problems: (a) breakdown of the insulation on the coils and the iron core material and (b) decrease in the electrical efficiency of the transformer due to higher resistance offered by the heated coils and iron core. The disastrous affect of excess dc component, and overheated coils, is shown in Figure 159. The coil shown is from a conventional ac welding power source that was not derated for current when gas tungsten arc welding.

The output characteristics of welding power sources specifically designed for the gas tungsten arc welding process are the same as those required for the shielded metal arc welding process. Volt-ampere curves for a conventional welding power source are illustrated in Figure 160. The open circuit voltage is relatively high at 80 volts. The maximum short circuit current of the unit is limited. The volt-ampere curve illustrated is typical for all power sources designed for use with the gas tungsten arc welding process, regardless whether manual or automatic process equipment is employed.

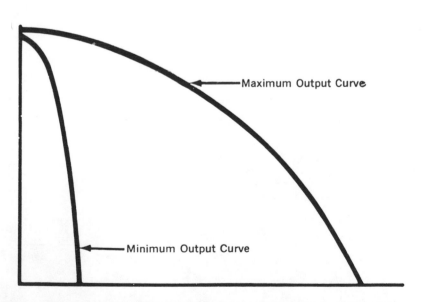

Figure 160. CONVENTIONAL POWER SOURCE VOLT-AMPERE CURVES.

The use of a particular kind of welding current, either ac or dc, will depend mostly on the type of metal to be welded, the shielding gas used, the welding technique employed, etc. Welding current selection is normally based on the type of metal to be welded.

Alternating current is usually used for the light metals such as aluminum and magnesium and their alloys. The basic reason for this selection is that ac provides the cleaning action necessary during the reverse polarity half-cycle.

There have been some excellent dc welding techniques developed in the past few years for gas tungsten arc welding of aluminum. Helium shielding gas is used with dcsp for this type of welding application. With these procedures aluminum up to at least one inch in thickness has been welded 100% in two passes.

Direct current, straight polarity is normally used for welding steel, low alloy steel, stainless steel, nickel and its alloys, copper and its alloys, refractory metals and other "hard" or dense metals. There is very little dcrp welding with the gas tungsten arc welding process.

Chapter 17

GAS METAL ARC WELDING

General Data.

The gas metal arc welding process had its inception in 1948. It was a natural development based on the concept of the gas tungsten arc welding process but with consumable electrodes instead of the non-consumable tungsten electrode.

The gas metal arc welding process requires a dc welding power source, some method of feeding and controlling the electrode wire (usually a feeder-control mechanism), a hose and cable assembly for the gas, wire and electrical power, and some type of gun or torch through which the electrode reaches the welding arc. Electrical contact is made at the barrel end of the gun, through a copper contact tube, to electrically energize the welding electrode. Some type of shielding gas is normally used with this welding process although the gas is not necessarily inert to the products of the weld zone. Welding speeds will vary according to the method of metal transfer used, the electrode type and diameter, the weld position and the type of metal being welded. In almost every instance gas metal arc welding is faster than shielded metal arc welding. If you were to categorize the gas metal arc welding techniques in the order of weld deposit speed per hour, they would probably line up as follows, the fastest being first listed and declining in speed of deposit as we go down the list.

1. **Spray Arc Transfer.** A solid wire method which is accomplished in an inert gas, usually argon or argon-rich mixtures.

2. **Gas Shielded Flux Core Electrode Welding.**
This is normally a globular transfer technique which may be as fast, or in some cases faster, than spray transfer.

3. **Buried Arc Transfer.** This is a very high speed welding technique for mild steel using CO_2 shielding gas only.

4. **Pulsed Current Transfer.** This is a type of spray transfer but much slower than true spray transfer methods. It is an all-position gas metal arc welding technique.

5. **Globular Transfer.** Although normally not an acceptable method of metal transfer for many applications, the globular transfer method may be used where spatter does not matter in the appearance of the final end product.

6. **Short Circuit Transfer.** This is the slowest of all the gas metal arc welding techniques because of its intermittent arc outages. Short circuiting transfer may be used in all welding positions with equal ease. Over a period of five minutes or more steady welding, short circuit transfer will beat any shielded metal arc welding electrode for deposition efficiency and inches of linear weld joint. This is based on the metal thickness being applicable for this method of welding.

(This list is compiled from data studied over a number of years by the author. I believe it reflects the present state of gas metal arc welding.)

Some of the trade names used for gas metal arc welding equipment reflects the type of metal transfer normally obtained with its use. Other trade names are for manufacturer identification only. According to the master chart of welding processes issued by the American Welding Society, the proper name for this process is **Gas Metal Arc Welding.** The abbreviation is GMAW.

The term "MIG" is sometimes used to identify the GMAW process. The initials stand for **M**etal **I**nert **G**as. The term originated in the early days of the process when the only shielding gases used were argon and helium, both of which are inert to the products of the weld zone. With the introduction of shielding gases with active elements, such as carbon dioxide (CO_2), the term MIG is not really applicable to the process. It is still used by many people in industry although it is no longer a technically correct name unless you are using a pure argon or helium shielding gas.

The gas metal arc welding process was patented in 1950. Part of the **process patent** states that the process is used with direct current, reverse polarity (dcrp). This is still basically true for all techniques of welding with gas metal arc processes regardless of the class and type of metal to be welded.

Alternating current has been tried for the gas metal arc welding process but with little success. Extremely high open circuit voltages have been necessary to maintain the welding arc. The main problem is the arc outage every 1/120th of a second (each half-cycle) which causes a loss of ionization in the shielding gas. High frequency is to no avail here for it gets to the wire drive motor and the extremely high voltage causes the motor to freeze and be destroyed. Research is still being conducted using various types of specially coated electrode wires, etc. At this writing there has been no significant development for using alternating current (ac) with the gas metal arc welding process.

Direct current, straight polarity is not normally used with the gas metal arc welding process although it has been found to be useful for very thin gauge steel welding with the short circuit method of metal transfer. Wire electrode melt rates are often twice as fast with dcsp as with dcrp. The penetration patterns with dcsp are very shallow with high crowned welds being the normal deposit configuration. With the electrode negative, it is the cathode in the welding arc. The electron flow is from the electrode to the work. The "cathode spot" may tend to wander at the electrode tip causing an erratic arc behavior. The weld metal transfer is globular instead of the spray transfer that would normally be expected **at the same amperage** with dcrp and argon shielding gas.

Specially coated electrode wires have been employed with dcsp to stabilize the welding arc. The rare earth coating used (usually Cesium) tends to decrease the melt rate of the electrode that would normally be expected with dcsp. Applications of the gas metal arc welding process and dcsp are presently limited to high speed welding of light gauge sheet metal and body shop applications where minimum penetration is required.

All of the commonly welded metals may be welded with the gas metal arc welding process. The specific technique of welding will usually determine if out-of-position welding can be accomplished. The term "out-of-position" welding means the arc is applied to a joint that is in other than the flat plane.

The gas metal arc welding process is designed to replace the shielded metal arc welding process in many applications. There are substantial differences in the two processes although they do have many similarities. There is some further discussion of this subject in the chapter entitled "Welding Processes Comparison And Use" later in this text.

Current density (the amperage per square inch of cross sectional area of the electrode) is the key factor in the gas metal arc welding process. In all cases there is the use of relatively high amperages with relatively small diameter electrodes. This combination provides high current density at the electrode tip. The shielded metal arc welding process employs relatively low current densities since the amperage is low for a specific electrode diameter. For example, one of the largest electrode diameters used with the gas metal arc welding process is 1/8″ diameter solid or flux core electrode. The welding amperage for this electrode may range up to 650-700 amperes in some conditions. With the shielded metal arc welding process a 1/8″ electrode will seldom operate satisfactorily above 120 amperes. Additional information on current density will be discussed in a subsequent portion of this text. There is a Current Density Chart in the appendix of this book.

Equipment Required.

The fundamental equipment necessary to operate with the gas metal arc welding process includes a welding power source (either conventional or constant potential), a wire feeder-control mechanism, the necessary interconnecting cable and hose assembly, and a gun or torch. A source of cooling water, if required by the gun, a proper shielding gas and the correct electrode wire complete the assembly. A line diagram is shown in Figure 161 indicating the required basic equipment for gas metal arc welding.

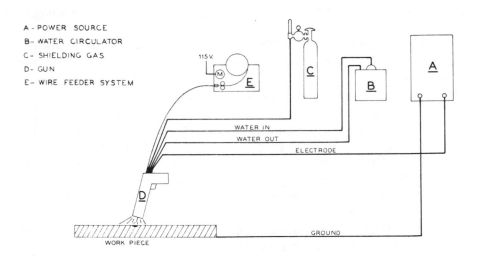

A - POWER SOURCE
B - WATER CIRCULATOR
C - SHIELDING GAS
D - GUN
E - WIRE FEEDER SYSTEM

Figure 161. BASIC GAS METAL ARC WELDING EQUIPMENT.

As the requirements of the welding industry have grown to meet exacting manufacturing specifications so has the development of welding equipment, including welding power sources, progressed. New and more versatile equipment has had to be designed and built to keep pace with the insatiable demands of the welding industry. The advance of technical developments has made some of the gas metal arc welding process equipment obsolete before it could even reach the welding market. It is estimated that welding has progressed more in the past twenty years than in all of the time previous.

Welding Power Source Development.

When the gas metal arc welding process was developed there was a limited selection of dc power sources available to the welding industry. The dc power sources then in use were designed and built for the shielded metal arc welding process. The output volt-ampere curves of this type of power source follow a negative, or sharply drooping, curve. Open circuit voltage for this type of welding power source is usually in the range of 72-80 volts. Maximum short circuit current will be approximately 150-175% of the amperage rating of the power source.

With the increased demand on the gas metal arc welding process to perform more and different applications came the need for welding power sources that could extend the versatility of the process equipment. Due to these demands the constant potential, or constant voltage, type welding power sources were developed.

Constant potential and **constant voltage** are synonymous when discussing power sources for welding. The constant potential type welding power sources have a relatively low maximum open circuit voltage. The maximum short circuit current, however, may be several times the value of the rated amperage of the power source. This is very plainly shown in Figure 162 which compares the conventional and the constant potential type welding power source volt-ampere curves.

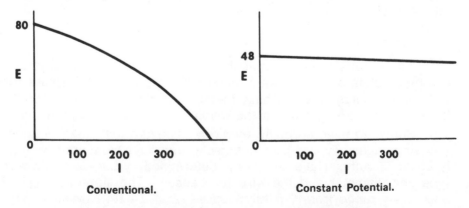

Figure 162. COMPARATIVE VOLT-AMPERE CURVES.

The chapter of this text entitled "Welding Power Sources For Gas Metal Arc Welding" has additional data concerning the development of the various types of welding power sources and their functions for the gas metal arc welding process.

Types Of Metal Transfer.

There are several types of metal transfer employed with the gas metal arc welding process. Although many times they are called "welding processes", the various types of metal transfer are actually a welding method or technique within the total process.

As an example, there are actually only two basic types of metal transfer with the gas metal arc welding process. They may be classed as the **gas shielded open arc method** and the **gas shielded short circuit method.**

The gas shielded open arc method of metal transfer considers that molten metal is separated from the welding electrode, moved across a physical air space within the arc column and deposited as weld metal in the joint to be welded. Terms such as "Spray Transfer", "Buried Arc Transfer", "Globular Transfer", and "Pulsed Current Transfer" are applied to the various gas shielded open arc methods of metal transfer.

The short circuit method of metal transfer deposits the molten weld metal by direct contact of the welding electrode with the base metal. Many trade names are presently in use for equipment designed for this type of welding. No matter what name is used for the equipment **the welding technique is exactly the same.** The short circuit method of metal transfer has many applications in the welding industry today. Some of the trade names to describe the process equipment include Short Arc, Dip-Matic, Micro-Wire, Dip-Transfer, Mini-Arc, etc.

Spray Transfer.

Spray transfer is accomplished by the movement of a stream of tiny droplets of molten weld metal from the electrode, across the welding arc column, to the base metal. The arc has a characteristic buzzing or humming sound if the welding condition is properly set.

Spray transfer normally employs relatively high load voltages and amperages. Typical load voltages would range from approximately 25 to 33 volts. The amperage is determined by the electrode wire type and diameter and the wire feed speed. The slope settings, if used, may range from 0-6 turns based on a 14 turn reactor. This is applicable only if such a device is part of the power source circuitry.

Spray transfer may be accomplished outside the values indicated, however, depending on the type and diameter of the electrode wire used. For example, spray transfer welding of aluminum with 0.030″ diameter electrode wire has been done very successfully at 23 load volts. The ranges indicated are typical of the process in most cases.

Shielding gases that are used with this technique of welding include argon, helium, argon-helium mixtures, argon-oxygen mixtures and helium-oxygen mixtures. An artists conception of spray type metal transfer is shown in Figure 163.

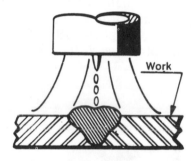

Figure 163. SPRAY TYPE TRANSFER.

In particular, notice the electrode tip and the manner in which it is pointed. This is typical of the spray transfer method of welding since the molten weld metal literally runs off the end of the electrode in an almost continuous stream.

Buried Arc Transfer.

The buried arc technique is truly an open arc welding method. Metal transfer occurs below the surface of the base metal. A crater is created by the arc force which contains the spatter inherent in open arc CO_2 welding of steel. The base metal acts as a crucible for the molten weld metal.

The basic concepts of buried arc welding with CO_2 are relatively simple. High amperages and high load voltages are the norm. Electrode wire diameters usually range through 0.045″, 1/16″ and 3/32″ diameters. For example:

0.045" dia. = 350-450 amperes @ 33-36 load volts.
1/16" dia. = 450-650 amperes @ 34-39 load volts.
3/32" dia. = 500-800 amperes @ 35-40 load volts.

These ranges are for use as guides only and are based on the experiences of the author over a period of years. The minimum steel plate recommended for this welding technique is 1/4" since this is a very deep penetration welding arc.

The buried arc welding technique is normally used for mild steel applications with CO_2 shielding gas. With the low alloy steels presently being used the electrode wire would have to be formulated with a chemical composition that would be compatible with the base metal when deposited in the weld joint.

Plate thickness will determine which diameter electrode wire you should use. Speed of travel can be quite high with this welding technique, especially if automatic welding equipment is used. The critical thing to watch is that travel speed doesn't outrun the solidification pattern of the weld metal deposit. This could cause plastic cold shuts, porosity and other weld defects.

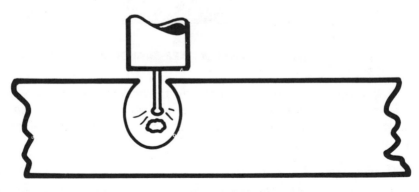

Figure 164. BURIED ARC TRANSFER.

There are several interesting points about the buried arc welding technique which is illustrated in Figure 164. The relatively high speed of welding can certainly be a cost savings to the manufacturer. The fact that all metal transfer occurs below the surface of the base metal is important. A crater, or cavity, is created by the arc force and all spatter, and most of the arc heat, is retained within the subsurface crater. This has two specific advantages: (1) the welding gun will operate at cooler temperatures because there is not the tremendous amount of reflected heat on the gun barrel and, (2) the electrode deposition efficiency is high because the spatter is trapped within the crater and used as filler metal.

Buried arc is a high speed, deep penetration process. Constant potential type welding power sources with a slightly negative volt-ampere curve are preferred for this process because they provide the fastest response to changing arc conditions. Also important is the fact that substantial amperage changes can be made with very little change in load voltage. The constant potential power sources used for the buried arc welding technique should have a rating of not less than 600 amperes, 40 load volts, 100% duty cycle. Ideally, the power source would have electric control of welding output so the welding operator has remote control of the power source at his work station.

Pulsed Current Transfer.

The pulsed current method of gas metal arc welding is a form of spray transfer. It is not as fast as normal spray transfer since the metal transfers from the electrode at specific intervals of time. The number of current pulses per second are usually based on the input primary power frequency. For example, 60 hertz power as the primary frequency provides either 60 or 120 pulses per second on present power sources. There is one droplet of metal transferred with each pulse. Almost any wire feeder-control system may be used with this welding technique although the constant potential type power source used is unique.

In one type of welding power source used for pulsed current welding there are two levels of current supplied. Both levels of current come from the same power source although from two separate transformers. The main transformer is a three phase system which supplies what is known as "background current". The objective of this amperage setting is to maintain the welding arc at all times without melting the electrode wire. It also keeps a steady flow of heat into the weld joint and maintains gas ionization. Background current is adjustable to suit the type and diameter of electrode wire used.

The second current level is the "pulse current". This is supplied from a single phase transformer which is connected to the main transformer through reactors. The pulse amperage is adjustable and should be set just above the spray threshold amperage for the specific type and diameter of electrode wire used. (It is important to know that **every electrode type and diameter has an amperage level above**

which it will automatically go into spray transfer in an argon-rich atmosphere).

The pulsed current transfer technique provides the capability to perform a modified type of spray transfer under controlled heat input conditions. With this welding method spray transfer can be accomplished in all welding positions.

Pulsed current metal transfer is unique in that only one droplet of metal transfers across the arc with each pulse. If the unit is set for 60 pulses per second there will be 60 droplets of metal moving across the arc every second. This is, of course, much slower than normal spray transfer which has a continuous stream of molten metal droplets moving across the arc. It is faster than the short circuit method of metal transfer.

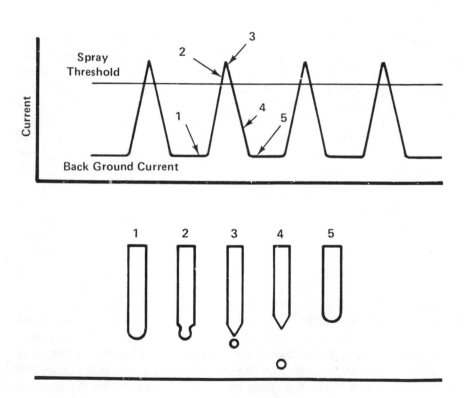

Figure 165. **PULSED CURRENT METAL TRANSFER.**

Globular Transfer.

As the name implies, globular transfer of weld metal is in the form of large irregularly shaped globules of molten electrode material transferring across the welding arc column. This technique of gas metal arc welding is typical of mild steel welding with CO_2 shielding gas. Globular transfer may also occur because of low arc voltage or low arc amperage when using other shielding gases.

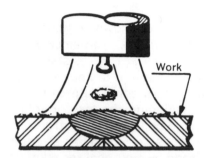

Figure 166. GLOBULAR TRANSFER.

The use of globular transfer is not normally a desirable condition because of the excessive spatter losses that occur with this welding technique. Figure 166 shows the molten ball bulging at the electrode tip which is typical and indicative of globular transfer welding.

Short Circuit Transfer.

The short circuit method of metal transfer is unique in that no molten metal transfers across the welding arc. Weld metal transfer occurs only when the electrode makes physical contact with the base metal. The actual number of short circuits per second that takes place will depend on the electrode type and diameter, the shielding gas used, the open circuit voltage, the load voltage used and the wire feed speed. With all other welding conditions constant, the number of short circuits per second will be directly influenced by wire feed speed. The faster the wire feed speed, the more short circuits per second. Keep in mind, however, that excess wire feed speed will cause the electrode to stub into the weld deposit.

As illustrated in Figure 167 short circuit transfer is a series of changing voltage and amperage values which are controlled by the pre-set welding condition.

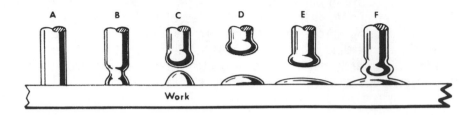

Figure 167. SHORT CIRCUIT TRANSFER.

The sequence of action shown for one short circuit cycle will follow a relatively stable pattern. The electrode wire makes physical contact as shown at point (A). The resistance of the electrode wire to electrical current flow causes it to heat and "neck down" as shown at point (B). Several factors contribute to the necking down of the electrode at point (B) including the loss of columnar strength in the electrode and the development of a very strong magnetic field around the electrode wire. The loss of columnar strength is due to the heat generated in the electrode by electrical resistance heating. The strong magnetic field is caused by the rapid increase in welding current value in the electrode when it short circuits to the base metal workpiece.

The action begins to pick up at point (C). At this point, the electrode has separated from the workpiece and an arc is established. Keep in mind that **the wire feed speed is constant** and does not change at any time in the sequence. **The electrode melt rate is faster than the wire feed speed at point (C).** At point (D) we have reached a point of equilibrium where the **electrode melt rate and wire feed speed are exactly equal.** The maximum load voltage is reached and the arc length will not increase beyond this level. It is apparent that the melt rate has steadily decreased from point (C) to point (D). The next step in the illustration shows the electrode progressing toward the workpiece at point (E). The melt rate of the electrode has slowed still more. At this point **the melt rate of electrode is slower than the wire feed speed.** A ball of molten electrode metal has formed at the electrode tip. At point (F) the electrode again short circuits to the workpiece, the weld metal is passed to the base metal by direct contact, and the cycle begins again. Consider that the number of short circuits per second may be within a range of 40-250 short circuits per second and you have some idea what is occuring with this welding technique.

All industrial class process equipment designed specifically for the short circuit method of metal transfer, regardless of the manufacturer, will push small diameter electrode wire (0.030"; 0.035"; 0.045" diameters) through a casing assembly to the welding torch or gun. Welding power is applied to the electrode wire as it leaves the nozzle end of the gun or torch. Electrical contact is made between the electrode wire and a copper contact tube located in the torch body.

The short circuit technique of gas metal arc welding is essentially a low amperage and low voltage welding method. The welding power sources used with this technique are constant potential type units with slope control either built-in as a fixed reactance or else variable through a range of slope settings. It is important to know that slope control may be achieved in constant potential type welding power sources by either electrical resistance or through a slope reactor device. This subject will be discussed further in the chapter entitled "Welding Power Sources For Gas Metal Arc Welding."

The slope settings are usually such that the maximum short circuit current of the power source is limited to a relatively low value. This precludes the possibility of explosive high current levels at the electrode tip which could cause excessive spatter in the welding arc. The use of slope control reactors also slows down the rate of response of the constant potential type power source to changing arc conditions.

Typical load voltages used with the short circuiting method of metal transfer will range approximately 15-21 volts. Slope settings will range from 6-14 electrical turns, based on a 14 turn reactor. (You will recall that an electrical "turn" is one full wrap of the conductor wire around the periphery of a coil). The exact slope settings will be determined by the type of metal to be welded and the diameter of the welding electrode. For example, slope settings for mild steel will normally be 6-10 turns based on a 14 turn reactor. For 0.030" diameter electrode wire 6 turns would probably be very good, using either welding grade CO_2 or 75% argon-25% CO_2 shielding gas. For 0.035" diameter electrode wire either 6 or 8 turns of slope would be acceptable. There would probably be more weld spatter with 6 turns of slope. For 0.045" diameter wire either 8 or 10 turns would be correct.

Slope settings for austenitic stainless steels welded with the short circuit method of metal transfer will be approximately 10-12 turns. An ideal setting for 0.035" diameter electrode wire is 10 turns of slope using the **tri-gas mix** of 90% helium-7 1/2% argon-2 1/2% CO_2. The bead will come out with very little build-up, especially in vee groove joints.

Short circuit transfer can be used on aluminum and its alloys but the slope setting must be fairly high, usually 12-14 turns of slope. This is considering that the aluminum is sheet metal of 1/8" thickness or less. For heavier thicknesses of aluminum it would be necessary to decrease the number of turns of slope in order to put sufficient heat into the greater metal mass which will act as a heat sink and steal the welding heat from the welding joint.

While these slope ranges are indicative of the slope values normally used they may not be exact and should be used as guides only when setting up welding conditions for short circuit welding.

Shielding gases that are normally used for short circuit transfer welding include argon, helium, welding grade CO_2, argon-oxygen mixtures and helium-oxygen mixtures. As noted, there is a special tri-gas mixture of 90% helium-7 1/2% argon-2 1/2% CO_2 that is designed for use with the austenitic stainless steels and short circuit metal transfer welding.

Amperage Control.

In using the gas metal arc welding process with constant potential type welding power sources, amperage is thought to be controlled by wire feed speed. The faster the wire feed speed the higher the amperage output of the power source. At the same time the amperage increases, the load voltage value decreases. Let us note right now that wire feed speed is only one factor in setting a welding condition with the gas metal arc welding process. The open circuit voltage and the slope setting on the power source (if used) actually determines how much welding amperage the power source can supply at the welding arc. The wire feed speed simply determines the balance of welding amperage and load voltage at the arc.

Amperage control is actually a function of the output characteristics of the constant potential type welding power source. As illustrated in Figure 168 the constant potential type welding power source provides **automatic amperage control.**

As the wire feed speed is increased in inches per minute, the power source increases the amount of amperage to the arc so the electrode will be melted. This will continue until the wire feed speed is increased to the point where the power source cannot put out enough amperage, with sufficient load voltage, to melt the electrode wire. The wire will stub into the base metal at this point. By "stubbing", we mean that the electrode wire will make solid physical contact with the base metal and not create or sustain an arc. The electrode wire will just turn red from resistance heating and pile up at the surface of the base metal being welded.

The welding set-up pictured in Figure 168 considers a fixed nozzle end distance from the base metal workpiece. The actual arc length is determined by the open circuit voltage setting and the slope setting on the welding power source. The welding condition is deemed correct for the material, electrode type and diameter, shielding gas, etc.

The volt-ampere curve, with the "X" at the point of equilibrium, reflects the welding condition set in the drawing. **(The point of equilibrium is where the wire feed speed and the melt-rate of the electrode are exactly equal).** According to the diagram the weld is progressing from left to right. The problem appears to be a tack that was not removed from the path of the weld. For this example we will assume that the tack weld deposit does not melt significantly.

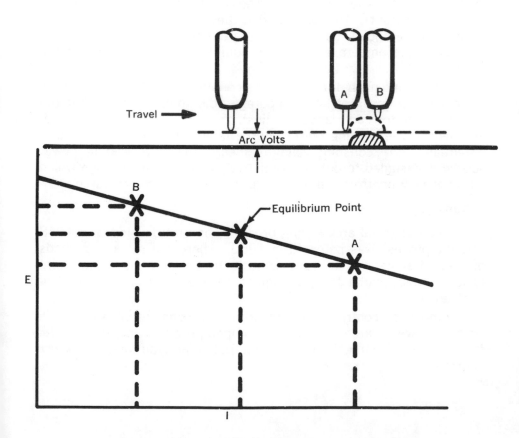

Figure 168. AUTOMATIC AMPERAGE CONTROL.

At the instant the welding arc reaches the tack (A) the arc length shortens. Since arc voltage is a function of arc length the arc voltage decreases momentarily. Looking at the volt-ampere curve we see the lower voltage at point (A) but a considerable increase in arc amperage. The increased amperage provides a **faster melt rate** for the electrode wire; in fact, faster than the wire is being fed into the weld joint. Remember: **wire feed speed remains constant!** The faster melt rate causes an increase in the arc length, with a corresponding slide **up the volt-ampere curve,** until the point of equilibrium is once again reached. At the point of equilibrium, of course, the melt rate of the electrode is exactly equal to the wire feed speed.

The back side of the tack is indicated as point (B) in the drawing. At this point the arc length increases momentarily. The increase in arc length also increases arc voltage. With the increase in arc voltage there is a simultaneous decrease in arc amperage. The melt rate of the electrode is less at lower amperage than the wire feed speed (which is constant). The result is that the electrode wire drives down toward the weld joint, decreasing the arc length and arc voltage and increasing the amperage, until the equilibrium point is once more established.

It is apparent that, although the constant potential type welding power source does not have amperage adjustment devices that the welding operator can manually adjust, there is a built-in current adjustment in the welding power source itself. The names "constant potential" and "constant voltage" means that the welding power source is designed to put out a relatively constant **arc voltage** which automatically controls the amperage level of the power source.

Summary.

The gas metal arc welding process was first developed in 1948 and the process was patented in 1950. There are several methods of metal transfer which are used with the process. It is important to know which metal transfer method to use for a specific welding application.

Amperage control is automatic in the constant potential type welding power source. There is no amperage output control device, for manual adjustment, on constant potential type welding power sources.

Chapter 18

SHIELDING GASES AND ELECTRODES

There are a number of shielding gases and shielding gas mixtures in common use with the gas metal arc welding process. Some of the gases and gas mixtures have a broad range of applications while others are restricted in their use. This chapter will discuss some of the shielding gases in general use for gas metal arc welding.

Argon.

Argon is a chemically inert gas that will not combine with the products of the weld zone. It has an ionization potential of 15.7 electron volts. The term "electron volts" has no particular significance here except that it defines the type of voltage used to determine ionization potential. (Ionization potential is the energy necessary to remove an electron from the gas atom. Removing the electron from the gas atom creates an ion with the atomic particles that are left. The ion, or electrically charged gas atom, provides a better electrical conductor for the welding current to follow from the electrode to the base metal workpiece).

Argon has low thermal, or heat, conductivity. The arc column is constricted with the result that high arc densities are present. Arc density is measured by dividing the arc current value by the cross sectional area of the arc column measured in square inches. The high density arc permits more of the available arc energy to go into the base metal workpiece as heat. The result is a narrow weld bead width with a deep central penetration of the weld deposit. The cross section described is typical of argon gas shielded weld deposits in aluminum and aluminum alloys.

Argon causes a more concentrated arc than any of the other commonly used shielding gases employed with the gas metal arc welding process. This is one reason that argon has the reputation of being a "cleaning gas" for removing surface oxides. It is actually the arc column concentration, and therefore the heat energy concentration, that causes the refractory oxides associated with aluminum and its alloys to loosen and disperse. The fact that the gas metal arc welding process uses dc reverse polarity (the cleaning half-cycle of polarity) probably helps in removing the surface oxides.

Argon normally has a purity of 99.995+% and is used as the shielding gas for many of the commonly welded metals. Its primary use is with the non-ferrous metals such as aluminum, magnesium, copper and their respective alloys.

In some gas metal arc welding applications argon does not provide the penetration characteristics needed for thicker metals. In many instances other shielding gases such as helium, oxygen or carbon dioxide (CO_2), are mixed with argon to obtain better arc characteristics and weld deposit configurations.

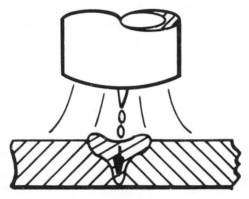

Figure 169. ARGON SHIELDED WELD IN ALUMINUM, GMAW.

A typical argon shielded gas metal arc weld in aluminum material is illustrated in Figure 169. It is important to remember that the drawing shows an **argon shielded gas metal arc weld.** (FOR GAS TUNGSTEN ARC WELDING THE CROSS SECTION OF THE WELD IS CONSIDERABLY DIFFERENT WITH ARGON SHIELDING).

Helium.

Helium is an inert gas and may be compared to argon in that respect. Helium has an ionization potential of 24.5 electron volts. It is lighter than air and has excellent thermal (heat) conductivity. The helium arc plasma and arc column will expand under the heat of the arc producing thermal ionization of the gas and reducing the arc density.

With helium there is a simultaneous change in arc voltage when the voltage gradient of the arc length is increased by the discharge of heat from the arc column. This means that the arc voltage is greater for a given arc length than would be the case with argon because some of the arc energy is lost in the arc column itself and is not transmitted to the work. Using helium shielding gas there will be a broader weld bead, with relatively shallower penetration, than with argon. (JUST THE OPPOSITE IS TRUE WHEN USING HELIUM SHIELDING AND THE GAS TUNGSTEN ARC WELDING PROCESS). This also accounts for the higher arc voltage, **for the same arc length,** obtained with helium as opposed to argon.

Helium is derived from natural gas wells. The natural gas is cooled and compressed. During the process the hydrocarbons are drawn off, then nitrogen, and finally helium. This is the process of liquifying the gases until, at —453° F., the helium is isolated.

Helium has been in short supply at times and limited industrial application as a shielding gas has been the result. Recent discoveries of natural gas deposits with high helium content have made helium gas more readily available to the general welding public and its increased use has been applied to many applications of gas metal arc welding.

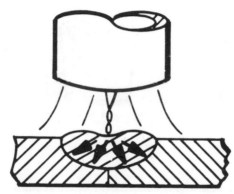

Figure 170. HELIUM SHIELDED WELD IN ALUMINUM, GMAW.

Figure 170 shows a typical cross section of a **helium shielded gas metal arc weld.** The reason for the difference in the gas metal arc welding deposit and the previously illustrated gas tungsten arc welding deposit lie in the fact that, with gas metal arc welding, heat plus molten metal is put into the weld deposit.

Helium is used primarily for the non-ferrous metals such as aluminum, magnesium, copper and their alloys. It is also used as a mixtured gas with other shielding gases.

Carbon Dioxide (CO_2).

Carbon dioxide is a compound gas which means that it is made up of more than one element. Where argon and helium have only one

— 299 —

gas atom to the molecule, carbon dioxide has two gas atoms to the molecule. Argon and helium are, therefore, monatomic while carbon dioxide is classed as a diatomic gas. Diatomic means that carbon dioxide has **two atoms per molecule** in its chemical composition. The primary constituents are carbon monoxide and atomic oxygen. **CO_2 is not an inert gas** such as argon and helium. In fact, CO_2 has a characteristic different than argon or helium which is its ability to dissociate and re-combine in the atmosphere. It is this factor that permits more heat energy to be absorbed in the gas passing through the welding arc. It also uses the free oxygen in the arc area to superheat the weld metal transferring from the electrode to the workpiece. As illustrated in Figure 171, CO_2 has a slightly wider arc column than argon but less than is apparent with helium.

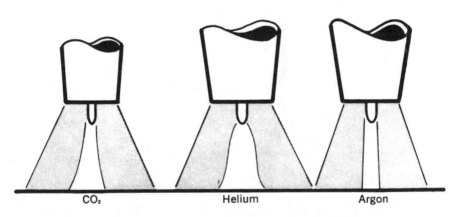

CO₂   Helium   Argon

Figure 171. COMPARATIVE ARC COLUMNS.

Care must be taken when using **open arc** CO_2 shielded welding techniques so that plastic cold shuts, gas entrapment and intrabead cracking do not occur. Such faults in the weld could be caused by the penetration characteristics of the arc or by the cooling of the globular transfer of molten metal across the arc. The use of welding grade CO_2 shielding gas and electrode wires with the proper deoxidizing elements will decrease the possibilities of such flaws occuring in the weld deposit.

CO_2 has a tendency to cause a spattering, unstable arc when used for **open arc transfer** of weld metals. Open arc means the molten weld metal is physically transferred across a space from the electrode to the base metal. The weld spatter may be contained to some degree by maintaining an extremely close arc; in fact, by burying the arc in the workpiece. This is the Buried Arc technique which has been discussed. With this welding technique, the arc literally creates a cavity in the base metal within which almost all of the spatter is used as filler metal in the weld. The buried arc welding technique is normally used for heavier thicknesses of steel plate where high speed welding is the criteria. This technique is usually used with CO_2 shielding gas and on those steel materials where the filler wire chemistry will provide sufficient mechanical and physical strengths in the weld deposit to comply with code requirements.

In most cases CO_2 is used only for mild steel welding applications. This is because the carbon dioxide (CO_2) gas breaks down under the heat of the welding arc into approximately 33% carbon monoxide by volume, **33% free oxygen** by volume, and the rest remains CO_2 since it is outside the arc heat zone. Carbon dioxide is also used in a number of shielding gas mixtures for other steel materials.

Argon-Oxygen.

Pure argon is an excellent shielding gas for the gas metal arc welding process because it permits the use of spray type metal transfer with all the commonly welded metals. When depositing flat or horizontal welds on steels or stainless steels, however, the "quick-freeze" characteristics of argon shielded welds does not permit the molten metal to "wet out" to the toes of the weld. This will invariably cause undercut at the edges of the weld bead. The drawing shown in Figure 172 notes the typical high crowned weld bead, with the attendant undercut, of an argon shielded weld on stainless steel.

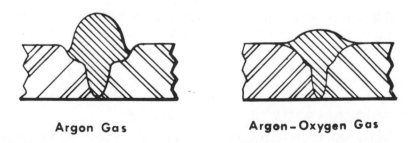

Argon Gas   Argon - Oxygen Gas

Figure 172. COMPARATIVE WELDS WITH ARGON AND ARGON-OXYGEN.

The tendency to undercut ferrous metal welds may be minimized or prevented by the addition of 1-5% oxygen to the argon gas. This mixture is commonly available in cylinders of argon plus 1%, 2%, 3% and 5% oxygen mixtures. The oxygen super-heats the weld metal transferring across the welding arc column. The super-heating effect **retards the cooling rate** of the molten weld metal by a few milli-seconds. In keeping the weld metal molten the few milli-seconds long-er, the deposited metal has the opportunity to flow out to the toes of the weld thus virtually eliminating the possibility of undercut. This also acts to control the weld bead profile by flattening the weld through movement of the liquid metal away from the weld bead cen-terline.

The natural question is, "When do I use 1% oxygen and when do I use 5% oxygen?". There is no set percentage to use because each application is different. As a guide we offer the following data. Keep in mind that the reason for the oxygen addition is basically to super-heat the molten metal transferring across the arc. The greater the volume of molten metal the more oxygen required for super-heating.

   a) Stainless steel thickness of $1/16''$-$1/8'' =$ argon $+ 1\% \ O_2$.
   b) Stainless steel thickness of $1/8''$-$3/16'' =$ argon $+ 2\% \ O_2$.
   c) Stainless steel thickness of $3/16''$-$1/4'' =$ argon $+ 3\% \ O_2$.
   d) Stainless steel thickness of over $1/4'' =$ argon $+ 5\% \ O_2$.

This is "rule of thumb" data only. Your specific application will de-termine the argon-oxygen mixture you should use.

Argon-Helium.

Argon-helium mixtures are usually used to obtain the best and most favorable characteristics of both gases. The addition of helium may be in the percentages of 20%-90% and is normally mixed by the user to suit his specific requirements.

The normal procedure is to purchase a cylinder of each type of shielding gas and use flowmeters and a mixing chamber to con-trol the percentages of each gas in the mixture. A relatively new de-vice called a gas proportioner is presently available to perform the function of mixing the shielding gases with more accuracy. The use of either type of device permits the user to select the exact welding characteristics he wants from the shielding gas.

Argon-helium mixtures are normally used where heavy sections of non-ferrous metals are to be welded. This would include mag-nesium, aluminum and copper. The heavier the metal thickness, the greater the percentage of helium in the gas mixture. Although the atoms of the two shielding gases do not combine, they do inter-mix by the venturi action of drawing off the gases.

— 302 —

Argon-CO₂.

Welding grade CO_2 does not provide the arc characteristics re-quired for welding some types of steels. In mild steel applications the problem may be excess spatter which can scar an exposed metal surface. In low alloy steels it would manifest itself by excessive oxidation of the alloying elements.

In such cases the use of a mixture of argon-CO_2 has usually eliminated the trouble. Some welding authorities believe the argon-CO_2 mixture should contain no more than 25% CO_2. Others are of the opinion that mixtures with up to 80% CO_2 are practical. The individual user should determine his requirements for his specific application.

Cost is the primary reason for using as much CO_2 as possible. The price of argon is normally quite a bit higher than the cost of CO_2 in most areas of the United States. By using a cylinder of each type of shielding gas (argon and CO_2) the mixture percentages may be varied by using flowmeters or a gas proportioner.

If a simple "wye" connection is used it is possible to have a quantity of one gas (normally the one with the greatest flow rate from the cylinder) reach the weld in an unmixed state. This would normally be followed by a smaller quantity of the other gas, in its unmixed state, reaching the weld. The result could be disastrous to the weld quality. Such dissociations of mixtured gases usually occur at the beginning of a weld.

Argon-CO_2 mixtures in cylinders normally sell for the same price as pure argon. The reason is the additional cost of mixing the gases to quite exact percentages. When the shielding gases are purchased in separate cylinders and mixed by the user the cost is more propor-tionately distributed according to the amount of each gas used.

Argon-CO_2 gas mixtures are employed for gas metal arc welding steel, low alloy steel and, in some few cases, for some thin-gauge stainless steel applications.

Helium-Argon-CO₂.

The tri-gas mixture of shielding gases was developed primarily for welding the austenitic stainless steels with the short circuit method of metal transfer. It is usually sold as a pre-mixed gas having a con-tent of 90% helium, 7 1/2% argon and 2 1/2% CO_2. The key factor is the use of a high helium content in the gas mixture. The helium tends to spread out the arc column thus providing greater heat input to the weld joint over a larger surface area. This action helps retain the heat in the weld joint a few milli-seconds longer since heat transfers much more slowly through hot metal than it does through cold metal.

The combination of gases imparts a unique characteristic to the weld. Using the tri-gas mixture, it is possible to make a weld with very little build-up of the top bead profile. The result is excellent where a high crowned weld is detrimental rather than a help. The welding of austenitic stainless steel pipe with the short circuit method of metal transfer has been accomplished with relative ease since the gas mixture retards the cooling rate of the weld metal long enough for it to wet out to the edges of the weld deposit.

The small addition of CO_2 has not shown the expected carbide precipitation at the grain boundaries of the weld metal. Apparently the time of reaction is so short, and the heat level in the weld maintained at such a low value, that the material passes through the critical range (1400° F.-800° F.) with little or no carbide precipitation occuring.

Other Gases.

Other shielding gases have been used in combination with argon and helium. For example, a trace of chlorine was added to argon in one experimental mixture for welding aluminum. It was thought that the chlorine would provide a fluxing action that would tend to eliminate porosity from the weld deposit. The basis for this opinion comes from the fact that chlorine is used as a purifying and fluxing agent for aluminum when it is held in the molten state. Massive purges of chlorine are injected directly into the molten aluminum (approximately 1,325° F. and silvery-red in color) while it is in the holding furnace and just before it is cast into ingots. At this temperature the chlorine removes the impurities from the aluminum. They float to the top of the molten metal and are removed as dross.

The use of argon-chlorine mixtures for gas metal arc welding was not commercially successful because of the inherent danger from the breakdown of the chlorine gas under the welding arc. The chlorine will decompose into a gas called phosgene which is a real health hazard.

Nitrogen has been used as a shielding gas, either in pure form or in combination with argon, for welding copper and copper alloys. This work was done in Europe where helium is not readily available. The arc is very stiff and harsh when using pure nitrogen. A mixture of approximately 70% argon and 30% nitrogen has been very successful for gas metal arc welding copper and copper alloys. The addition of argon stabilizes the welding arc and creates less turbulence in the welding puddle.

Summary Of Section.

The selection of the proper shielding gas for any specific application will depend on several variables. The type and thickness of the metal to be welded, the joint design, method of metal transfer, welding position, etc. will all materially affect the final shielding gas selection.

Further data on shielding gases is contained in the various data charts in the appendix of this text.

Welding Wire For Electrodes.

Any discussion of electrodes for the gas metal arc welding process must be prefaced by some comment on the area to be covered in the discussion. It is not the function of this text to compare one brand name with another. We will discuss some of the constituents in the chemical composition of some electrodes to explain how they affect the weld deposit.

Alloys And Deoxidizers.

It is very seldom that chemically pure metals are used by the welding industry. While pure aluminum is used in some areas of manufacture it is not considered a structural material. Pure iron, of which there are only **a few pounds in existence,** is simply a laboratory curiosity. Most other pure elements fall in this same category. The reason for this is that there are always at least trace elements of impurities in most "pure" metals.

It is for the reasons stated that most welding electrodes are always an alloy of two or more elements. For example, low carbon steel is an alloy of iron and carbon with fractional percentages of other elements, some of which are impurities. In most cases, the impurities are sulphur and phosphorus. Some of the elements added to electrodes serve as alloying agents while others function as deoxidizers and scavengers of unwanted elements that could damage the weld. In many cases, the added element performs both the scavenging and alloying function.

Some of the elements serve as deoxidizers in one material, alloys in another and yet are the prime metal in still other alloys. Figure 173 shows how aluminum, a versatile metal, may be used as a deoxidizer in steel, as a primary base metal and as an alloying element in aluminum bronze.

To further illustrate that many elements are used as alloys we list those used in some of the aluminum alloys:

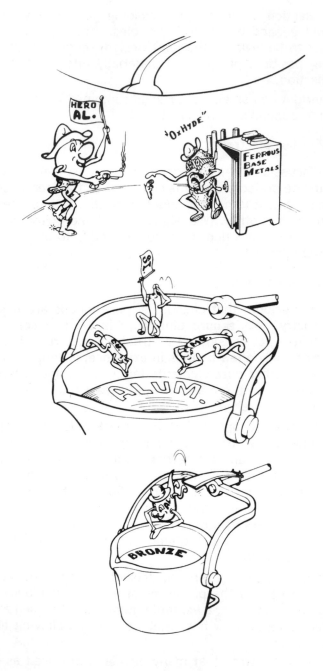

Figure 173. ALUMINUM, A VERSATILE METAL.

Copper, silicon, manganese, iron, zinc, chromium, nickel, lead, titanium, and bismuth.

When considering all the steels presently available, it is difficult to list all the alloying elements, and deoxidizing elements, that might be used in their manufacture. Most of the same elements are also used in the manufacture of electrodes for welding. The type of alloying element, and the percentage of content in the electrode wire, will be determined by the type of steel to be welded and the shielding gas used.

The reason for considering the shielding gas is that some shielding gases cause severe oxidation losses of alloying elements across the arc. A case in point is CO_2 which is used for welding steel. Carbon dioxide (CO_2) breaks down under the heat of the arc into approximately 33% by volume carbon monoxide, 33% by volume atomic oxygen and the balance will remain CO_2 because it is outside the influence of the arc heat and energy.

It is well known that both silicon and manganese are subject to high oxidation losses in the CO_2 shielded welding arc. The percentage of alloy transfer may be improved by the addition of more silicon or more manganese to the electrode chemistry. Even the addition of a larger percentage of **either** of the elements will decrease the percentage losses of **both** elements across the arc.

The type of electrode used will have a definite effect on the transfer efficiencies of the alloying elements and the deoxidizer elements. Transfer efficiencies and shielding gases are shown for several steel electrodes:

    1. Solid electrode wire in argon-oxygen gas shield is the most efficient. (Normally spray type metal transfer).

    2. Solid electrode wire in CO_2 shielding gas is next best.

    3. Flux cored electrode wire in CO_2 is the least efficient.

Deposition Rates.

Deposition rates will depend on several variables in the welding condition. The type and diameter of electrode wire, type of welding power source used, the shielding gas used, the method of metal transfer, and the position of the weld joint all have an effect on the deposition rate.

Data is provided in charts in the appendix of this book for solid electrode wire calculation and deposition rates. It is a simple matter to put the applicable values in the formulas and obtain the correct answer. The charts should be used as guides only since local conditions will determine the reasonably exact deposition rates that may be expected from any specific electrode wire.

Flux Cored Electrodes.

In recent years an entirely new concept of welding electrode has been developed for gas metal arc welding. Called flux cored electrodes, they incorporate a granular flux as the core of the electrode. It is fabricated by several methods although the result is essentially the same.

The use of flux for protecting the molten weld metal while it is cooling and solidifying is well known. It has been employed for years with the submerged arc welding process and the shielded metal arc welding process. The use of flux with the gas metal arc welding process has been considered feasible only within the past few years. It was not until about 1954 that serious thought was given to developing the process for gas metal arc welding.

There are many reasons for the welding industry to promote the use of flux cored electrodes for gas metal arc welding. The use of CO_2 shielding gas with this type of electrode provides the fast, clean weld deposits required for welding mild steel, low alloy steel and medium carbon steels.

Some of the benefits of fabricated flux cored electrodes are:
a) Continuous welding (no electrode changing).
b) Almost total elimination of electrode stub loss.
c) Greater operator arc-on time; more production.
d) Less damage to electrodes during shipping and handling.
e) Deeper penetration welds.

The manufacturing process for flux cored electrodes is relatively simple. Each step in the fabrication of the electrode is highly important so quality control is rigid and a very important factor. The electrode is made from steel strip (called "skelp") which is run through a set of forming dies. In manufacturing electrodes with a simple tubular construction the first step is to form a "U" shape as shown in Figure 174. Flux is added from a hopper. The amount of flux is carefully measured and regulated to maintain the integrity of the electrode. The partially formed electrode is then fed through additional roll stations to complete the forming and closing operations.

Flux cored electrodes are normally fabricated in 3/16" diameters and reduced to the final electrode diameter desired in succeeding operations. Fabricated electrodes are presently commercially available in diameters of 0.045", 1/16", 5/64", 3/32", 7/64" and 1/8".

Figure 174 shows the steps in manufacturing one type of flux cored electrode.

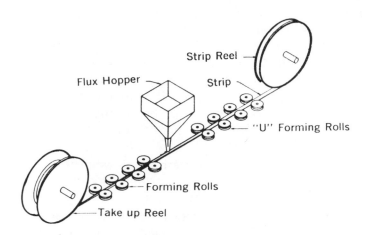

Figure 174. FABRICATION FLUX CORED ELECTRODES.

It is interesting to note that some of the flux cored electrodes are designed for single pass operation only while others are designed for multi-pass operation. The single pass electrodes **require** some dilution of the weld metal deposit and the base metal to achieve their physical characteristics. The selection of the proper fabricated electrode for a specific application will be determined, to some degree, by the requirement for one or more welding passes.

The American Welding Society has issued specifications for both solid electrode wires and flux cored electrode wires for gas metal arc welding. Copies of these specifications may be obtained by writing to the following address:

The American Welding Society
2501 NorthWest 7th St.
Miami, Florida 33125

Summary.

Welding costs include many variables that add up to a total figure per manufactured item. The object of using gas metal arc welding, with either solid electrode wire or flux cored electrode wire, is to cut net welding costs by faster welding speeds, higher deposition rates and more welding operator arc time than is possible with the shielded metal arc welding process. Most fabricated electrodes are designed for use with low and medium carbon steels and low alloy steels. There has been some extraordinary advances in stainless steel flux cored electrodes, as well as hard surfacing electrodes, in recent years.

Some flux cored electrodes are designed for use without benefit of externally supplied shielding gas. They are termed "self-shielding" flux cored electrodes. The weld deposits obtained with this type of electrode have shown a tendency to have lower physical and mechanical properties than if a gas shield had been employed.

Most flux compositions in use with flux cored electrodes today are non-hygroscopic. The term "non-hygroscopic" means the flux will not retain moisture. The shelf life of flux cored electrodes is quite long because of this factor.

The small amount of slag covering the weld bead is easily removed. Its purpose, of course, is to protect the molten weld metal through the cooling and solidification cycle.

Applications of the flux cored electrodes include heavy construction, bridges, shipyard fabrication, foundries and almost any other heavy weldment assembly problem. Deposition rates will depend on the electrode diameter and other variables but it is not uncommon to obtain 18-25 pounds of deposited weld metal per hour.

Chapter 19.

WELDING POWER SOURCES FOR GAS METAL ARC WELDING

Welding power normally used for the gas metal arc welding process is **direct current, reverse polarity.** When the process was first developed in 1948 the only dc welding power sources available were the rotating armature units classed as motor-generator (MG) sets. In 1950 the transformer-rectifier type welding power source was first presented to the welding public. It was also in 1950 that the process patent was issued for the gas metal arc welding process.

Both the electro-mechanical rotating type welding equipment and the static transformer-rectifier type equipment were designed for the shielded metal arc and gas tungsten arc welding processes. Although inherently different in their design concepts both types of welding power sources have the same general output characteristics. The conventional, or "drooper", class power sources normally have 70-80 volts open circuit voltage and a predictable maximum short circuit current. The value of the maximum short circuit current should be approximately one and one-half times the rated amperage output of the power source according to NEMA Standard EW-1. A volt-ampere curve for a typical three range, dc output power source is shown in Figure 175.

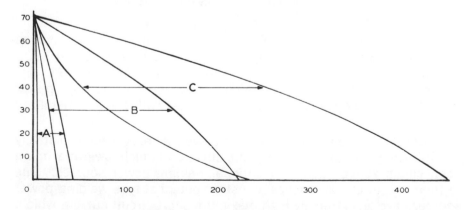

Figure 175. **CONVENTIONAL DC POWER SOURCE VOLT-AMPERE CURVE.**

The development of the transformer-rectifier type welding power source was a great improvement for the gas metal arc welding process because of the faster rate of response of the units to changing arc conditions. The motor-generator (MG) welding power sources have an inherently slow rate of response to changing arc conditions due to the mechanical inertia of the rotating mass of iron and copper conductor wires. This inertial effect is termed the "flywheel effect".

To illustrate the point, a conventional transformer-rectifier power source will have a response time two to three times faster than a conventional motor-generator set. One of the benefits of faster response time is better arc starts with less possibility of cold-lapping occuring at the beginning of the weld.

A comparison of motor-generator and transformer-rectifier power source **response times** is shown in Figure 176. The response time curves are typical for the two types of welding power sources having comparable ampere output ratings at 60% duty cycle.

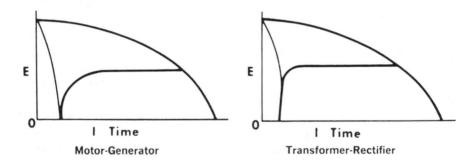

Figure 176. RELATIVE RESPONSE TIME CURVES.

In the time period from 1948 to 1953 the only dc power sources available were the conventional, or constant current, type. In 1953 the first constant potential type welding power sources were displayed to the welding public. Also called constant voltage power sources, they were expressly designed for use with the gas metal arc welding process.

Constant potential type welding power sources have a relatively flat volt-ampere output characteristic curve and much **lower maximum open circuit voltage** than conventional welding power sources. Due to their output characteristics, constant potential type welding power sources have an extremely high maximum short circuit current which, if the power source is allowed to reach, will literally destroy the unit by burning the insulation and short circuiting the coils and core of the main transformer.

The lower open circuit voltage is very suitable for the gas metal arc welding process because of the smaller diameter electrodes that are used. The maximum electrode diameter in present use is 1/8"— but it is used at amperage ranges of 600-700 amperes. The lower open circuit voltages, with the relatively small diameter electrodes, will permit "run-in" arc starts while "scratch" starts are required for initiating the arc on most metals with the shielded metal arc welding process and conventional welding power sources. Low open circuit voltage will also help to reduce the possibility of high current surges at the beginning of a weld.

Conventional Welding Power Sources.

Since conventional, or constant current, welding power sources were the first used with the gas metal arc welding process they will be discussed before proceeding to the constant potential type welding power sources. It is interesting to note that, because of the limited maximum short circuit current and the "drooping curve" output characteristics of conventional power sources, the first wire feeder systems used for gas metal arc welding were arc voltage controlled for speed. A signal lead was normally attached to the workpiece. As the arc length became shorter, for example, the arc voltage became lower in value and the wire feed speed would slow down accordingly. If the arc voltage became higher, due to a longer arc length, the wire feed speed would increase. It was not the most satisfactory arrangement but it was necessary for the conventional type welding power sources. When constant potential type welding power sources were developed the wire feeder manufacturers went to a constant speed drive motor for wire feeders.

The conventional transformer-rectifier type welding power source is normally a three phase transformer with either a selenium or a silicon rectifier to provide dc welding power output. This type of welding power source has loose magnetic, and inductive, coupling in the primary-secondary coil relationship of the main transformer. Loose magnetic coupling means that a physical air space exists between the primary coil and the secondary coil. This may also be termed "loose inductive" coupling of the coils.

Air is, of course, a good resistor to electrical current flow. In addition, **it has poor magnetic permeability.** With the air space between the primary and secondary coils of the main transformer, the magnetic field created in the main transformer iron core does not transfer power efficiently to the secondary coil.

It is the physical air space between the primary and secondary coils of the main transformer that provides the limited maximum short circuit current and the negative, or drooping, volt-ampere output curve characteristic. It is also one reason that relatively high open circuit voltage is required. A lower open circuit voltage (for example, 50 volts) would provide less maximum short circuit current for a given volt-ampere curve. This is illustrated in Figure 177 by comparative curves at different open circuit voltages. The lower open circuit voltages obviously limit the operating range of the power source.

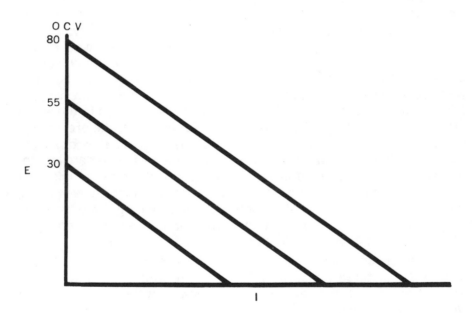

Figure 177. COMPARATIVE VOLT-AMPERE CURVES AT DIFFERENT OPEN CIRCUIT VOLTAGES.

Most conventional welding power sources are designed for the shielded metal arc welding process and gas tungsten arc welding process. The output characteristics of this class of welding power source make it possible to have some small voltage and amperage adjustments made at the welding arc by changing the arc length and, therefore, the arc voltage and amperage relationship. In the illustration, Figure 178, the only variable is the physical arc length. **Remember that arc length determines arc voltage.** The greater the air space distance between the electrode tip and the base metal workpiece surface, the greater the electrical resistance between the two points. Since voltage is a force, it is logical that more voltage would be required to push the arc amperage across the greater arc length. Conversely, as the arc length decreases the voltage requirement decreases and a lower arc voltage results.

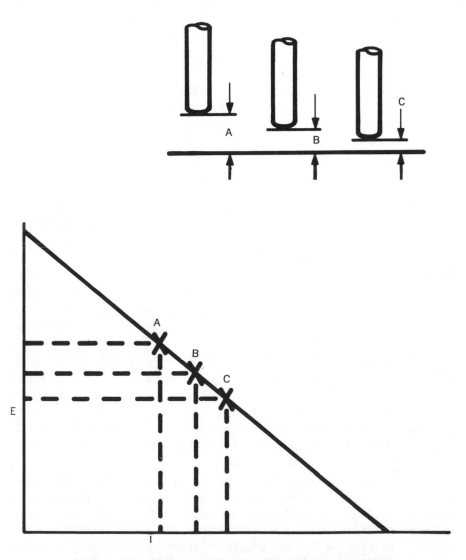

Figure 178. RESULTS OF ARC LENGTH ADJUSTMENT.

The "X" at point "B" indicates the pre-set voltage and amperage normally used for the welding condition. The "X" at point "A" appearing higher on the volt-ampere curve indicates higher arc voltage and lower arc amperage. The "X" appearing lower on the volt-ampere curve at point "C" shows lower arc voltage and higher welding current. The welding operator may control both welding amperage and arc voltage, to some degree, by moving the shielded metal arc welding electrode tip away from, or closer to, the workpiece to be welded.

There is a certain amount of amperage and voltage control option to the welding operator when using conventional power sources and certain welding processes. Much of the success of the welding application depends on the physical skill and dexterity of the person operating the welding equipment.

When considering the conventional type of welding power source for gas metal arc welding the metal aluminum is invariably mentioned. Aluminum was the first metal to be welded with the gas metal arc welding process.

Aluminum has a relatively low melting point (1218° F.) and consequently a relatively high melt rate as an electrode. The metal cannot tolerate high current surges at the electrode tip bcause of the explosive force that would be generated. The result would be excessive spatter and poor weld quality. This combination of factors makes it almost mandatory that the power sources used with the metal aluminum, and the gas metal arc welding process, have a limited maximum short circuit current.

It was a stroke of good fortune that the power sources available in 1948 were just what was needed to weld the metal. They were, and are, excellent power sources for spray transfer welding and globular transfer welding with the process.

It is interesting to note that the recommendations of the manufacturers of the first wire feeder-control equipment stated that 3/16" was the minimum thickness of aluminum that could be welded with the gas metal arc welding process. Many of the new models of wire feeder-controls and the newer models of constant potential type welding power sources are capable of welding much thinner gauges of aluminum with today's techniques.

Conventional welding power sources may be used with the fabricated flux cored electrodes presently on the market. This is particularly true of the hard surfacing electrodes. The wire feeder-control system should be a voltage sensitive unit with a variable speed drive motor. The maximum welding current applicable seems to be about 400-450 amperes. In the opinion of most Welding Engineers, fabricated flux cored electrodes should be used with constant potential type welding power sources. Deposition rates, weld quality and weld appearance are substantially better with the constant potential type unit.

In conclusion of this section of the text we can say that **for gas metal arc welding processes only,** the conventional type welding power source is not considered the best unit available by most welding authorities. The preference of the welding industry is for constant potential (constant voltage) type welding power sources where the gas metal arc welding process is the only process used.

Constant Potential (Constant Voltage) Power Sources.

The terms "constant potential" and "constant voltage" can be confusing if not properly defined. "Potential" and "voltage" are synonymous in electrical terminology. There really is no such thing as a constant potential, or constant voltage, welding power source. If a constant potential power source were truly constant potential the arc voltage would be the same at any amperage output as it is at open circuit voltage. This cannot be because of the internal resistance of the power source circuitry. Since this is not the output characteristic of constant potential power sources we will refer to them as **constant potential type** welding power sources.

The first constant potential type welding power sources had a relatively flat volt-ampere output characteristic. There was some voltage drop per hundred amperes of output but usually not more than a volt and a half drop. The maximum open circuit voltage was approximately 48-50 volts with an adjustable open circuit voltage capability. These power sources had relatively unlimited maximum short circuit current which could reach several thousand amperes. This is why all manufacturers of welding power sources have recommended that **large stick electrodes not be used** with constant potential type welding power sources. If the stick electrode makes short circuit contact with the base metal, and it will eventually because of the globular type transfer, the response of the power source is so fast that it will reach maximum short circuit current before the electrode can be broken free. In this situation there would be several thousand amperes flowing through transformer coils designed to carry a few hundred amperes. The resistance heating in the coils would be almost instantaneous, the insulation would be destroyed and the coils would short circuit to the main transformer iron core. This would be evidenced at the time by a quantity of smoke and a brilliant blue electrical flash from the interior of the power source. You have just blown the main transformer of the power source!

At first there was no concern for "slope" control in the constant potential type welding power source. The characteristic relatively flat output curve was thought to be the answer to all problems of gas metal arc welding. A typical constant potential power source output volt-ampere curve is illustrated in Figure 179.

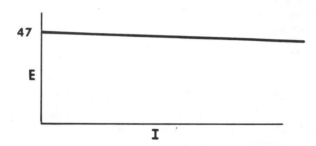

Figure 179. CONSTANT POTENTIAL VOLT-AMPERE CURVE.

Slope was, of course, a later development in the evolution of constant potential type welding power sources for gas metal arc welding.

The constant potential type power source has a much different main transformer design than the conventional type welding power source. The primary-secondary coil relationship has very tight magnetic, and inductive, coupling. This means that the primary and secondary coils of the main transformer are frequently wound on one coil together. (Interleaving and interwinding is also done. The cost of such coil configurations is considerably more and it is usually only used on coils carrying 750 amperes or more).

The main transformer of a typical constant potential type welding power source is illustrated in Figure 180. Note the close relationship of the primary and secondary coils. This coil configuration develops close magnetic and inductive coupling.

In the case of most constant potential transformer coils **either** the primary or the secondary coil may be wound to the outside of the total coil. The exact placement of primary and secondary coils would depend on the type and rating of the power source being built. Constant potential type welding power sources rated at 500 amperes and over may have a range switch for selection of a specific open circuit voltage range. The **primary** coil would normally be wound at the outer periphery of the total coil on this type of power source.

Most constant potential type power sources have one wide voltage range if they are rated under 300 amperes output at 100% duty cycle. No open circuit voltage range switch is required. In this case it is normally the **secondary coil** which is wound to the outside of the coil.

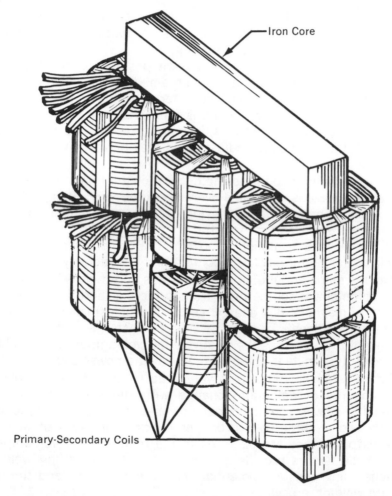

Figure 180. A TYPICAL CONSTANT POTENTIAL TRANSFORMER.

There is good reason for the two different coil winding methods. Many constant potential type welding power sources have mechanical controls for setting open circuit voltage. The control is achieved by moving a set of carbon-type brushes over the coil face which, incidentally, has had the insulation removed from the face surface for this purpose.

There are some constant potential type welding power sources which have totally electric control. In this class of power source it is the **load voltage** that is set and controlled. Open circuit voltage is a fixed value and does not change.

On any transformer type welding power source open circuit volage is controlled by the number of **effective electrical turns** in the secondary coil. (Effective electrical turns are those turns of the coil through which current is actually flowing).

If the primary coil is in the outside position on the total coil assembly, fine voltage adjustment is obtained by moving the carbon type brushes over the coil surface where the insulation has been removed. The voltage range switch is tapped off the secondary coil. This provides the various ranges, usually three to five in number, within which the fine voltage adjustments may be made with the sliding brushes.

If the secondary coil is to the outside of the total coil assembly open circuit voltage is controlled over one wide range. It is controlled by the carbon contact brushes moving over the bare coil surface. As noted, the coil surfaces over which the contact carbon brushes move have all the surface insulation removed. The coil surface is normally silver plated for better electrical contact and conductivity.

The magnetic and inductive coupling is very tight with constant potential type power source main transformers. There is only a very small loss of magnetic flux energy; certainly much smaller than would be in the conventional "drooper" welding transformer. It is this design concept that provides the relatively flat volt-ampere output characteristic of the constant potential type welding power source.

Figure 179 shows there is very little voltage drop for considerable amounts of amperage change with the constant potential type welding power source. Most power sources of this type will have only two or three volts drop per hundred amperes of welding current output if they are operated in the relatively flat volt-ampere mode. These factors provide greater versatility for setting welding conditions using constant potential type power sources and the gas metal arc welding process.

Constant potential type welding power sources have no facility for manual amperage control, either standard or remote. On the contrary, they supply welding current as required to maintain a preset arc voltage condition. The actual setting of arc voltage depends on the open circuit voltage, the slope setting, and the amount of wire feed speed set on the feeder-control system.

For example, the welding operation shown in Figure 181 has no value shown for voltage or amperage. The shape of the volt-ampere output curve would be determined by the open circuit voltage and the slope setting, if used, on the power source.

At the point where the wire feed speed and the melt-rate of the electrode are exactly equal (the equilibrium point) there is a solid "X". The arc voltage is indicated by the line going to the vertical ordinate of the volt-ampere curve. The welding current is shown on the horizontal axis of the curve.

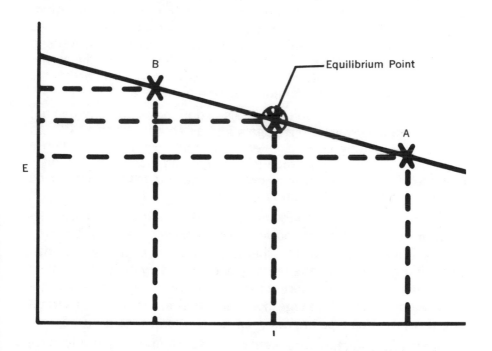

Figure 181. AUTOMATIC CURRENT CONTROL WITH CONSTANT POTENTIAL TYPE POWER SOURCES.

The key point to remember is that arc voltage is determined by arc length. Other factors affect arc voltage to some degree but it is arc length that is the final determining factor. Any variation of arc length will cause the arc voltage to change.

At point "A" in the illustration there is indicated a lower arc voltage and a higher welding amperage than at the equilibrium point. This could only occur if the arc length has been shortened while welding. Momentary shortening of the arc length could be caused by a sudden movement by the welding operator, running over a tack weld, etc. The other variables in the welding condition would normally remain the same.

The result of the lower arc voltage is higher welding amperage at the arc. At this instant of time the welding power source is furnishing considerably more welding current than it did at the point of equillibrium although the **wire feed speed remains the same.** The electrode melt-rate is increased as a result. The higher melt-rate increases the physical arc length and raises the arc voltage. In this manner the welding condition returns to the equilibrium point pre-set on the welding power source and the wire feeder-control system. **Important:** As long as none of the other welding condition variables are changed the welding power source will produce output amperage and welding voltage along the plotted volt-ampere curve. This factor causes the volt-ampere relationship to slide up, or down, the volt-ampere curve according to the physical conditions of the welding arc.

Of course, if the arc voltage is raised due to the arc lengthening, the state of equilibrium is again lost as shown at point "B". In this case the welding current is less than that required to sustain a proper melt-rate in the electrode for its type and diameter. The electrode therefore drives down, shortening the arc length, until once again the point of equilibrium is reached.

From an electrical standpoint, it should be noted that all conventional, or constant current, power sources have inherently high impedance transformer design. This impedance is basically provided by the air space between the primary and secondary coils of the main transformer. The volt-ampere curve has a steep negative angle between the open circuit voltage point and the maximum short circuit current point. This type of power source is often called a "drooper".

Constant potential type welding power sources naturally have a relatively flat volt-ampere output curve. They also have much lower open circuit voltage than conventional welding power sources. These characteristics are achieved by using a **low impedance** transformer design. This means there is **no air space between the primary and secondary coils** of the main transformer. This type of welding power source was designed especially for the gas metal arc welding process.

Of the two types of welding power sources the constant potential type units will reach welding voltage and amperage without the delay in time involved with conventional type units. Constant potential type welding power sources are preferred for the gas metal arc welding process in almost every type of application. Constant potential type power sources with some means of slope control are mandatory for the short circuit method of metal transfer. All methods of gas metal arc welding **metal transfer** may be achieved with some type of constant potential power source and proper gas metal arc welding equipment. Conventional power sources are applicable only for spray or globular transfer.

Voltages.

There are three types of voltage with which we are concerned in welding. They are Open Circuit Voltage (OCV), Load Voltage and Arc Voltage.

Open circuit voltage is determined when the welding power source is energized but under no welding load. The voltage measurement is made with a voltmeter at the output terminals of the power source.

Load voltage is also measured at the output terminals of the power source but while the unit is actually under welding load. Load voltage includes the total voltage load affecting the power source and includes the voltage drop through the welding cables, the ground connection, etc.

Arc voltage is measured only at the welding arc. It can actually be set only at the welding arc during welding operations. Special voltmeters are used to determine arc voltage. For most welding applications it is load voltage that is read on the meters of the power source and classed as "arc voltage" for procedure writing.

It is open circuit voltage that is normally set on constant potential type power sources having mechanical control. The load, and arc, voltages are determined by other variables in the welding condition. Although the methods of adjusting open circuit voltage may vary for different models of constant potential type power sources the results will be the same as long as the only variable physically changed is **open circuit voltage.**

There are some constant potential type welding power sources now available which have electric control of voltage. These units have a **fixed open circuit voltage.** Usually it is the load voltage that is set on this type of equipment. There are a number of reasons for having electric control of a power source but the main reason is to achieve remote control of the output of the unit by the welding operator at his work station.

To illustrate this point, a volt-ampere curve for a typical constant potential type welding power source is shown in Figure 182. The curve shows a relatively flat volt-ampere output characteristic. For this example we will assume a minimum slope setting in the welding condition.

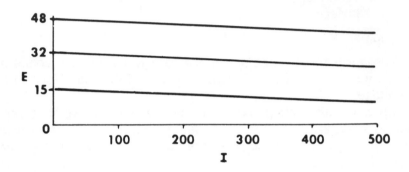

Figure 182. VARIABLE OPEN CIRCUIT VOLTAGE CONDITIONS.

The open circuit voltage range shown in the illustration is 15-48 volts. When changing open circuit voltage from one place to another on the vertical axis of the graph there will be parallel line development of the curves and projections.

If the voltage drop through the secondary circuit is 4 volts per hundred amperes of output, and the open circuit voltage is set at 48 volts, the **load voltage** at 300 amperes would be 36 volts. To clarify the proposition:

> Open circuit voltage = 48 volts.
> Load voltage @ 100 I = 44 volts.
> Load voltage @ 200 I = 40 volts.
> Load voltage @ 300 I = 36 volts.

The working rule that can be applied here is, "The voltage drop per hundred amperes of welding current output remains the same regardless of the open circuit voltage setting of the constant potential welding power source". This is illustrated in Figure 182. The rule holds true when the only variable physically changed is open circuit voltage.

Slope.

The term "slope" may be confusing to the person who is trying to understand the workings of some constant potential type welding power sources. Many people in the welding industry associate the term slope with the gas tungsten arc welding process. The terms "upslope" and "downslope" mean an increase or a decrease in the welding current values when welding with gas tungsten arc. In particular the term "sloping off" indicates a decrease in the welding current at the end of a gas tungsten arc weld for the purpose of crater filling to reduce the possibility of crater cracking.

When discussing slope for the gas metal arc welding process it must be understood that **it is the shape of the static volt-ampere curve to which we refer.** As noted in the preceding section about voltage the relatively flat volt-ampere curve has very little slope.

There are actually two methods of putting slope into a constant potential type welding power source. It may be done by either putting resistors in the power source circuit or putting a reactor in the secondary ac portion of the circuit. We will examine both methods and see what the results of each do to the power source welding current output.

Resistance slope control will modify the shape of the static volt-ampere curve but it does nothing else. Usually the resistors are some type of highly resistive metal such as ni-chrome (a nickel-chromium alloy) or something similar. This type of slope control is most often used on motor-generator (MG) sets. It is normally a three tap slope system with minimum, medium and maximum settings indicated on the three position range switch. Resistance slope control is fine for an MG set because the unit has inherently slow response time to changing arc conditions.

Reactor slope control is another thing entirely and it is the system used by manufacturers of most static welding power sources having slope control. In this case, slope is actually caused by impedance and is usually created by the addition of a substantial amount of inductive reactance to the power source circuitry. The electrical device used for slope control, in most constant potential type power sources, is **a fourteen turn slope reactor.** (Remember: an electrical turn is one wrap of the conductor wire around the coil periphery. A fourteen turn coil would have fourteen physical wraps of the conductor wire around the iron reactor core). More or less slope in the welding power source circuitry could be controlled by varying the amount of reactance in the circuit. Keep in mind that **a reactor inherently opposes change in the welding circuit.**

We have used some terms in the forgoing discussion which are explained in the Common Terms chapter. Some of the terms have been used in previous sections of this text but their use in this section may be somewhat different. The term **impedance** is a good example.

Impedance means to slow down but not stop a quantity of something. In this case it is something electrical. There are just two possible values in an electrical circuit which may be impeded and they are voltage and amperage. Since voltage is a force which causes

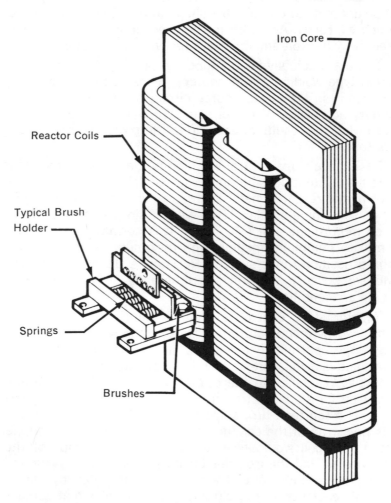

Figure 183. SLOPE CONTROL REACTOR, THREE PHASE.

amperage to flow in a circuit, **but which does not flow itself**, the amperage flowing in the circuit must be the impeded quantity.

To explain how impedance occurs we must investigate the **reactor as it is used for slope control.** A reactor consists of one or more current carrying coils placed around an iron reactor core. The core material is thin gauge electrical steel, insulated on both sides, and is the same quality and thickness as that used for transformer cores. A variable reactor iron core and coil assembly is shown in Figure 183. The slope control illustrated is typical of some mechanically controlled constant potential type welding power sources. It is usually called "variable slope control" because of the sliding brushes on each coil.

The coils illustrated have fourteen electrical turns and are especially designed for slope reactors. One side of the coil insulation has been removed by machining after which the bare copper conductor is polished and silver plated. Carbon contact brushes move over the face of the coil to provide more or less reactance in the welding power source circuit. It is the number of **effective turns** in the coil that determines the impedance factor of the reactor.

You may have noticed that there were six coils in the main transformer shown in Figure 180. There are also six coils illustrated in Figure 183 which is the slope reactor. Since both units operate in three phase alternating current circuits, the logical question is, "Why six coils in these components?" The answer is quite simple. The coils that are one above the other on a single leg of the transformer and reactor cores are in electrical **series.** This means they function as one coil in each phase. The reason for series stacking of the coils is design convenience. One single coil would make the power source twice as wide as it is thus requiring more floor space. Actually, it operates better with the coils in the series configuration.

Some of the following data has been applied in other sections of this text. Although the fundamental principles remain the same **the results** of their application to constant potential type welding power sources produces substantially different amperage output characteristics.

The Reactor For Slope Control.

Since the reactor is normally found in the secondary portion of the ac circuit of a welding power source it has welding amperage flowing through its coils. A magnetic field is developed in the reactor iron core around which the coil is located. Remember that magnetic field strength depends on three things:

1. The mass and type of iron in the core.
2. The number of effective electrical turns in the coil.
3. The alternating current value flowing in the coil.

Considering the foregoing data it is logical to assume the magnetic field strength could be varied in one of several ways:

1. Changing the amount of iron in the core.
2. Changing the number of turns effective in the coil.
3. Changing the amount of current flowing in the coil.

If a step-by-step evaluation of the reactor components were undertaken we could determine their function and the reaction of the other parts of the welding circuit.

In Figure 184 there is illustrated a profile view of a typical slope reactor coil and a volt-ampere curve. We will use these three devices to illustrate **how** the reactor works, **what** it does to the output amperage and **why** it does these things in the circuit.

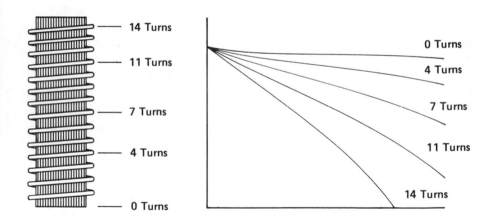

Figure 184. SLOPE CONTROL CONCEPTS.

Consider that the coil profile is just one of the six coils shown in Figure 183. The arrow is pointing to the section of the coil which has had the insulation removed. We will arbitrarily begin with **4 turns of slope** in the reactor.

The welding current flows in the reactor coils **(only four turns are effective although this is a 14 turn coil).** The coil is physically located around one leg of the reactor iron core where a magnetic field is created. This occurs as the current level builds up in the ac sine wave trace. **Energy** is used to create the magnetic field but it is not used up. It is, for the moment, stored in the magnetic field. When the current value reaches maximum amplitude, or strength, the magnetic field strength is at maximum too. Now the current value decreases, following the sine wave trace as it goes to zero each half-cycle. As the magnetic field collapses the energy stored in the magnetic field must go someplace. It goes to the only place it can go which is the reactor circuit coil where it originated in the first place. **This is termed self-inductance of the coil.**

Recalling the electrical function of a transformer we know when power was induced into the secondary coil it produced a voltage that directed force **counter in direction** to the impressed primary ac voltage. A similar counter voltage (counter emf) is created in the slope control reactor. This counter voltage opposes the flow of welding current in the reactor coils which has the effect on the volt-ampere curve shown in Figure 184 at **4 turns of slope.**

If we now set the slope at **7 turns,** half of the total reactor coil is effective in the welding circuit. As current builds up in strength there will be created a magnetic field. This time, however, the magnetic field is stronger because there are **more effective turns** in the reactor coil. The energy used to build the stronger magnetic field is stored momentarily in the magnetic field and, when the current goes to zero as it does each half-cycle, **the energy comes back on the reactor coil as a counter voltage.** This time it is stronger counter voltage because it took more energy to build the stronger magnetic field. The volt-ampere curve shows the result at **7 turns of slope**; a steeper slope to the curve and lower maximum short circuit current.

Progressing to a higher slope value of **11 turns** we have approximately three-fourths of the reactor coil now effective in the welding circuit. The magnetic field will be much stronger than before. Considerably more energy is used to create the magnetic field and that energy is again stored in the magnetic field as it increases to maximum strength. When the current level peaks and then moves toward zero, the magnetic field collapses and the energy is returned to the reactor coil as counter voltage; an impedance holding back the flow of current. The result is shown in the volt-ampere curve **(11 turns)** as a very steep slope, or negative curve, with limited maximum short circuit current.

If we put the entire **14 turns of slope** into the reactor there will be maximum magnetic field strength in the reactor iron core. Maximum energy will be used to create the magnetic field and, when the current goes to zero (as it does each half-cycle of ac input) there will be **maximum counter voltage, or impedance, to the flow of welding current.** The volt-ampere curve will be almost like a conventional power source with very limited maximum short circuit current, and a steep negative curve, as shown in Figure 184.

It is apparent there is a chain reaction to the function of a reactor for slope control. Current flows in the reactor coil wrapped around the reactor iron core. The energy used to create the magnetic field is momentarily stored in the field. When the magnetic field collapses each half-cycle the stored energy is returned to the reactor circuit as a counter voltage which is impressed on the coil where it originated. The counter voltage is called the **impedance factor.**

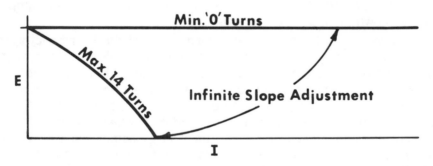

Figure 185. RESULTS OF VARIABLE SLOPE CONTROL.

The results of introducing various amounts of slope are shown in Figure 185. For the "0" slope setting the volt-ampere output characteristic is relatively flat. By adding slope turns on the reactor, up to 14 turns, the slope curve becomes progressively more negative (steeper) in angle and the maximum short circuit current becomes more limited in value.

There are two major results when slope turns of a reactor are added to the welding circuit. They are:

1. The maximum short circuit current is limited. (This is apparent in Figure 184 when you look at the volt-ampere curve).

2. The addition of slope turns **slows the rate of response** of the welding power source to changing arc conditions.

You may recall the point made in a previous statement that **the reactor inherently opposes change in the welding power circuit.** This means that, as more of the reactor coils are added to the circuit through additional slope turns, the response of the power source slows down proportionately. In Figure 186 we have illustrated two response conditions. While hypothetical, they do tell the story of response time rather graphically.

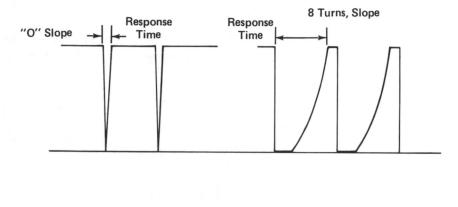

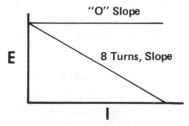

Figure 186. RELATIVE RESPONSE TIMES WITH SLOPE.

For example, if there is "O" slope in the welding power source we would not expect much of a short circuit transfer welding condition because the electrode, as it approached the base metal workpiece, would cause the arc voltage to drop. With the relatively flat volt-ampere characteristic, this would literally slide the welding volt-ampere position out to a prohibitively high amperage. The result would be explosively high currents at the welding electrode tip which would cause excessive spatter and poor penetration in the weld. The response time would be very fast.

By adding the correct amount of slope for the electrode wire type and diameter there is a slowing down of the response time of the power source. This is illustrated as 8 turns (typical for mild steel) of slope in the reactor controlled power source. But look at the results!

The response time is slowed to where an actual short circuit can be made between the electrode wire and the base metal weld. Transfer of the molten weld metal is accomplished during this short circuit. But the rate of rise of welding current is much slower than it is at "0" turns of slope. The time at full arc amperage and voltage is also much less than it was before. The result is **limited heat input** to the weld area and the capability of welding in all positions with the short circuit method of metal transfer.

Response time is the actual time it takes the welding power source to get from dead short circuit amperage and voltage to welding voltage and amperage.

Although the slope settings may change the shape of the volt-ampere curve drastically as more reactance is added to the welding circuit the **open circuit voltage remains the same.** It becomes apparent that the welding conditions can be changed substantially by adjusting the slope settings on a constant potential type welding power source.

When setting a welding condition on a constant potential type power source and wire feeder-control system, change only **one variable at a time.** Adjusting both variable slope and variable voltage at the same time will make setting the welding condition extremely difficult (and unlikely).

Some gas metal arc welding techniques require very little slope in the welding power source. Spray transfer is a good example. Other techniques, such as short circuit metal transfer, need the added reactance, or impedance, in the welding circuit to function properly. A very rapid power source response to changing arc conditions would cause the small diameter electrode wires used (0.030"; 0.035"; 0.045" diameters) to literally explode at the electrode tip. This would be shown as spatter along the weld edges.

For the short circuit transfer method of gas metal arc welding the slower rate of response to the short circuit connection between the electrode and the base metal workpiece is actually accomplished by the addition of reactance in the welding power circuit. The more reactance in the circuit the slower the rate of response. This permits the short circuit of the electrode to the base metal without the explosive burst of current at the electrode tip.

There is a rule of thumb which can be applied here for slope reactor circuits. It is:

a) THE HIGHER THE SLOPE NUMBER, THE LOWER THE MAXI-MUM SHORT CIRCUIT CURRENT WILL BE.

b) THE HIGHER THE SLOPE NUMBER, THE SLOWER WILL BE THE RATE OF RESPONSE OF THE POWER SOURCE TO CHANGING ARC CONDITIONS.

In truth, if the average welding man would remember this there would be no problem in understanding slope control for gas metal arc welding!

The Stabilizer Or Inductor.

There is a control device available on some models of constant potential type welding power sources called the **inductor** or **stabilizer**. It normally regulates the amount of stabilization in the dc portion of the power source welding circuitry. The control may be either built in the power source or furnished as an external control system. It is usually an optional item when purchasing a constant potential type welding power source.

The inductor, or stabilizer, control serves only one function which is to permit a slight changing of the rate of response of the power source without effectively changing the shape of the output volt-ampere curve. In the opinion of many welding authorities the inductor or stabilizer control has limited use with the gas metal arc welding process. The small amount of inductance achieved with the inductor control is relatively unimportant when you consider the total inductance already in the power source circuit.

Points To Remember:

1. A **slope reactor** is located in the **secondary ac** portion of the power source circuitry. The slope reactor will change **both** the power source **rate of response and limit the maximum short circuit current.**

2. An **inductor,** or **stabilizer,** is always in the **dc portion** of the power source circuitry and will **only limit rate of response** of the power source.

3. Resistance slope control will **only limit the maximum short circuit current of the welding power source.**

Gas Metal Arc Spot Welding.

Gas metal arc spot welding has been developed as an industrial tool for fabricating all weldable metals. The NEMA Class 1 power sources used with this process normally have a 600 ampere output rating although some units have been developed for sheet metal work that have output ratings as low as 150 amperes, 60% duty cycle. Such units are for light metal work.

In many cases the top plate thickness of the pieces to be welded requires a power source with higher open circuit voltage than is usual for the gas metal arc welding process. The high open circuit voltage is necessary to provide the power to penetrate the top plate in minimum time. Some constant potential type welding power sources are designed for this type of application.

A typical gas metal arc spot weld is illustrated in Figure 187. The weld interface is almost vertical through the top plate changing into a modified parabola in the lower plate. Depth of penetration is determined as a time function after the welding condition is initially set up.

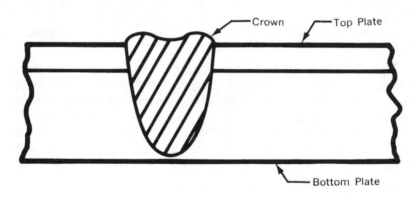

Crown   Top Plate

Bottom Plate

Figure 187. TYPICAL GAS METAL ARC SPOT WELD.

Applications of gas metal arc spot welding include tacking structural members in a weldment, manufacture of articles that must be welded from one side only, joining thick to thin plates, etc. The welding technique is fast and clean with no slag residue.

Most applications of gas metal arc spot welding are with mild steel and low alloy steel materials although the technique is certainly usable with other metals.

Constant potential type power sources are usually used in the flat volt-ampere characteristic when gas metal arc spot welding. A special spot control is normally employed with a standard wire feeder-control and gun system. The gun uses specially designed nozzles which permit flat, lap or fillet joints. The control system includes a timing device which counts welding time in cycles.

Constant Potential Power Sources And Air Carbon-Arc Gouging.

One of the newer applications of the constant potential type welding power sources is for air carbon-arc gouging and cutting of metals. Here is the way it works and why this type of power source can be used without problems **if used correctly.**

You may recall that we said not to use large diameter shielded metal arc electrodes with the constant potential type welding power source because they would **short circuit to the base metal** and the power source would destroy itself. Now we say that you can use various sizes of carbon electrodes with no harm to the power source. Why one and not the other type of electrode?

The key to the whole operation is that **carbon sublimes** when it reaches its melting point. This means that it goes from the solid state as an electrode to the gaseous state without the intermediate liquid state. The volt-ampere curve illustrated in Figure 188 shows what happens.

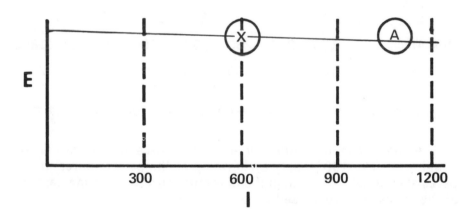

Figure 188. VOLT-AMPERE CURVE FOR CARBON ARC GOUGING.

All carbon electrode diameters have a minimum-maximum amperage range within which they will operate normally. For example, the amperage range for 3/8″ carbon electrodes is 350-600 amperes. The "X" with a circle around it shows the normal operating amperage and voltage for a given carbon electrode. If the welding operator has too much angle on the electrode it would usually tend to short circuit, build up current and then vaporize and create a crater before it stubs again. This is typical of the air carbon-arc gouging technique using a constant current welding power source. With the constant potential type welding power source, however, the electrode doesn't stub. As the electrode gets closer to the workpiece surface, the arc voltage decreases. When the arc voltage decreases, the amperage increases substantially as shown in the illustration, point "A". The tremendous increase in amperage is almost instantaneously available at the electrode tip which **vaporizes** and cannot stub.

Many foundries have switched to high amperage constant potential type welding power sources and the air carbon-arc process for removing risers, gates, pad washing, etc. Very often, the same power source has also been set up with a flux core wire feeder-control system so the welding operator can make casting repairs as soon as the air carbon-arc gouging is completed. It is a very efficient operation.

The following chart shows the recommended amperage ranges for various diameters of electrodes for the air carbon-arc process.

Electrode Diameter, inches	5/32	3/16	1/4	5/16	3/8	1/2	5/8	3/4	1
Min. Amps.	90	150	200	250	350	600	800	1200	1800
Max. Amps.	150	200	400	450	600	1000	1200	1600	2200

Figure 189. RECOMMENDED AMPERAGE RANGES, CARBON ELECTRODE.

A safe recommendation for a power source for air carbon-arc gouging is: **The 100% duty cycle rating of the power source in amperes should not be less than the maximum current carrying capacity of the carbon electrode used for the application.**

Chapter 20

WELDING PROCESSES COMPARISON AND USE

This chapter discusses the comparison of various facets of the gas metal arc welding process to other welding processes. Frequently when the gas metal arc welding process is being considered for an application the intent is to replace another welding process already in use. Normally the process in use is shielded metal arc, gas tungsten arc or submerged arc. To properly evaluate the processes for selection and use it is necessary that comparisons be made to determine which welding method will do the job **at the lowest net cost to the user.**

Such evaluations will involve the considerations of labor costs, equipment costs, electrode and shielding gas costs, time in material preparation costs, actual arc time costs, weld cleaning time costs, location of the welding job, etc. The result of the evaluation must show the gas metal arc welding process will perform the operation at lower net cost to the user or there is no reason to make the change.

Accumulating the facts for a survey of a users shop welding requirements takes time and effort. The results of knowing the proper equipment and processes are being applied to the job makes any effort worthwhile.

For this section of the text we will consider the comparative data for shielded metal arc and gas metal arc welding processes. Gas tungsten arc and submerged arc are not evaluated although some of the data is applicable.

Joint Preparation.

For many years there have been typical weld joint designs used with the arc welding processes. Many of the designs were developed especially for the shielded metal arc welding process. Due to the inherent bulk of the shielded metal arc welding electrodes (because of the flux coatings) the opening in the base metal had to be large enough to accomodate them. The root face of the joint (the portion of the root with a vertical face) had to be no more than about 3/32" because the penetration of even the deepest-digging mild steel electrode (E-6010) is only about 1/8″ maximum into the base metal. The penetration depth would be somewhat less for alloy electrodes.

It was deemed good practice to open up the joint and have as much interface as possible between the deposited metal and the base metal. While this concept has been modified to some degree in the past few years it is not uncommon to find vee butt joint designs with an included groove angle of 75 degrees or more. The drawing in Figure 190 shows a typical weld joint for the shielded metal arc welding process having 75 degrees included groove angle. In section "B" of the same drawing there is a typical joint design for the gas metal arc welding process. It shows a 45 degree included groove angle.

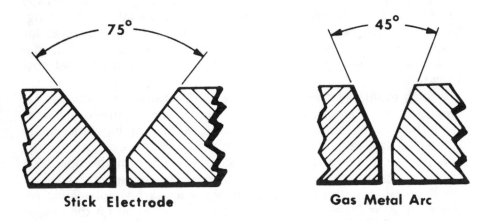

Stick Electrode   Gas Metal Arc

Figure 190. VEE BUTT GROOVE ANGLE COMPARISON.

The vee butt joint design shown in "A" has a relatively thin root face (or "land") area. As stated before, the root face for shielded metal arc welding is normally no more than 3/32″ where full 100% penetration of the weld joint is desired. With a greater root face area the penetration of the electrode into the base metal would probably be less than 100%.

In any type of shielded metal arc weldment the joint design should be considered for the most economical use of filler metal. The actual joint design will depend on several factors including material type and thickness, position of welding, welding process and technique used, etc. For example, 3/8″ steel plate would normally require some type of joint preparation regardless of the position of welding. The type of joint design will depend on the factors previously mentioned as well as the methods available for performing the preparation. The most frequently used edge preparation process for steel plate is the oxy-acetylene (oxygen-fuel gas) cutting torch which may be manually held and operated or be fully automatic equipment. Edge beveling and square cutting of shapes in steel are accomplished with relative ease with this type of equipment.

For shielded metal arc joints requiring 100% penetration a square butt joint welded from both sides may be used for steel thicknesses up to 1/4". For best results the root opening should be approximately 1/16"-3/32". (The root opening is the physical distance the root faces of the two pieces are apart). Thicknesses of steel in excess of 1/4" require some form of joint preparation for welding with the shielded metal arc process.

Although the same factors are applicable to gas metal arc welding the joint designs may be considerably different. For example, it is not uncommon for penetration patterns in mild steel 1/4" plate to achieve 100% penetration using the buried arc technique and CO_2 shielding gas.

Figure 190 illustrates the relative filler metal requirements for shielded metal arc and gas metal arc welding joint designs. It is obvious that more filler metal is needed to complete the shielded metal arc weld joint. This means greater heat energy input per linear inch of weld with the attendent possibility of locked in stresses and part distortion. Since there is less filler metal placed in the gas metal arc weld joint there is less possibility of weldment distortion.

A narrow weld joint groove means that less base metal must be removed as scrap. The cash value of scrap metal is normally about 25% the original cost of the base metal. By not removing the base metal (or removing less of it) for gas metal arc welding there is a substantial savings in both base metal costs and filler metal electrode costs.

Welding Electrodes.

When discussing electrode cost per pound it is not uncommon to find that shielded metal arc welding electrodes are about half the price per pound of gas metal arc welding electrode wire for the same type of welding applications. It is not the **purchased price per pound** of welding electrodes that should be of concern to the consumer. It is the **deposited pound cost** of any filler metal that is the important factor. An examination of comparative deposition efficiencies of the **two electrode types will help clarify the point.**

The losses by weight of electrode materials are substantial when using shielded metal arc welding electrodes. If the facts are based on 5/32" diameter electrodes the losses will be approximately as follows:

Stub end loss (based on 2" rod stub ends) 17%
Spatter and flux coating losses 27%

The total is 44% loss by weight of **the purchased pounds of electrodes.** As noted the figures are based on two inch rod stub ends which are seldom achieved. Most fabricators and contractors, when estimating electrode efficiencies, calculate no more than 35-40% of the purchased pounds of electrodes for shielded metal arc welding will be deposited as filler metal. Some estimates range as low as 30% deposition efficiencies.

There are electrode losses with the gas metal arc welding process but they are minimal. For example, we will consider a welding operator using 0.035" diameter solid steel electrode wire. The application is mild steel base metal using CO_2 shielding gas and the short circuit method of metal transfer.

Let us assume that the welding operator cuts the equivalent of 1/2" from the end of the electrode wire each time he completes a weld. The assumed quantity of 1/2" is probably twice as much as he would normally cut off in actual welding practice.

The number of inches of electrode wire per pound is shown in Data Chart 2 of the book appendix. Finding mild steel at the left of the chart, we move to the right of the column headed "0.035" diameter. There we find there are 3,650 inches per pound of 0.035" diameter electrode. Based on the assumed loss of 1/2" of electrode wire each time a weld is completed, it is apparent that the **welding operator can make 7,300 individual welds** before there is a pound of waste electrode wire.

In view of the forgoing data the cost per pound of deposited weld metal, including the cost of shielding gas, is usually less with the gas metal arc welding process.

Weld Cleaning.

Although many people outside the manufacturing and fabrication of weldments consider the actual welding time as the major factor in welding cost considerations, it isn't. The greatest cost of shielded metal arc welding is the cleaning and finishing of the parts **after** the welding operation is completed. This fact has been proven in numerous surveys of costs in the welding industry.

There is very little requirement for weld cleaning and finishing with the gas metal arc welding process. With the exception of flux cored electrodes there is no flux, and therefore, no slag cleanup involved. If the welding procedure and process are correctly set up there should be very little spatter from the welding operation. Chipping, grinding and finishing will be at a minimum.

Operator Training.

The need to be trained in the gas metal arc welding process often is a concern of welding operators. Most of them are craftsmen who are competent in one or more of the existing welding processes. Some are reluctant to try learning to use the gas metal arc welding equipment because it is new to them and they are afraid they might fail to master it properly. Their fears are unfounded.

It is true that shielded metal arc and gas tungsten arc welding processes require much time and practice to learn. Many apprentice programs require some welding courses through each of the four years of work and study. The time element required to learn the gas metal arc welding process is much less than with the other two processes mentioned. Welding operators who are proficient in other welding processes can be trained to use the gas metal arc welding process in a matter of hours. Unskilled operators have been trained to use the process in forty to sixty working hours of shop time.

The relative ease of training welding operators to use the gas metal arc welding process is simple to explain. Much of the control required for welding with the process is incorporated into the constant potential type welding power source and the gas metal arc welding equipment. The only basic variables the welding operator must watch are the gun angle relative to the workpiece, the speed of travel and the gas shielding pattern. The welding operator will be able to master the more difficult gas metal arc welding techniques as more welding experience is gained with the process.

Fast Controlled Welding.

One of the outstanding features of the gas metal arc welding process is the **elimination of weld starting and stopping due to changing electrodes.** Many shielded metal arc weld defects are caused by slag inclusions, crater cracking, cold lapping, etc., which occur when shielded metal arc electrodes are changed. The listed defects are certainly not the only reasons for weld failures but they are detrimental to quality welding. Decreasing the number of weld starts and stops will limit the possibility of such defects occuring.

The gas metal arc welding power source and equipment controls several variables which permit the welding operator to weld as fast as he is physically capable of moving and seeing. (At about 40 inches per minute the human eye can no longer focus on the arc area and the weld seam). The welding operator has the opportunity to set the welding condition on the gas metal arc welding feeder-control and the constant potential type power source before beginning the actual weld. The welding condition will remain stable until open circuit voltage, wire feed speed, slope, gas flow rate or some other variable is physically changed. This permits the welding operator to concentrate on the weld path, the weld puddle and the welding technique he is using. He has high current density at the electrode tip which allows him to obtain controlled penetration, controlled bead dimensions and a stable arc plasma and arc column. It is these added features that result in high speed, high quality gas metal arc welds.

Current Density.

Current density may be defined as the amperage per square inch of cross sectional area of the electrode. At a specific amperage, for example, there would be much higher current density with 0.030" diameter electrode than with 0.045" diameter electrode. The explanation is the cross sectional area of the two electrodes as shown:

0.030" dia. wire = 0.00071 square inches of area.

0.045" dia. wire = 0.00160 square inches of area.

The current density may be easily calculated for any diameter electrode. Dividing the welding amperage by the electrode cross sectional area in square inches results in amperage per square inch of electrode. For 0.030" diameter electrode the problem would appear as follows:

$$0.00071 \text{ inches}^2 \overline{) \quad 100 \text{ amps.}} \quad = 140,845 \text{ amps/inches}^2.$$

where 100 is the value of the welding current. The current density would be approximately 141,000 amperes per square inch of electrode.

Each electrode type and diameter has a minimum and maximum current density capability. If the current density is too high for the electrode type and diameter there is the possibility of electrode burnback into the contact tube. Low current density could cause stubbing of the electrode into the work. The term "stubbing" means the electrode has short-circuited to the workpiece and there is no welding arc present.

A condition called "roping" may occur in conjunction with stubbing. In a roping condition the electrode becomes red hot due to the short circuit condition with the base metal. The heating of the electrode is caused by the resistance of the electrode wire to current flow. The electrode will continue to feed from the gun and, since it is in a semi-plastic state, will coil or "rope" on the surface of the base metal. There is no welding arc present, of course.

Arc Energy.

A high current density arc has more arc energy concentrated at one point, the electrode tip. For example, where gas metal arc welding has high current density, shielded metal arc welding does not. The gas metal arc column is sharp and incisive. With shielded metal arc welding the arc column is relatively soft and widespread. The relative arc columns, and deposit configurations, are illustrated in Figure 191.

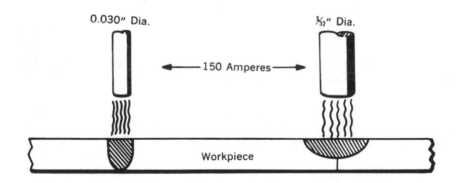

Figure 191. DEPOSIT COMPARISON AT THE SAME AMPERAGE.

It is logical that the weld deposit made with shielded metal arc will have a wider, shallower and more parabolic shape than a deposit made with gas metal arc welding. In this circumstance, the welding speed of the shielded metal arc welding process has to be slower, with more heat applied per linear inch of weld, than the gas metal arc process.

Since there is a greater volume of molten metal deposited, per linear inch of weld, with the shielded metal arc welding process, the heat input differential of the two welding processes is considerable. There is greater penetration with the gas metal arc welding process because of the higher current density at the electrode tip. It is logical that the bead-to-width ratio will be less with gas metal arc than with shielded metal arc welding. Comparative weld deposits are shown in Figure 192.

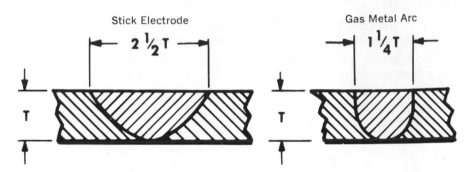

Stick Electrode

Gas Metal Arc

Figure 192. WELD DEPOSIT COMPARISON.

A normal shielded metal arc weld on 1/4″ steel plate would have a top bead width of approximately 2 1/2—3 "T" where "T" is the plate thickness. Using the gas metal arc welding process on the same material (with CO_2 shielding gas) the top bead width would be about 1-1 1/2 "T".

The gas metal arc welding process permits narrower joint preparation, faster welding speeds and a corresponding decrease in the cost per pound of deposited filler metal.

Other Benefits.

Other benefits of the gas metal arc welding process include less heat affected zone (HAZ) width, less possibility of grain growth in the base metal and less distortion of the weldment due to thermal expansion and contraction. The heat energy is applied to the weld instead of dissipating into the base metal. Due to the speed of travel, and the fact of welding heat being concentrated in a small area of the workpiece, there is less heat energy put into the work per linear inch of weld when using the gas metal arc welding process. This assumes a comparison to shielded metal arc welding when welding a given thickness of material over a prescribed distance of weld bead length.

Gas Metal Arc Welding Applications.

It has been estimated that approximately 80% of all welding performed today is on some type of steel material. Probably 75%

of the total steel welded is classed as low carbon or mild steel. In the past few years the gas metal arc welding process has successfully competed for the right to be used in place of other processes for a good share of this work. The reason for its wide acceptance by industry is because it welds faster, cleaner and more efficiently where applicable. The process has found applications wherever metal working and metal joining industries operate.

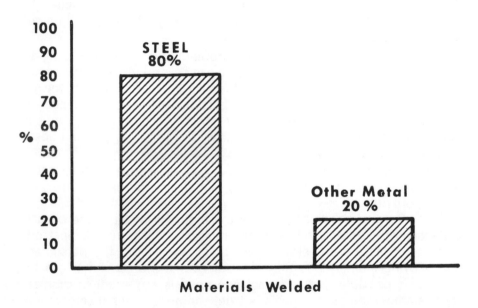

Figure 193. METALS WELDED BY GENERAL PERCENTAGES.

Steel Welding.

The short circuit method of metal transfer is designed primarily for welding materials 1/4″ thickness or less. Either welding grade CO_2 or argon-CO_2 gas mixtures are employed with this process.

Applications for the short circuit method of metal transfer include all types of sheet metal fabrication, angle iron frames, pipe welding, out of position work, etc. Some nominal voltages and amperages used with various diameters and types of electrodes are listed in the appendix of this book.

Spray transfer welding of steels is normally accomplished with argon-oxygen shielding gas mixtures. Spray transfer of steel, using solid electrode wire and argon-oxygen shielding gas, is a high speed method of welding. It is normally only done in the flat or horizontal welding positions.

Porosity.

In any type of welding there is always a percentage of failures for one reason or another. A weld defect can cause re-work of the part and, in some cases, can actually cause the part to be scrapped. Since re-work is always expensive (at times it can run five to ten times the cost of the original weld) it is to be avoided whenever possible. Some of the weld defects that can cause re-work are slag inclusions, lack of penetration, lack of fusion and porosity. Probably the single most common defect is porosity.

Porosity is one of the recurring problems faced when welding any metal with any welding process. There are several general rules which may help to minimize porosity problems when welding with the gas metal arc welding process. The suggestions made do not include all the possible solutions for eliminating porosity occurrences. Due consideration to the points made will help decrease the possibilities of porosity in welds.

1. Welding speeds that are too fast may cause either partial or complete loss of the shielding gas pattern in the arc area and cause porosity.

2. Current densities that are too high will often cause porosity because of the excessive heat of the molten metal from the electrode. In some cases there is severe oxidation loss of both the deoxidizers and alloying elements as they move across the arc column. In this case, it is possible that the electrode used is too small in diameter for the amperage used. If this is true, the next larger diameter electrode should be tried. If the smaller diameter electrode **must** be used it will be necessary to reset the welding condition variables.

3. The shielding gases used with the gas metal arc welding process must be the right type for the metal being welded. The correct flow rate must also be used or unsatisfactory results will occur. It is imperative to have clean, dry shielding gases. Argon and helium of commercial grade have purities of approximately 99.995+ percent.

4. It is important that the welding electrode be maintained in the center of the shielding gas flow pattern. If the electrode wire is off center it can cause erratic arc behavior and porosity.

In all steel welding there is the possibility of a silicate residue on the surface of the weld bead. This is particularly noticeable when welding steels having a high silicon content or when using double or triple-deoxidized electrode wires having high silicon content. The residue appears normally as a glassy brown substance that is somewhat difficult to remove. It should be cleaned off between multiple passes or when the weldment is to be painted or plated. Removal may be accomplished with chipping tools or by sandblasting.

Aluminum.

The use of aluminum has increased greatly since World War II. The metal is light in weight yet has the capability of being alloyed to achieve strengths comparable to mild steel.

Aluminum has excellent electrical and thermal conductivity. It is factors such as these that make aluminum a highly desirable metal for many applications. Its high strength-to-weight ratio has made it an invaluable material for aircraft and missile industries. Not the least of its attributes is its low weight, neat appearance and ability to be formed into almost any shape or configuration desired.

Aluminum and most of its alloys may be welded by any of the gas metal arc welding techniques. It does take some skill to weld aluminum out-of-position with the short circuit method of metal transfer. Argon and helium, as well as mixtures of the two gases, are used for shielding aluminum welds made with the gas metal arc welding process.

A consideration of prime importance when welding aluminum is the cleanliness of the base metal. Aluminum forms a heavy refractory oxide at the metal surface which may cause irregularities in the weld if not removed prior to welding. Aluminum oxide has a melting point in excess of 4,000° F.

Pre-weld cleaning may be either mechanical or chemical. A power wire brush may be used for mechanically removing the surface oxides but a heavy metal scraper blade will accomplish the job faster and with more certain results. Be sure the wire brush is really removing the oxide if this is the cleaning method selected. It often happens that the oxide is only being polished instead of removed.

Chemical cleaning is another method used for removing surface aluminum oxides. A system of cleaning solutions are often used in the following sequence:

1. Sodium hydroxide bath (approximately 150° F.).
2. Hot water rinse.
3. Chromic acid bright dip.
4. Hot water rinse.
5. Hot air dry (optional).

The important thing to remember is that aluminum oxides must be removed from the welding zone prior to welding.

Aluminum plate that has been sheared presents a problem unique to the material. The action of shearing aluminum is approximately 35% shear and 65% controlled tear of the metal. The sliding action of the oiled shear blade past the metal edge will usually result in some fold-over of the material. There is a definite possibility that oil and other foreign particles will be entrapped in the edge of the metal. A sheared edge should always be degreased and at least 0.020″ removed from the sheared surface **before welding** with any process.

Spray transfer of aluminum is a fast welding technique. Most welding is done in the flat or horizontal positions although it is not unusual to weld out of position with spray transfer and relatively fine diameter electrode wires. The gun is normally pointed in the direction of travel so that gas shielding is assured. The method is called forehand welding.

Summary.

It is apparent that the gas metal arc welding process has much to offer the welding industry in the way of faster, cleaner welds for many applications. It is not a panacea for all the ills of the world of welding so please don't get caught up in that idea! As every welding process, in its time, has made a contribution to the advancement of welding technology, so will the gas metal arc welding process.

Chapter 21

ROTATING TYPE WELDING POWER SOURCES

General Data.

Rotating type welding power sources, such as motor-generators and various types of engine driven units, have been used by the welding industry for many years. As a matter of fact, motor-generator welding power sources were some of the first electric welding power sources built. The engine driven welding power sources were instrumental in freeing the profession from the shop and giving mobility to this most useful craft.

An important thing to remember is that **all rotating welding power sources are some type of electro-mechanical device.** This means they are a composite unit which uses mechanical energy to generate electrical energy for welding. This is the same principle used in power generation plants all over the world.

All generator and alternator units produce a maximum total amount of secondary kilowatts. If you recall that volts times amperes equals watts, and divided by 1,000 equals kilowatts, the rest is easy. The rule, as given earlier in this text, is that for a specific amount of total power, "As the voltage increases the amperage will decrease proportionately. Conversely, as the voltage decreases, the amperage must rise proportionately".

If there is a relatively high open circuit voltage at some particular setting on a welding power source there must be a relatively limited amount of maximum short circuit current at the same time. This is typical of a number of models of motor-generator or engine driven welding power sources on the market today. When the power source is used for small diameter electrodes, such as 1/16" and 3/32", a check of open circuit voltage at the output terminals of the unit will show something in a range of 100-105 volts.

For normal operations with 1/8", 5/32" and 3/16" diameter electrodes, the open circuit voltage is usually lower (in a range of 75-80 volts) but the maximum short circuit current of the power source is considerably more. This is shown in the several volt-ampere curves illustrated in Figure 194.

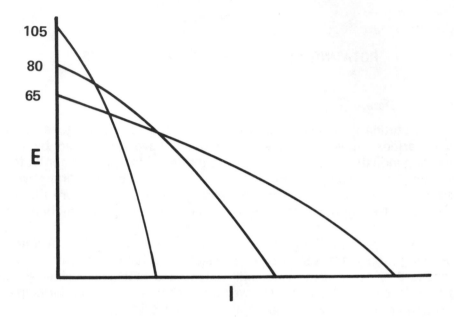

Figure 194. TYPICAL VOLT-AMPERE CURVES FOR A VARIABLE OPEN CIRCUIT VOLTAGE MG SET.

Setting the rotating type electric welding power source for larger electrodes will produce open circuit voltage that is further decreased in value (possibly ranging in value from 55-70 volts) and maximum short circuit current that is quite high.

The question is, "Why the different open circuit voltages and maximum short circuit current levels for different diameters of electrodes used with shielded metal arc welding?" The answer is one of logic.

Consider, for example, the small electrode diameters of 1/16" and 3/32". If you measure the cross sectional area of the electrode core wire, in square inches, and compare that to the cross sectional area of the **electrode flux,** in square inches, you will find **there is more flux than core wire cross section.** This makes it difficult to initiate the welding arc with most small diameter electrodes unless there is relatively high open circuit voltage at the welding power source output terminals.

The normal, or most commonly used, electrode diameters for shielded metal arc welding are 1/8″, 5/32″ and 3/16″. **In most cases these electrodes will have greater core wire cross sectional areas than they have flux cross sectional area.** They do not require the extra high open circuit voltage to initiate the welding arc but they do need more welding amperage from the power source. This is shown in the volt-ampere curves in Figure 195.

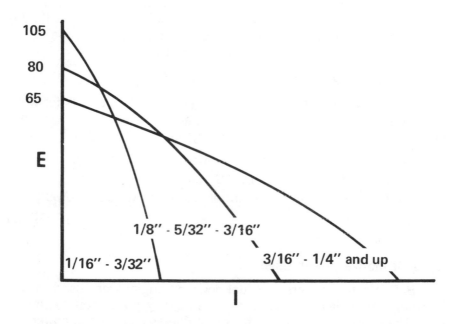

Figure 195. VOLT-AMPERE CURVE AND ELECTRODE DIAMETER CORRELATION.

Of course, the larger diameter electrodes need even less open circuit voltage but much more welding amperage to operate correctly. This is illustrated in the volt-ampere curves.

The "voltage control" normally located on the control panel of constant current type rotating power sources used with the shielded metal arc welding process is basically an **open circuit voltage control,** regardless of what it may be called by the manufacturer of the unit.

The flux coatings on some of the various shielded metal arc welding electrode classifications may require relatively high open circuit voltage for arc initiation and stabilization. For example, AWS Class E-6010 welding electrodes have a tendency to "pop out" if open circuit voltage is too low at the power source. This can occur while actually welding with the electrode particularly with the 1/8" diameter, and smaller, electrodes. This problem has been minimized to some degree by the addition of iron powder in relatively small amounts to the electrode flux mixture. For this type of electrode the iron powder addition is normally not more than about 10% by volume of flux. The iron powder addition to the flux will increase the fluidity of the weld puddle to some extent so this type of electrode is not necessarily applicable to all types of welding applications.

Types Of Welding Generators.

There have been three types of commonly used designs for welding power generators over the years. Each has its place in the history of electric arc welding. The three generator types are:

DC Generator = Normally DC welding output only.
AC Generator = Normally AC welding output only.
AC Alternator = Can be either AC or DC welding output.

Each of these electro-mechanical devices for generating welding power will be explained in detail in this section of the text.

Types Of Motive Power Used.

Both electric motors and fuel-powered engines are used to turn the rotor assemblies of the rotating type welding power source generators. The basic criteria for motor or engine size and horsepower rating is that the electric motor or fuel-powered engine be capable of **permitting the generator to reach full power output.**

Electric Motors.

The electric driving motors used for motor-generator type welding power sources are normally **ac induction type motors.** The NEMA EW-1 Standard for Electric Arc Welding apparatus states in paragraph EW-1-3.05-B, "Alternating current induction motors driving dc generator and dc generator-rectifier arc welding machines shall be **three phase** and shall have voltage and frequency ratings in accordance with the following:

60 hertz = 200, 230, 460 and 575 volts.
50 hertz = 220, 380 and 440 volts."

Fuel Powered Engines.

There are a variety of engine types and sizes used for portable engine driven welding power sources. Both liquid and air cooled engines are employed for specific power source applications. Many smaller engine driven units of less than 250 amperes output rating use air cooled engines. Most larger engine driven power sources have liquid cooled engines.

A variety of fossil fuels may be used for running the engines of welding power source generators. Gasoline is probably the most popular because it is readily available in most areas and is used for other engine driven equipment besides welding power sources. Diesel fuel is very popular in some areas because of its lower cost and low pilferage losses. Some Federal laws will only permit diesel fuel for engines in certain specific applications. A good example of this restriction is the use of diesel engines for welding power sources on off-shore drilling rigs. Diesel fuel has a much higher flashpoint than gasoline and is considered much less dangerous particularly as a fire hazard.

Propane is used in some applications of engine driven welding power sources. It is less expensive and cleaner burning than gasoline but it requires a special carburation system.

The DC Generator Design Concept.

The dc generator design concept considers that the rotor assembly is comprised of a through shaft, two end bearings to support the rotor and shaft load, an armature which includes the laminated armature iron core and the current carrying armature coils, and a commutator. **It is in the armature coils that welding power is generated.**

The stator is the stationary portion of the generator within which the rotor assembly (the armature) turns. In this design **the stator holds the magnetic field coils** of the generator. The magnetic field coils have a small amount of dc voltage and amperage applied to maintain the necessary continuous magnetic field required for power generation. The dc amperage is normally no more than 10-15 amperes and very often is less.

In electric power generation there must be relative motion between a magnetic field and a current, or electric, field. It makes no difference which type of field is in motion as long as there is **relative motion** between the two. In the dc generator it is **the armature that is the current, or electric, field.** The magnetic field coils are in the stator. The armature turns within the stator, and its magnetic field system, and welding current is generated.

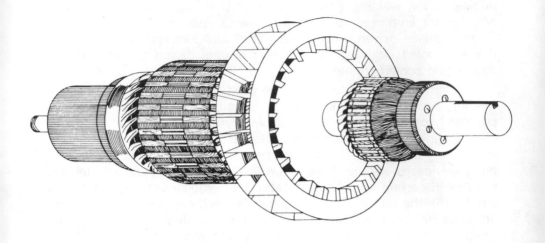

Figure 196. A TYPICAL DC ARMATURE ASSEMBLY.

The current that is generated in any welding power generator is alternating current (ac). The alternating current is carried to the copper commutator bars through electrical conductors from the armature coils. The conductors are soft-soldered to the individual commutator bars. The commutator bars may be considered as terminals, or "collector bars", for the generated alternating current from the armature.

The commutator is a system of copper bars that are placed concentric to the centerline of the rotor shaft. Each copper bar of the commutator has a machined and polished top surface upon which carbon type contact brushes ride to pick up each half-cycle of the generated alternating current. The purpose of the commutator is to reverse every other half-cycle to form a unidirectional current which is called direct current (dc). Each of the copper bars of the commutator is insulated from all the other copper bars.

The carbon type contact brushes actually pick up each half-cycle of generated alternating current and direct it into a conductor as direct current. It may be said that the brush-commutator arrangement is a type of mechanical rectifier since it does change the generated alternating current (ac) to direct current (dc). Most of the carbon type brushes used for this purpose are an alloy of carbon, graphite and small copper flakes.

The dc generator is so-called because it has the commutator-brush arrangement for changing ac to dc welding power. Normally the dc generator is a three phase electrical device. Three phase systems provide the smoothest welding power of any of the electromechanical welding power sources.

DC generators may be either electric motor driven or fuel-powered engine driven units. The engine driven power sources are especially applicable for field welding operations.

The AC Generator Design Concept.

The AC generator design concept is similar to the dc generator with the major exception that **it has no commutator and brush arrangement** for changing the generated ac to dc. AC generators do have a rotor assembly which consists of the armature iron core and armature coils, through shaft and bearings and, something different in place of the commutator, **slip rings.** (Slip rings are solid brass parts that are machined and polished and fitted concentric to the shaft centerline). It is still in the armature coils, located on the rotor assembly, that welding power is generated.

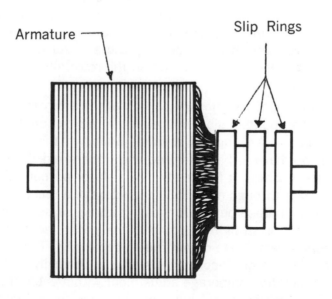

Armature —   Slip Rings

Figure 197. A TYPICAL AC ROTOR ASSEMBLY.

The stator is the stationary portion of the generator in which the magnetic field coils are located. As the rotating armature moves within the magnetic field of the stator electric welding power is generated. The generated power is normally single phase alternating current.

The generated electrical power is carried by conductors to the slip rings. The alternating current is picked up by carbon type contact brushes from the slip rings and conveyed by conductors to some type of control device such as a reactor. The alternating current is then directed to the output terminals of the welding power source. The welding power output is alternating current, normally 60 hertz frequency. AC generators may be driven by either electric motors or some type of fuel powered engine.

The AC Alternator Design Concept.

The AC alternator design concept has both the rotor and the stator as do the other two types of electric power generators. There is, however, a marked difference in the design characteristics of this unit. The rotor consists of a through shaft, **end bearings at one end of the shaft** and a coupling plate for direct connection of the rotor to the engine driving it at the other end. The rotor assembly has the magnetic field coils instead of the armature coils as the other generators have. There are brass slip rings through which small amounts of dc voltage and amperage are brought to the magnetic field coils for excitation purposes. In many designs there is some portion of the exciter circuit power system located on the rotor although this is not a requirement.

The stator is the stationary portion of the alternator. It has the armature coils wound in slots in the armature iron core which is also the basic stator. Welding power is generated in the armature coils but it does not have to be picked up and transferred through carbon type contact brushes. It is already in the stator portion of the circuit and is conveyed through electrical conductors to a reactor for division into various amperage output ranges. In those cases where the unit is either ac/dc, or straight dc, output, the generated welding power goes through a rectifier to obtain the dc capability. The rectifier may be either selenium or silicon. A rheostat is normally in the control circuit to obtain fine amperage adjustment within each of the output ranges. The rectifier does only one thing which is to change the generated alternating current to direct current.

MAGNETIC
ROTOR—FIELD COILS

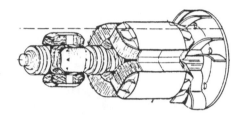

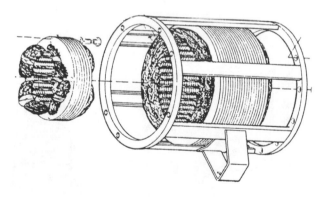

STATOR—ARMATURE COILS

Figure 198. A TYPICAL AC ALTERNATOR SYSTEM.

This type of welding power source is normally an engine driven unit and totally portable for field operations. In some cases the welding power sources of the AC alternator design also have the capability to function as auxiliary 60 hertz electric power plants. For this purpose the generator, or alternator, is a single phase unit.

Most engine driven welding power sources used for auxiliary power generation are four pole systems. If the engine welding speed is above 1800 rpm it must be slowed to 1800 rpm to obtain 60 hertz, 115 or 230 volt electrical power for proper operation of ac tools, lights, and other electrical equipment.

Comparison: Rotating Armature And Rotating Field Coil Designs.

To better understand the design concepts, and the differences, of both the rotating armature and rotating field generating systems we will examine the two types of units as shown in Figure 199.

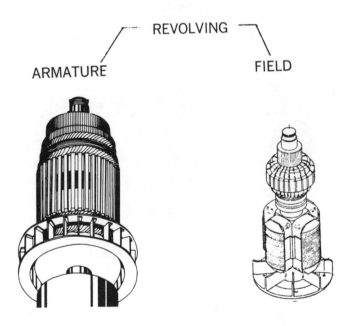

Figure 199. COMPARATIVE ROTORS.

In looking at the **rotating armature design** first you see the massive iron core, the armature coils and the rather large commutator assembly. The mass and weight are considerable. The stator portion is also quite massive although its most important function is to house the magnetic field coils of the generator.

By comparison **the rotating magnetic field coil design** rotor looks small. You see the four magnetic field coils and their iron core material. (This is, of course, a four pole electrical system). The stator holds the armature coils in their slots. The stator iron is the core material for the armature.

The interesting point is that **both units** illustrated are rated at 200 amperes welding output, DC. The duty cycle rating is different, however, since the **rotating armature unit is rated at 200 amperes, 60% duty cycle** and the **rotating magnetic field coil unit is rated at 200 amperes, 100% duty cycle.** (Remember: duty cycle is based on a ten minute period of time. At 60% duty cycle the rotating armature type power source can operate at rated amperage (200 amperes) for six minutes out of ten. The unit must idle and cool the other four minutes of the cycle.)

The rotating magnetic field power source may be operated **continuously** at, or below, its rated 200 amperes of welding current output. There is no cooling problem, no overheating which could result in job downtime and possible damage to the power source.

The logical question is, "How can this be when the one unit— the rotating armature type and design—is so much larger and heavier than the rotating magnetic field coil system?" Fortunately there is a reasonable and logical answer to the problem and the question.

Consider that all welding amperage on these units is generated in the armature coils. It makes no difference to the power source if the armature coils are in the stator or the rotor. They still carry welding amperage as generated. You will recall that **all electrical conductor materials have some measure of electrical resistance to current flow.**

In the rotating armature generator the armature coils are located on the rotor assembly **which rotates within the stator assembly.** There is no way the armature coils can be cooled adequately and so, as the generated welding current flows through them, the coil conductor temperature increases due to the electrical resistance of the conductor wires. Since the thermal (heat) energy cannot be dissipated it heats the iron core material on the rotor as well as the armature coils. The result is, while the generator is producing the proper amount of amperage for its rating, that amperage cannot all get through the conductor wires because of the resistance heating in the circuit.

The effects of the resistance heating in the rotor assembly will show up in the welding output after a very few minutes of operation. It is not uncommon for the welding operator to return to his power source and readjust the amperage output control to a higher setting so he can continue to weld with the same diameter electrode. If you were to ask him why he is adjusting the output control he will probably tell you, "My machine got cold and I need more current to weld!"

As a matter of fact, his power source components **became hot due to the resistance heating** and the lack of adequate cooling. The additional electrical resistance in the armature coils, the iron core material, the conductors, the commutator bars and probably the electrode leads used up a considerable amount of the output kilowatts from the generator. The energy just never had a chance to reach the welding arc.

In the rotor assembly of the rotating magnetic field coil system the only electrical current flowing is a small amount of dc necessary for the excitation of the magnetic field coils. At maximum this may amount to 12-15 amperes; hardly enough to cause an overheating problem in the magnetic field coils or any other portion of the rotor assembly.

The armature coils carrying the generated welding current are located in the stator which is wide open to the cooling air. The result is that the armature coils have a better opportunity to remain relatively cool while under the welding output load. There is less electrical resistance and not nearly the decrease in welding current output that is found in the rotating armature design. Very seldom does a welding operator have to adjust his rotating magnetic field type power source upward for current output when welding with the same diameter electrode.

An item of general interest should be inserted here. All of the power generating turbines used in power generating plants the world over are the rotating magnetic field coil design. As a matter of fact, the efficiency of rotating magnetic field coil design power generators is very high.

Paralleling Electro-Mechanical Power Sources.

When two or more welding power sources are placed in parallel operation the amperage of the two units is additive while the load voltage remains the same as it was for one power source. When paralleling electric motor-generator units the situation is time consuming and difficult. The very nature of paralleling the output amperage requires that the excitation of the two units be in phase, the phasing of the two electric motors be in phase and the output power generated in the armature be in phase. This gets to be a little rough unless you are a highly skilled and qualified electrical man.

One of the basic problems that is encountered in paralleling dc generators, either electric motor driven or engine driven, is that of balancing the output amperage of the two units being paralleled. If they are not almost **exactly the same amperage output** there is a tendency for the unit with the higher amperage output to feed current back into the other power source. This current goes through the parallel connection, the welding power source output terminals, through the brushes and the commutator and dissipates as heat energy in the armature coils. Of course, the armature coils are already operating at elevated temperatures. Many rotating armature type welding power sources are burned out every year because precautions were not taken to keep this imbalance from occurring when paralleling two or more of the MG power sources.

Those rotating magnetic field coil design power sources which have dc welding output achieve that characteristic by putting the generated ac power through a rectifier in the circuit. The rectifier may be either silicon or selenium. Rectifiers, of course, permit current to flow in one direction only, in this case, **out of the welding power source.**

It is a simple matter to put two or more such power sources in parallel operation. Even if the welding amperage output is not exactly the same on both units, it is not possible for one power source to become the slave of the other since the possibility of current feedback into **either** of the rectifier type units is absolutely stopped by the rectifiers themselves. There is no way current from one power source can get into the armature coils of the other unit.

Auxiliary Power Plant Operation.

Many of the AC and AC/DC engine driven welding power sources have capability for use as an auxiliary power generating plant. This is often a blessing to the contractor in the field who has no other source of electrical power for operating electric tools, lights or other electrical equipment. If the generator is a four pole system, and most of them are, the engine should be run at 1800 rpm when the unit is used as a power plant. The reason is that 1800 rpm produces 60 hertz, 115 or 230 volt power. This is the voltage and frequency required for electric tools that operate from alternating current only. This does not include the "universal electric tools" which may operate from either ac or dc power.

Any speed of rotation that is higher or lower than 1800 rpm will cause either higher or lower frequency in hertz per second. For example, if the unit were operated at 1500 rpm the hertz per second frequency would be low and the ac tool could stall and possibly burn out the electric motor. If the rotation speed of the generator was high, say 2500 rpm, the frequency in hertz per second would be very high and the overspeed would again burn up the electric tool.

Summary.

Engine driven welding power sources are a necessity for the many field welding applications presently being performed. There is a difference in the design of various engine driven units. The rotating armature design has been relatively unchanged for many years. Until about 1950, the rotating armature type welding power source, either electric motor driven or engine driven, was the only dc welding power source available to the welding industry.

Rotating magnetic field coil design power sources are smaller and have better electrical efficiencies than rotating armature units. The rotating magnetic field coil design is similar to the design of turbines used in power plants all over the world.

Some engine driven power sources may be used for auxiliary power generation. To obtain 60 hertz frequency, the engines must turn the rotor at 1800 rpm in a four pole electrical system.

Which unit should be specified for a particular application? That is up to the user to decide. Logic and experience will soon tell which type of power source requires the least maintenance, will have higher operating capabilities and lower cost per ampere of output.

Chapter 22

SPECIAL WELDING PROCESSES

Introduction.

In this text we have discussed the Gas Welding method, Shielded Metal Arc Welding, Gas Tungsten Arc Welding and Gas Metal Arc Welding. All of these welding processes have broad commercial applications in the welding industry. There are some other welding processes, however, which have many uses but, because of specific limitations, cannot be used for general overall welding applications. It is the intent of this chapter to briefly describe some of these welding processes for you.

The welding equipment employed is often rather sophisticated compared to the "everyday" welding process equipment previously mentioned. The welding power sources are often modifications of standard units although in some cases special power sources are required. In many cases it is the **process equipment controls** that are the key to the successful function of the equipment for welding.

Submerged Arc Welding.

Submerged arc welding really came into its own as a welding process during the second World War. Hundreds of miles of ship deck seams were welded with this process. Submerged arc welding may be either semi-automatic or full automatic.

The necessary equipment includes a high amperage rated welding power source, either ac or dc, some type of electrode wire feeder system with drive rolls, some type of welding head, and a flux hopper system for holding the flux and bringing it to the weld area.

The bare electrode is fed through the drive roll system to the head tube assembly where the electrode is energized with welding current. The welding current flows from the welding power source, through welding cables to the head assembly contact tube, through the welding electrode and the flux blanket to the base metal being welded. The welding electrode is solid bare wire and is continuously fed during the entire welding operation. It is usually provided in reels, spools or pay-off paks of considerable weight.

The actual welding arc is shielded by a blanket of granular flux which covers the weld joint. The flux has a low melting temperature; considerably lower than the melting point of the metal to be welded. At room, or ambient, temperature the flux is considered to be a non-conductor of electrical current. When in the molten state, however, the flux is an excellent conductor of electric current. The flux covers the molten weld puddle as well as the base metal immediately adjacent to the joint. The main purpose of the flux is to protect the molten weld puddle from atmospheric contamination although it does have some other functions as well. It insulates the welding heat so that relatively narrow, deep penetration welds can be made.

Initiating the welding arc may be accomplished in several ways. The flux, being a non-conductor when cold, makes arc starting difficult. In some instances a small portion of steel wool is placed between the electrode tip and the base metal. When welding current is applied the steel wool melts almost at once. This melts a small amount of the granular flux which permits actual welding arc initiation to take place. Other methods used are the touch start where the electrode is actually brought into contact with the base metal and high frequency energy which is often impressed on the electrode wire to improve arc initiation.

The continuously fed, consumable electrode is normally low carbon steel for most welding applications. It may be supplied in pay-off pak drums or as coils of wire.

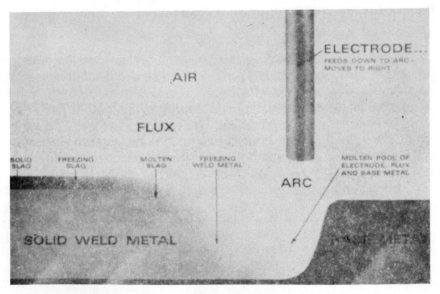

Figure 200. SUBMERGED ARC WELDING.

The granular flux is deposited along the weld joint. Its flow is directed so that flux is slightly ahead of the welding puddle at all times. The electrode wire melts as it goes through the molten flux, combining with the molten base metal to make the weld. After the weld has cooled and solidified, the flux will retain its molten condition where it is next to the weld metal deposit. When the flux has cooled and solidified, the excess flux is picked up with an industrial type vacuum cleaner, screened and then re-used. The slag residue on the weld becomes brittle and is easily removed.

The submerged arc welding process is adaptable for either semi-automatic or automatic operation. Semi-automatic systems are used where mobility of equipment is essential, mainly because of short welds. Fully automatic techniques and equipment are used where long seams can be brought to the work station and welded. This is a much more economical method of welding with the submerged arc welding process.

The nature of the weld puddle (very large volume and fluid) plus the use of granular flux dictates that submerged arc welding be done in the flat position. Welding may be done on materials down to approximately 0.060″ thickness. Using multiple passes, metals of several inches of thickness have been welded with submerged arc.

In fully automatic operations two or more electrodes may be used to make the weld. They may be positioned side by side or in-line along the longitudinal axis of the weld joint. In this type of operation both electrodes normally melt into the same weld puddle for faster deposition rates and higher travel speeds.

Some applications will not permit the massive weld puddle inherent with two or more electrodes melting in the same weld. Many times the electrodes are put in-line as before but far enough apart so that two separate weld puddles are formed, and two separate welds made, as a multi-pass weld with one traverse of the welding head.

Advantages And Disadvantages.

Some of the advantages of submerged arc welding are:
a) Less joint preparation normally.
b) No welding helmet required for the operator although safety glasses are recommended. The flux shields the arc.
c) Deoxidizers and alloying elements, if any, are normally contained in the welding flux. The electrode wire is less expensive than that used for gas metal arc welding.
d) Welding speeds and deposition rates are high compared to some other welding processes.

e) Submerged arc welding may be used in exposed areas where wind is a factor.

Some of the disadvantages of submerged arc welding are:

a) The welding flux can contaminate the work and cause weld porosity if proper procedures are not used.
b) The process equipment is bulky and normally not highly mobile. The work must be brought to such equipment.
c) The solidified molten flux must be removed after every welding pass.
d) Although used for metals as thin as 0.060", submerged arc welding is not normally recommended for metals under 1/8" thick. Even this thin a metal will require weld fixtures for reproducible weld quality.

Submerged arc welding is an established welding process used basically for making heavy weldments. There are several manufacturers of submerged arc welding process equipment.

Electron Beam Welding Process.

The electron beam welding process is relatively new to the welding industry. The principle is reasonably simple in actual concept although the equipment necessary to perform welding is somewhat expensive.

The electron beam is always generated in a hard vacuum within a vacuum chamber. In many applications the welding is also done within the vacuum chamber. Recent developments have made it possible to create the electron beam energy in the vacuum chamber while the work to be welded may be at atmospheric pressure.

Generation Of The Electron Beam.

The idea of the electron beam is that a controlled, narrow beam of electrons (the fundamental unit particle of negative electricity) is caused to impinge on the surface of the base metal at a high velocity. There are several major parts to the process equipment. The key part is the cathode ray tube, commonly called the electron beam gun. This unit emits the electrons at greatly accelerated velocities. Electronic and optical equipment focus the beam and prevent it from becoming widely divergent at the work surface to be welded. This is important since the success of the electron beam welding process depends on the size of the beam focal point at the work surface.

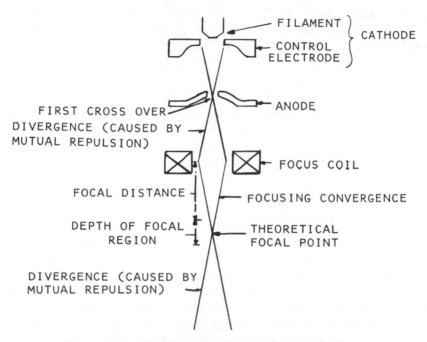

Figure 201. TYPICAL ELECTRON BEAM SYSTEM.

The extremely high density of the electron beam is dependent on the amperage and voltage of the electron beam power and the beam focal point size. The power density is calculated in watts per square inch (Watts/inch^2).

Electron Beam Welding.

When the welding equipment is set up properly the actual electron beam weld follows a rather simple pattern. The power density of the electron beam is so great that it will vaporize a small hole in the weld joint. The very thin layer of molten base metal completely surrounds the hole made by the electron beam. As the work, or the electron beam, is caused to move laterally along the joint the molten metal flows in behind the hole. As the intense localized heat is moved away the molten metal solidifies and completes the actual weld.

Metals That Can Be Welded With The Electron Beam.

All the known metals that are commonly welded with other processes may be welded with the electron beam technique. In addition, some metals previously considered unweldable have been joined by the process. Beryllium has been welded using high, or hard, vacuum equipment. **NOTE:** Beryllium is hazardous material to work with in any manner. **All applicable safety rules and standards for working beryllium must be complied with.**

Certain precautions must be taken when welding some of the nickel, aluminum, high strength steels, columbium alloys, etc. Normally the manufacturer of the material or alloy will be able to provide information on the weldability of the material with electron beam equipment. Another excellent source of information is the Technical Department, American Welding Society.

Ultrasonic Welding.

The ultrasonic method of metal joining is unique in that no specific heat application is made to the metal to be welded. Instead, the bond between two solid metal surfaces is made using vibratory energy supplied by transducers of some type. A **transducer** is a device which changes electrical high frequency energy to mechanical vibratory energy in this application. In a sense, a transducer is nothing more than a device which receives one type of energy and transmits that energy in another form to some type of receiver.

The specific metal surfaces to be joined are brought together under relatively low pressure. The ultrasonic energy is then brought to the area to be welded. It is considered that the ultra high frequency of vibration probably causes the surfaces at the joint interface to reach a compatible resonance. This permits the grains of metal at those surfaces to slip and intermingle (diffuse) sufficiently to create a sound, cohesive metallurgical bond.

Many different metals have been joined by ultrasonic welding. Usually the metal thicknesses are of the foil variety with most welds made in the foil thickness range of 0.001-0.040" for the harder, more dense metals. Low density metals such as aluminum and aluminum alloys have been successfully welded with this process in thicknesses up to 0.100". The limitations on thickness apply only to the thinnest member of the weld joint which, incidentally, **must be placed next to the vibratory welding tip.**

Plasma Arc Welding.

Plasma arc welding is relatively new since the first commercially used equipment was introduced in 1963. The principle of plasma arc welding is similar to tungsten arc welding as far as most of the equipment is concerned. The torch that is used does have some significant differences in characteristics and configurations.

"Plasma" may be defined, for welding, as follows: "A gas that has been heated to the temperature where thermal ionization takes place". Such ionization of the gas makes it an electrical conductor of excellent quality. The actual plasma contains both positively charged ions plus essentially an equal number of free electrons. Other sub-atomic particles such as neutrons are also found in plasma arc columns.

Plasma Production.

There are at least two methods for producing the plasma arc column. In one case, heat is produced by current flowing through a gaseous column between a non-consumable tungsten electrode and the base metal workpiece. This method is called **transferred arc plasma welding.**

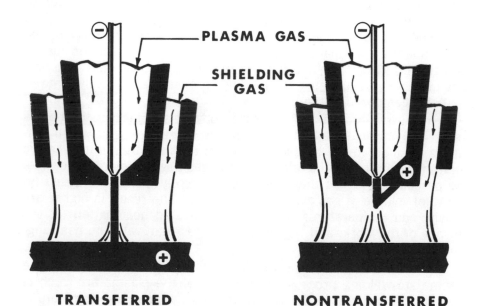

TRANSFERRED   **NONTRANSFERRED**

Figure 202. PLASMA ARCS, TRANSFERRED AND NON-TRANSFERRED.

A second method is termed **non-transferred plasma arc welding.** In this method the plasma arc is initiated between the non-consumable tungsten electrode and the metal cup, or nozzle, containing the rather restricted orifice (hole) necessary for this process. In all cases there must be a gas medium to ionize. In many applications the plasma gas is argon, an inert gas.

An ion is simply a gas atom that has had one or more electrons removed and which is, therefore, electrically charged positive. The electrically charged, ionized gas is called plasma when it is actually conducting the welding current.

The gas plasma is, of course, capable of conducting current. Consider that the ionized plasma is electrically charged positive. The negatively charged electrons, which constitute the flow of electrical current, are emitted from the non-consumable tungsten electrode. The electrons are attracted to the plasma as a carrying vehicle since similar electrical charges repel but **dissimilar electrical charges attract**. There is, then, an ideal situation for current flow through the gas plasma. It is considered that the plasma gas maintains its temperature because of velocity and the electrical resistance of the gas to current flow. The actual continuous ionization of the plasma gas is due to thermal ionization caused by the heat of the welding arc. The gas flows through a constricted hole, or orifice, in the output nozzle (cup) of the plasma torch.

There is, of course, some plasma in all welding arc columns. The amount of plasma is limited in volume since there is no means to restrict the nozzle orifice size and diameter. For example, the nozzles used for gas tungsten arc welding torches are relatively large in diameter compared to the plasma arc torch nozzle orifice.

In plasma arc welding a nozzle body of special construction is used. The nozzle has a restricted size and diameter orifice which constricts the diameter of the plasma stream flowing through it. The arc column is centered in the orifice. As the gas flows through the orifice, the percentage of gas that is ionized is greatly increased, based on the total volume of gas available in the arc. The result is higher arc density and temperatures, a more concentrated heat pattern at the surface of the work, and higher arc voltages than with the gas tungsten arc welding process.

The constriction of the welding arc is a unique feature of the plasma arc welding process. The design of the plasma arc torch is the controlling factor in the arc constriction. Since the plasma arc is actually more constricted it has more directional flow capability and is not easily disturbed by magnetic fields or other conditions that would disturb a conventional gas tungsten arc weld. It is logical that a constricted plasma arc will have much higher current densities than a standard gas tungsten arc welding arc column.

Some Plasma Arc Process Fundamentals.

As noted before, the plasma arc welding process employs a specially designed electrode holder, or "torch". It also is the conduit for the gases used with the process. Non-consumable tungsten electrodes are used. Although any of the commonly available types of tungsten electrodes may be used, thorium-bearing electrodes containing 2% thoria are normally preferred for plasma arc welding.

The plasma arc torches usually have two gas flow systems. Each system has its special job to do to make the process function correctly.

The inner gas column is the plasma gas which is normally argon. An inert gas, argon will not contaminate the tungsten electrode in any manner, including oxidation. It is the inner gas system that supplies the orifice gas for plasma.

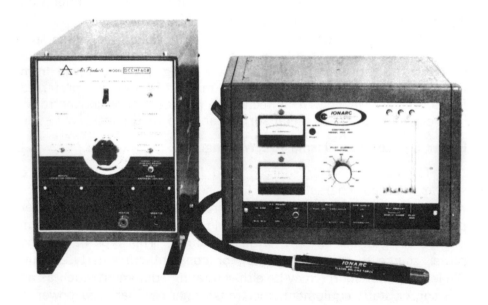

Figure 203. TYPICAL PLASMA ARC TORCH SYSTEM.

The outer gas system is called the shielding gas. It serves the same function as any shielding gas in that it keeps the outside atmosphere away from the weld area. Normally this gas will also be argon although some other gases, or gas mixtures, have been used for special applications.

The output orifice is, of course, in the end of the gas nozzle. The orifice-to-work distance is not extremely critical although it should be held to a distance between 1/8″ and 1/4″ with 3/16″ being the median distance used.

Types Of Arc Used With Plasma Arc Welding.

There are two basic types of welding arcs used with the plasma arc welding process. They are the **transferred arc method** and the **non-transferred arc method.** Transferred arc welding is done with the welding arc established between the tungsten electrode and the base metal workpiece. It is probably the most used of the plasma arc welding processes.

The non-transferred arc plasma welding technique is done with the plasma arc established between the tungsten electrode and the inner wall of the orifice nozzle body. There is some tendency for erosion of the nozzle inner wall surface with this technique. It is not a serious problem since the materials used for the nozzle inner wall are designed to prevent the erosion.

When welding with the transferred arc plasma technique, the welding heat comes directly from the welding arc. With the non-transferred arc plasma technique, the welding heat is derived from the plasma only. The transferred arc method is normally preferred for plasma arc welding. This technique provides more concentrated heat energy to the base metal surface and the weld joint.

Power Sources For Plasma Arc Welding.

The welding power sources used for plasma arc welding are similar in output characteristics to those used for gas tungsten arc welding processes. Usually they are three phase transformer-rectifier power sources with dc output only. Most plasma welding is accomplished using direct current, straight polarity. This type of welding power source is the conventional, or constant current, type unit. While the power source may be either rotating equipment, such as an MG set, or static equipment, the transformer-rectifier type power is preferred by most plasma welding manufacturers and users.

The actual welding current used with plasma arc welding may range from 0.10 ampere to 500 amperes. With the greater heat concentration of plasma arc welding, power sources are normally rated no higher than 500 amperes. Actual welding currents will seldom exceed 400 amperes.

Low ampere welding with the plasma arc process is considered to range from 0.10 ampere to 100 amperes approximately.

High ampere welding with plasma arc is normally considered to be between 100 amperes and 500 amperes. As previously noted, seldom are amperages exceeding 400 amperes used.

Summary of Section.

Plasma arc welding permits a highly concentrated arc plasma to impinge on the base metal surface to be welded. The high current density present provides deep penetration welds with low amperages.

The electrodes used are non-consumable tungsten electrodes. They may be any of the various classes of tungsten although 2% thoria bearing tungsten is usually preferred for most operations.

Power sources used for plasma arc welding are constant current output with a "drooping" volt-ampere output characteristic.

Laser Welding.

The laser welding process is probably the newest of the new welding processes. The word "laser" is derived from the first letters of the following words: "Light amplification by stimulated emission of radiation". The laser welding process has the potential to perform work similar to the electron beam process. Deep penetration welds, with minimum heat affected zone, are characteristic of the electron beam welding process. As previously noted, however, the electron beam must be created in a high vacuum chamber. This is not so with the laser. By its very nature, it retains its beam through normal atmosphere without serious diffusion of the beam diameter. The laser has the ability to provide extremely high beam power densities at the surface of the base metal to be joined. The beam has the additional advantage of being able to operate through any transparent medium such as air, gas and even specific liquids.

Generation Of The Laser Beam.

The device used for generation of the laser beam in some cases is a ruby crystal in rod form. The ruby is made up of aluminum oxide with a small amount of chromium oxide in solution in the material. Normally, the ruby is exposed to high intensity flash lamps, such as Xenon flash lamps. The high intensity radiation from these lamps excites the chromium atoms to high energy levels. These atoms lose their high energy almost immediately through heat dispersion. As the chromium atoms drop back to their natural state there is an evolution of radiation in the form of red light. The ends of the ruby rod are mirrored with silver so that the light reflects back and forth along the ruby rod. One end of the ruby rod has a small hole in the silver plating through which an in-line, coherent light beam issues. This is the laser beam used for welding. There is very little diffusion of this beam even over considerable distances. The beam can be directed and manipulated easily with rather simple optical systems which bend and focus the beam as necessary for the welding application.

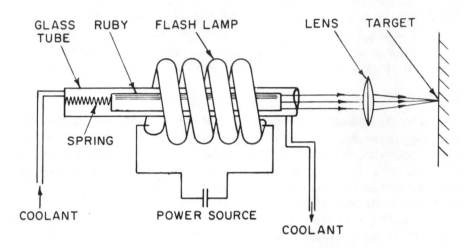

Figure 204. LASER EQUIPMENT.

Other types of laser generating systems include the continuous gas lasers, YAG (yttrium-aluminum-garnet) lasers and other systems still in development. Most ruby or YAG lasers are of relatively short duration pulses since much of the Xenon flash radiation used for exciting the solid state generators (for example, the ruby rod) is not the correct wave length for excitation of the transition element atoms (chromium). The Xenon flash lamps used create substantial heat as well as light and they must be pulsed because of heat build-up in the lamps themselves. As a result, most laser welding has been on very thin materials of about 0.025″-0.030″ thickness.

Summary Of Section.

Laser welding is being accomplished although most applications are in the micro-weld areas such as electronic components.

There have been some significant results indicated for deep penetration laser welding. This technique will probably very closely approximate the electron beam process in that a vaporized hole of very small diameter will be made in the weld joint and moved along the joint at a speed that will permit the molten metal to flow in behind the hole and become a homogenous weld upon solidification of the liquid metal.

Since it is difficult to measure power levels of the laser beam with accuracy, control systems have not been developed that will produce consistent results. This could change in a matter of days considering the amount of research that is being done in the field of laser welding. The problem of surface vaporization of the base metal is common, especially at the higher levels of power input. This may be overcome in the future by correlating the speed of weld travel to the power input as has been done with the electron beam welding process.

One of the exciting factors about laser welding is its potential for making welds comparable to electron beam welds in atmosphere. There is no requirement for the high vacuum apparatus with laser as there is with electron beam welding.

Once again it appears that the welding industry is on the threshold of new techniques for joining metals and other elements. With the special welding processes briefly discussed here, and some others still being developed or in the process of invention, the sky is the limit for people in welding today—and tomorrow!

EPILOGUE

This book has been written for you that you might be able to glean from its pages some ideas of what has gone before in the welding field. The fundamental information and natural laws stated will not change. "For every action there is an equal and opposite reaction". It is true in welding—and it is true in life and work.

There is much opportunity in the welding profession and it is getting more so all the time. Welders, Welding Technicians, Welding Engineers, Welding Supply Salesmen, Welding Consultants, Welding Foremen—there is no end to the occupations a person can achieve within the welding industry.

This is no easy touch profession. If you will work and study there is a good and rewarding life for you in welding. If you are looking for the easy way, this is not the profession for you. The income possibilities in various welding jobs are virtually unlimited—except by you—and what you will put into the work.

If this writing has made you understand more about welding and the equipment used with the various processes it is a successful book. Don't accept everything you read or hear as gospel, though. Ask questions—think—reason out the problems—and you, too, will find satisfaction that only a job well done can bring.

God bless you all and may He walk with you all your lives.

E. R. Pierre

Appendix

DATA CHART 1

APPLICABLE ELECTRICAL LAWS

Coulombs Law.

"Charged bodies attract, or repel, each other with a force that is directly proportional to the product of the charges on the bodies and inversely proportional to the square of the distance between them".

Newtons Law (Gravity).

"Every object attracts every other object with a force that is directly proportional to the product of the masses of the objects and inversely proportional to the square of the distance between them".

Ohms Law.

"The value of the current, in amperes, in any electrical circuit is equal to the difference in potential, in volts, across the circuit divided by the resistance in ohms of the circuit".

Lenz's Law.

"The induced emf of any circuit is always in such a direction as to oppose the effect that produces it".

DATA CHART 2

WIRE CALCULATION DATA

In calculating welding costs it is necessary to know the cost of the welding wire. A helpful figure is the number of inches in a pound of a specific type and diameter electrode. This can be calculated by the formula:

$$\text{Inches per pound} = \frac{\text{Linear inches per cubic inch}}{\text{Pounds per cubic inch}}$$

Linear inches per cubic inch is a constant value regardless of the material. Pounds per cubic inch is the unit of density of the material. These constants are shown in table form:

Table I

Diameter of Wire (solid). Decimal	Fraction	Linear inches per cubic inch.
0.020	. . . .	3,180
0.025	. . . .	2,190
0.030	. . . .	1,415
0.035	. . . .	1,040
0.040	. . . .	796
0.045	. . . .	629
0.062	1/16	332
0.078	5/64	208
0.093	3/32	148
0.125	1/8	81

Table II

Wire Type	Pounds per cubic inch.
Magnesium	0.063
Aluminum	0.098
Aluminum Bronze (10%)	0.275
Stainless Steel (4xxx)	0.280
Mild Steel	0.285
Stainless Steel (3xxx)	0.290
Silicon Bronze	0.308
Copper-Nickel (60-40)	0.320
Nickel	0.321
Deoxidized Copper	0.325

DATA CHART 2 (cont.)

With these constants several useful calculations can be made. As an example, suppose 1/16" diameter aluminum electrode wire is being fed at 300 inches per minute and the welding time of the joint is ten minutes:

FROM TABLES I AND II—

$$\text{Inches per pound} = \frac{332 \text{ (table I)}}{.098 \text{ (table II)}} = 3{,}390 \text{ inches per pound}$$

Length of wire used = 300 in. per minute X 10 minutes = 3,000

$$\text{Weight of wire} = \frac{3{,}000}{3{,}390 \text{ in./lb.}} = 0.885 \text{ lb.}$$

If the aluminum wire cost $1.30 per pound the cost of the deposited metal is $1.15.

The chart below lists the number of inches per pound of the various wire types and diameters.

INCHES PER POUND—MATERIAL

Fraction Decimal	0.020	0.025	0.030	0.035	0.045	1/16 0.062	5/64 0.078	3/32 0.093	1/8 0.125
Mg.	50500	34700	22400	16500	9990	5270	3300	2350	1280
Al.	32400	22300	14420	10600	6410	3382	2120	1510	825
Al. Br.	11600	7960	5150	3780	2290	1220	756	538	295
SS-400	11350	7820	5050	3720	2240	1180	742	528	289
Mild St.	11100	7680	4960	3650	2210	1160	730	519	284
SS-300	10950	7550	4880	3590	2170	1140	718	510	279
Si. Br.	10300	7100	4600	3380	2040	1070	675	480	263
Cu-Ni	9950	6850	4430	3260	1970	1040	650	462	253
Ni.	9900	6820	4400	3240	1960	1030	647	460	252
DO-Cu.	9800	6750	4360	3200	1940	1020	640	455	249

Mg. = Magnesium
Al. = Aluminum
Al. Br. = Aluminum Bronze
SS-400 = 400 series Stainless Steel
Mild St. = Mild Steel
SS-300 = 300 series Stainless Steel
Si. Br. = Silicon Bronze
Cu-Ni = Copper-Nickel
Ni. = Nickel
DO-Cu. = Deoxidized Copper

DATA CHART 3

CURRENT DENSITY CALCULATION CHART

Wire Diameter	Decimal Equivalent	Area Inches^2	Wire Diameter	Decimal Equivalent	Area Inches^2
0.020	0.020	0.00031	3/16	0.1875	0.0276
0.025	0.025	0.00051	13/64	0.2031	0.0324
0.030	0.030	0.00071	7/32	0.2187	0.0376
0.035	0.035	0.00096	15/64	0.2344	0.0431
0.045	0.045	0.00160	1/4	0.2500	0.0491
3/64	0.047	0.00173	17/64	0.2656	0.0553
1/16	0.0625	0.00307	9/32	0.2812	0.0621
5/64	0.0781	0.0048	19/64	0.2969	0.0692
3/32	0.094	0.0069	5/16	0.3125	0.0767
7/64	0.109	0.0093	21/64	0.3281	0.0845
1/8	0.125	0.01227	11/32	0.3437	0.0928
9/64	0.1406	0.0154	23/64	0.3594	0.1014
5/32	0.1562	0.0192	3/8	0.3750	0.1105
11/64	0.1719	0.0232			

Current Density.

Current density is calculated by dividing the electrode area, in square inches, into the welding amperage value used. The result is the amperage per square inch of electrode. For example:

7/64" dia. wire = $0.0093 \overline{)\ 500\ \text{amps}}$ = 53,762 amps/inches^2.

DATA CHART 4

SHIELDING GASES AND THEIR APPLICATIONS

Shielding gases have been discussed in this text. The chart shown here is a guide that may be used for the selection of the proper shielding gas for a specific application.

Shielding Gas	Chemical Behavior	Uses and Usage Notes
Argon	Inert	Welding almost all metals except steels.
Helium	Inert	Al, Cu, Mg and alloys.
Argon + Helium (Various percentages)	Inert	Al, Mg, Cu and alloys. Greater heat, quiet arc.
Argon + Chlorine (Trace chlorine)	Essentially Inert	For Al—to minimize porosity.
Nitrogen (N_2)	Reducing	For Cu, very powerful arc.
Argon + Nitrogen (25-30% N_2)	Reducing	For Cu, powerful but smoother arc than N_2.
Argon + 1 — 5% Oxygen	Oxidizing	Stainless and alloyed steels, carbon steels. Use deoxidized electrode wire.
Argon + CO_2	Oxidizing	Plain carbon and low alloy steels.
CO_2	Oxidizing	Plain carbon steels. Use deoxidized electrode wire.

Data for this chart taken from the Welding Handbook, Sixth Edition, with permission of the American Welding Society.

DATA CHART 5

PROPERTIES OF METALS

PROPERTIES OF METALS

Element	Symbol	Melting Point °F.	Coefficient of Exp. °F.	Elect. Cond. % Pure Cu	Lb/Cu. In.
Aluminum	Al	1218	0.0000133	64.9	0.098
Antimony	Sb	1167	0.00000627	4.42	0.239
Beryllium	Be	2345	0.0000068	9.32	0.066
Bismuth	Bi	520	0.00000747	1.50	0.354
Cadmium	Cd	610	0.00000166	22.7	0.313
Chromium	Cr	2822	0.0000045	13.2	0.258
Cobalt	Co	2714	0.00000671	17.8	0.322
Copper	Cu	1981	0.0000091	100.0	0.323
Gold	Au	1945	0.0000080	71.2	0.697
Iron	Fe	2795	0.0000066	17.6	0.284
Lead	Pb	621	0.0000164	8.35	0.409
Magnesium	Mg	1204	0.0000143	38.7	0.063
Mercury	Hg	−38	0........	1.80	0.489
Molybdenum	Mo	4748	0.00000305	36.1	0.368
Nickel	Ni	2646	0.0000076	25.0	0.322
Platinum	Pt	3224	0.0000043	17.5	0.774
Selenium	Se	428	0.0000206	14.4	0.174
Silver	Ag	1761	0.0000105	106.00	0.380
Tellurium	Te	846	0.0000093	....	0.224
Tin	Sn	450	0.0000124	15.0	0.264
Tungsten	W	6170	0.0000022	31.5	0.698
Vanadium	V	3110	0........	6.63	0.205
Zinc	Zn	787	0.0000219	29.1	0.258

DATA CHART 6

METRIC CONVERSION TABLES

The unit of length of the metric system, the meter, is intended to be one ten-millionth part of the distance from the equator to the pole. The meter is divided into ten parts called decimeters. Each decimeter is divided into ten equal parts known as centimeters and each centimeter is divided into ten millimeters. A millimeter is one thousandth part of a meter.

The metric units of length, mass and capacity are sub-divided decimally using the Latin prefixes of deci-, centi-, and milli-. The Greek prefixes deka, hecto, kilo and myria are used to indicate the multiplication of the units by ten.

U. S. to Metric	Metric to U. S.

Linear

1 inch = .0254000 meters	1 meter = 39.3700 inches
1 foot = .304800 meters	1 meter = 3.28083 feet
1 yard = .914400 meters	1 meter = 1.09361 yards
1 mile = 1609.35 meters	1 kilometer = .62137 miles
= 1.60935 kilometers	

Square

1 sq. inch = 6.452 sq. centimeters	1 sq. cm. = .1550 sq. inches
1 sq. foot = 9.290 sq. decimeters	1 sq. meter = 10.7640 sq. feet
1 sq. yard = .836 sq. meters	1 sq. meter = 1.196 sq. yards

Cubic

1 cu. inch = 16.387 cu. cm.	1 cu. cm. = .0610 cu. inches
1 cu. foot = .02832 cu. meters	1 cu. meter = 35.314 cu. feet
1 cu. yard = .765 cu. meters	1 cu. meter = 1.308 cu. yards

Capacity

1 fluid ounce = 29.57 milliliter	1 centiliter = .338 fluid ounces
1 quart = .94636 liters	1 liter = 1.0567 quarts
1 gallon = 3.78544 liters	1 hectoliter = 26.417 gallons

AMERICAN WELDING SOCIETY

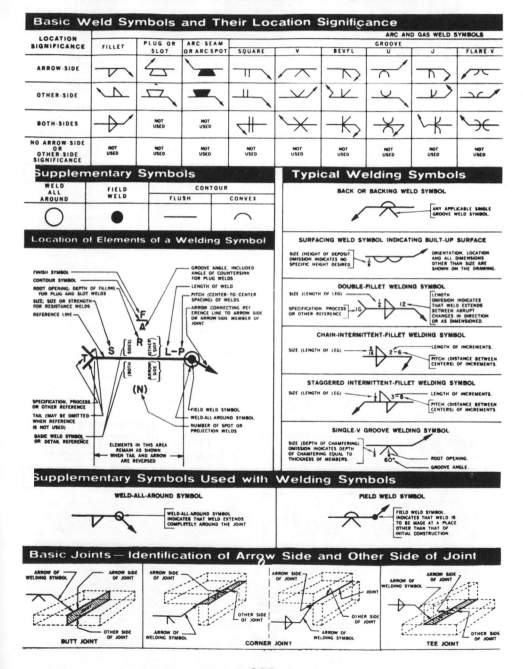

STANDARD WELDING SYMBOLS

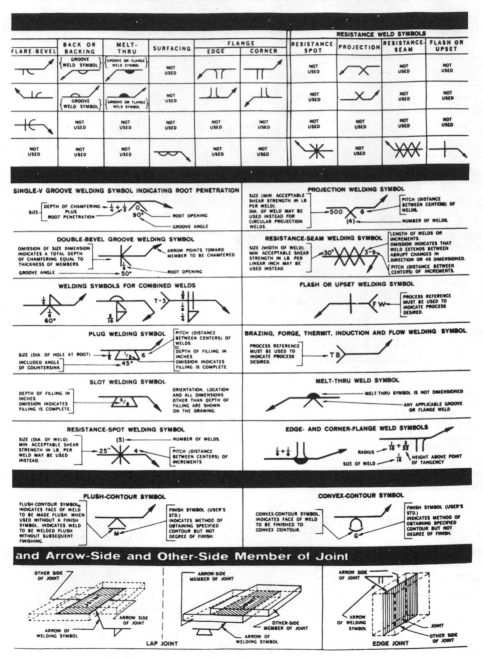

DATA CHART 8

F = FAHRENHEIT

C = CENTIGRADE

TEMPERATURE CONVERSION

$C = 5/9 \ (F-32)$
$F = 9/5 \ C \ +32$

$0°C = 273.16°K$
$0°F = 459.688°R$

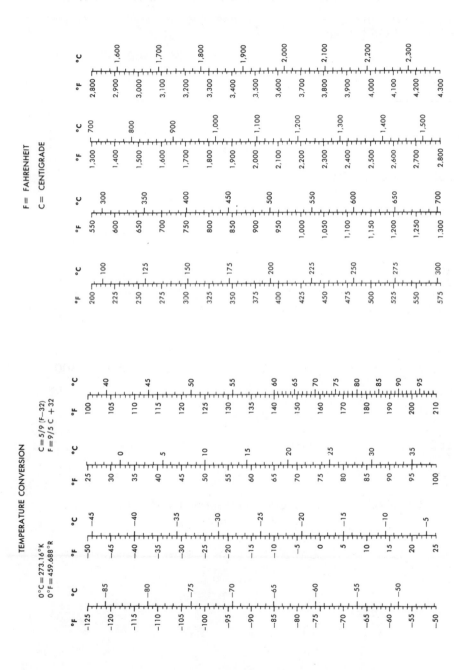

WELDING CABLE SIZES, AT 4 VOLTS DROP

Copper Welding Leads.

Amps.—	100	150	200	250	300	350	400	450	500	600
Feet From Terminals										
50'	2	2	2	2	1	1/0	1/0	2/0	2/0	3/0
75'	2	2	1	1/0	2/0	2/0	3/0	3/0	4/0	
100'	2	1	1/0	2/0	3/0	4/0	4/0			
125'	2	1/0	2/0	3/0	4/0					
150'	1	2/0	3/0	4/0						
175'	1/0	3/0	4/0							
200'	1/0	3/0	4/0							
225'	2/0	4/0								
250'	2/0	4/0								
300'	3/0									
350'	4/0									
400'	4/0									

The welding cable cover should be oil and moisture resistant.

The data above is based on the use of direct current.

Aluminum Welding Leads.

Amps.—	100	150	200	250	300	350	400	450
50'	2	2	1/0	2/0	2/0	3/0	4/0	
75'	2	1/0	2/0	3/0	4/0			
100'	1/0	2/0	4/0					
125'	2/0	3/0						
150'	2/0	3/0						
175'	3/0							
200'	4/0							
225'	4/0							

The data above is based on direct current and 4 volts drop

DATA CHART 10

MECHANICAL PROPERTIES OF
MILD STEEL AT ELEVATED TEMPERATURES*

Temp. °F.	Ultimate Strength psi	Yield Point psi	Elastic Limit psi	Modulus of Elasticity psi	Elongation in 2 inches %	Reduction of area %
100	54,000	39,000	36,000	30,700,000	48	68
200	56,000	43,000	33,000	29,000,000	43	66
300	58,000	46,000	28,000	27,300,000	39	65
400	60,000	47,000	22,000	25,600,000	37	64
500	63,000	46,000	16,000	23,900,000	37	64
600	66,000	44,000	14,000	22,200,000	40	66
700	60,000	40,000	12,000	20,500,000	44	68
800	52,000	37,000	10,000	18,000,000	49	76
900	43,000	32,000	8,000	17,100,000	55	84
1000	34,000	27,000	7,500	15,400,000	60	90
1100	25,000	21,000	5,000	13,700,000	64	94
1200	19,000	17,000	3,500	12,000,000	68	96
1300	14,000	13,000	2,000	10,300,000	72	98

Welding induces residual stresses in the weldment. These stresses may be reduced by post-weld thermal stress relief. The residual stress remaining in the weldment after thermal stress relief will depend on the rate of cooling. Uneven cooling of the part from stress relief temperatures may set up additional stresses and cancel the effects of the thermal stress relief cycle. After a part is stress relieved, the rate of cooling should be constant and consistent with the type of material being worked. Stress relieving temperatures, and cooling rates, should be based on manufacturers recommendations and appropriate regulatory bodies standards.

* American Welding Society Welding Handbook, 6th Edition.

DATA CHART 11

TYPICAL STRESS RELIEVING TEMPERATURES*

Material	Soaking Temperature, °F.
Carbon Steel	1100-1250
Carbon—½% Moly steel	1100-1325
½% Chrome—½% Moly steel	1100-1325
1% Chrome—½% Moly steel	1150-1350
1¼% Chrome—½% Moly steel	1150-1375
2% Chrome—½% Moly steel	1150-1375
2¼% Chrome—1% Moly steel	1200-1375
5% Chrome—½% Moly (Type 502) steel	1200-1375
7% Chrome—½% Moly steel	1300-1400
9% Chrome—1% Moly steel	1300-1400
12% Chrome (Type 410) steel	1350-1400
16% Chrome (Type 430) steel	1400-1500
1¼% Mn—½% Moly steel	1125-1250
Low-alloy Chrome-Ni-Moly steel	1100-1250
2%-5% Nickel steels	1050-1150
9% Nickel steel	1025-1085

* American Welding Society Welding Handbook, 6th Edition. The charted temperatures are to be used as a guide only. Actual stress relieving temperatures used will depend on local codes and established procedures.

DATA CHART 12

AMERICAN NATIONAL STANDARDS INSTITUTE.

MILLER OR ANSI ONLY

ACCEPTED CIRCUIT DIAGRAM SYMBOLS

CONTROL RELAY COIL — (CR₁) —

TIME DELAY RELAY COIL — (TD₁) —

WELD CONTACTOR COIL — (W) —

GAS SOLENOID COIL ⟋⟍ GS₁

WATER SOLENOID COIL ⟋⟍ WS₁

CONTROL OR TIME DELAY RELAY CONTACT (N.O.- *NORMALLY OPEN) ⊣⊢ CR₁ OR TD₁

CONTROL OR TIME DELAY RELAY CONTACT (N.C.- *NORMALLY CLOSED) ⊣⊬ CR₁ OR TD₁

*NORMAL IS DE-ENERGIZED POSITION

WELD CONTACTOR ⊣⊢ W
(OR RELAY CONTACT)   CR₁

MALE PLUG OR RECEPTACLE ⟶➤ RC₁

FEMALE PLUG OR RECEPTACLE ➤ RC₁

REACTOR OR STABILIZER   Z
⎓⎓ᒬᒬᒬᒬᒬ

CHOKE   RFC   RFC
⎓ᒬᒬᒬᒬᒬ⎓   ⎓ᒬᒬᒬᒬᒬ⎓
AIR CORE   FERRITE CORE

IRON CORE TRANSFORMER   T₁   CURRENT TRANSFORMER ⊣ᒬᒬ⊢ CT

AIR CORE TRANSFORMER   T₁

SATURABLE REACTOR   ᒬᒬᒬᒬᒬ
     OR   ═══════ SZ OR MA
MAGNETIC AMPLIFIER   ᒬᒬᒬᒬᒬ

CIRCUIT BREAKER

THERMAL OVERLOAD   MAGNETIC OVERLOAD

— 391 —

DATA CHART 12 (cont.)

FIXED RESISTOR	R_1
RHEOSTAT	R_1
POTENTIOMETER	R_1
POTENTIOMETER CONNECTED AS A RHEOSTAT	R_1
FUSE	F_1
THERMAL OVERLOAD	OL_1
SOLID STATE RECTIFIER (HALF WAVE)	SR_1 − + A.C.
SOLID STATE RECTIFIER (FULL WAVE)	D.C. SR_1 + − D.C. A.C.
CAPACITOR	C_1
GROUND	GND
SWITCH, PUSH BUTTON (NORMALLY OPEN)	S_1
SWITCH, PUSH BUTTON (NORMALLY CLOSED)	S_1
SWITCH, SINGLE-POLE, SINGLE-THROW	S_1
SWITCH, SINGLE-POLE, DOUBLE-THROW	S_1
SWITCH, DOUBLE-POLE, SINGLE-THROW	S_1
SWITCH, DOUBLE-POLE, DOUBLE-THROW	S_1
TERMINAL POINT	L_1
TERMINAL STRIP	1 2 3 4 TE_1
PILOT OR INDICATOR LAMP	PL-1
ROTARY SWITCH	S_1

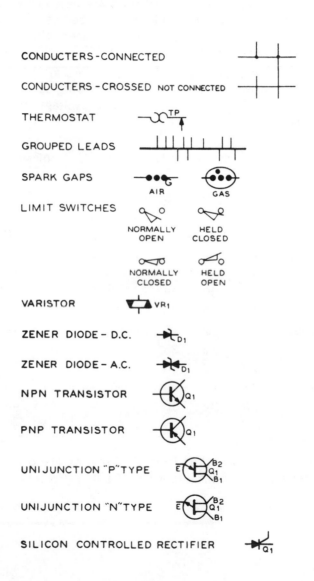

CONDUCTERS - CONNECTED

CONDUCTERS - CROSSED NOT CONNECTED

THERMOSTAT

GROUPED LEADS

SPARK GAPS

LIMIT SWITCHES

VARISTOR

ZENER DIODE - D.C.

ZENER DIODE - A.C.

NPN TRANSISTOR

PNP TRANSISTOR

UNIJUNCTION "P" TYPE

UNIJUNCTION "N" TYPE

SILICON CONTROLLED RECTIFIER

TYPICAL ARC VOLTAGES FOR THE
GAS METAL ARC WELDING PROCESS

The arc voltages presented in the chart are based on a plus or minus 10% value. The lower voltages would normally be used for thinner metals at low amperages. For heavier materials the higher voltage and amperage values would be applicable.

We stress that the values given in the chart are for reference purposes only and do not necessarily provide the best welding conditions. The chart is intended to provide a reasonable beginning voltage for you to use in setting the welding condition you need.

MATERIALS	SPRAY ARC-1/16" WIRE					DIPMATIC 0.035" WIRE			
	A	He	A + He	A + O	CO_2	A	A + O	A + CO_2	CO_2
Aluminum	25	30	28	—	—	19	—	—	—
Magnesium	27	31	30	—	—	19	—	—	—
Plain C steel	—	—	—	28	30	17	18	19	20
Low alloy steel	—	—	—	28	30	17	18	19	20
Stainless steel	24	—	—	26	—	18	19	21	—
Nickel	26	30	28	—	—	22	—	—	—
Ni-Cu alloy	26	30	28	—	—	22	—	—	—
Ni-Cr-Fe alloy	26	30	28	—	—	22	—	—	—
Copper	30	36	33	—	—	24	22	—	—
Cu-Ni alloy	28	32	30	—	—	23	—	—	—
Si Bronze	28	32	30	28	—	23	—	—	—
Al Bronze	28	32	30	—	—	23	—	—	—
Phos. Bronze	28	32	30	28	—	23	—	—	—

Data for this chart taken from the Welding Handbook, Sixth Edition with permission of the American Welding Society.

DATA CHART 14

GENERAL RECOMMENDATIONS FOR THE GAS SHIELDED METAL ARC PROCESS

The recommendations made in this chart are for filler metals and shielding gases.

Base Metal	Shielding Gas	Specific Alloy To Be Welded	Filler Metal	Elect. Diameter	Current Ranges
Aluminum and its alloys	Pure Argon or He-Ar mixture	1100 3003, 3004 5050 5052 5154, 5254 5083, 5084, 5456 6061	1100, 4043 4043 4043, 5554 5554, 5154 5554, 5154 5556, 5356 4043, 5556	0.030 0.045 1/16 3/32 1/8	50-175 90-250 160-350 225-400 350-475
Magnesium and its alloys	Pure Argon or He-Ar mixture	AZ31B, 61A, 81A, ZE10XA ZK20XA AZ31B, 61A, 63A, 80A, 81A, 91C, 92A, 100A, AM80A, ZE10XA XK20XA AZ63A	AZ61A AZ92A AZ63A	0.045 1/16 3/32	220-280 240-390 330-420
Copper	Pure Argon or He-Ar mixtures	Deoxidized Copper	Deox. Cu Si, 0.25% Sn, 0.75% Mn, 0.15%	1/16	300-470
Copper-Nickel alloy	Pure Argon	Cu-Ni alloy 70-30 90-10	Ti dexo. 70-30 90-10	1/16	250-300
Plain low Carbon steel	CO₂ A + CO₂ A + O	Hot or cold rolled sheet ASTM A7, A36 A285, A373 or equivalent	Deox. plain C steel	0.030 0.035 0.045 1/16 5/64	50-150 75-230 100-300 300-450 300-500
Low alloy steel	A + O A + CO₂	Hot or cold rolled sheet	Deox. low alloy	0.030 0.035 0.045 1/16	50-150 75-230 100-350 300-450
Stainless steel	Argon + O	302, 304 321, 347 309, 310 316, etc.	Elect. to match base metal	0.030 0.035 0.045 1/16	75-150 100-160 140-310 280-350
Nickel and Nickel alloys	Argon or He-Ar mixtures	Nickel Monel Inconel	Ti deox. to match base metal	0.035 0.045 1/16	100-150 150-260 200-400
Bronzes	Argon He + Ar mixtures A + O	Mn Bronze Al Bronze Ni-Al Bronze Sn Bronze	Al Bronze Al Bronze Al Bronze Phos. Bronze	1/16 5/64	225-300 275-350

Data for this chart taken and modified from the Welding Handbook, Sixth Edition, Section Two, published by the American Welding Society with permission.

DATA CHART 15

PROTECTIVE LENS FOR WELDING, COLOR CHART
RECOMMENDED SHADES

OPERATION	SHADE NUMBER
SOLDERING	2
TORCH BRAZING	3 OR 4
OXYGEN CUTTING UP TO 1 IN.	3 OR 4
1 TO 6 IN.	4 OR 5
6 IN. AND OVER	5 OR 6
GAS WELDING UP TO 1/8 IN.	4 OR 5
1/8 TO 1/2 IN.	5 OR 6
1/2 IN. AND OVER	6 OR 8
SHIELDED METAL-ARC WELDING	
1/16, 3/32, 1/8 & 5/32 IN. ELECTRODES	10
TIG WELDING (NONFERROUS)	
MIG WELDING (NONFERROUS)	
1/16, 3/32, 1/8 & 5/32 IN. ELECTRODES	11
TIG WELDING (FERROUS)	
MIG WELDING (FERROUS)	
1/16, 3/32, 1/8 & 5/32 IN. ELECTRODES	12
SHIELDED METAL-ARC WELDING	
3/16, 7/32 & 1/4 IN. ELECTRODES	12
5/16 - 3/8 ELECTRODES	14
ATOMIC HYDROGEN WELDING	10 TO 14
AIR CARBON-ARC WELDING, GOUGING, CUTTING	14

DATA CHART 16

Typical Gas Metal Arc Welding Conditions

c P250 TS .

Spray Transfer

Wire Type	Slope	Gas
Steel	0-4	$Ar + O_2$
Flux Core M.S.	0-2	CO_2
Aluminum	0-6	Ar or He or Mixture
Stainless Steel	0-4	$Ar + O_2$
Cu; Cu Alloys	0-2	Ar + He

Short Circuit Transfer

Wire Type	Slope	Gas
Mild Steel	6-10	CO_2 or $Ar + CO_2$
Stainless Steel	10-12	He-A-CO_2
Aluminum	12-14	Ar

Flux Core Electrode Conditions

Wire Dia.	Amperes	Load Volts
1/16"	240	24-25
5/64"	350	26-27
3/32"	460	30-31
7/64"	575	33-34
1/8"	650	35-36

All conditions and settings are approximate. The tables are to be used as a guide only.

DATA CHART 17
FRACTIONS, DECIMALS, MILLIMETERS

		DECIMALS	MILLIMETERS	MILLIMETERS	DECIMALS	
	1/64	.0156	0.3969	13.0969	.5156	33/64
1/32		.0313	0.7938	13.4938	.5313	17/32
	3/64	.0469	1.1906	13.8906	.5469	35/64
1/16		.0625	1.5875	14.2875	.5625	9/16
	5/64	.0781	1.9844	14.6844	.5781	37/64
3/32		.0938	2.3813	15.0813	.5938	19/32
	7/64	.1094	2.7781	15.4781	.6094	39/64
1/8		.125	3.1750	15.8750	.625	5/8
	9/64	.1406	3.5719	16.2719	.6406	41/64
5/32		.1563	3.9688	16.6688	.6563	21/32
	11/64	.1719	4.3656	17.0656	.6719	43/64
3/16		.1875	4.7625	17.4625	.6875	11/16
	13/64	.2031	5.1594	17.8594	.7031	45/64
7/32		.2188	5.5563	18.2563	.7188	23/32
	15/64	.2344	5.9531	18.6531	.7344	47/64
1/4		.250	6.3500	19.0500	.750	3/4
	17/64	.2656	6.7469	19.4469	.7656	49/64
9/32		.2813	7.1438	19.8438	.7813	25/32
	19/64	.2969	7.5406	20.2406	.7969	51/64
5/16		.3125	7.9375	20.6375	.8125	13/16
	21/64	.3281	8.3344	21.0344	.8281	53/64
11/32		.3438	8.7313	21.4313	.8438	27/32
	23/64	.3594	9.1281	21.8281	.8594	55/64
3/8		.375	9.5250	22.2250	.875	7/8
	25/64	.3906	9.9219	22.6219	.8906	57/64
13/32		.4063	10.3188	23.0188	.9063	29/32
	27/64	.4219	10.7156	23.4156	.9219	59/64
7/16		.4375	11.1125	23.8125	.9375	15/16
	29/64	.4531	11.5094	24.2094	.9531	61/64
15/32		.4688	11.9063	24.6063	.9688	31/32
	31/64	.4844	12.3031	25.0031	.9844	63/64
1/2		.500	12.7000	25.4000	1.000	1

— 398 —

DATA CHART 18
HARDNESS CONVERSION TABLE
FOR CARBON AND ALLOY STEELS

ALL VALUES ARE APPROXIMATE

Brinell Hardness Number (Carbide Ball)	Rockwell Hardness Numbers					Tensile Strength	
	C Scale	A Scale	15N Scale Superficial	B Scale	30T Scale Superficial	1000 Lb./Sq. In.	kgf/mm²
—	66	84.5	92.5	—	—	—	—
722	64	83.4	91.8	—	—	—	—
688	62	82.3	91.1	—	—	—	—
654	60	81.2	90.2	—	—	—	—
615	58	80.1	89.3	—	—	—	—
577	56	79.0	88.3	—	—	313	220
543	54	78.0	87.4	—	—	292	205
512	52	76.8	86.4	—	—	273	192
481	50	75.9	85.5	—	—	255	179
455	48	74.7	84.5	—	—	237	167
443	47	74.1	83.9	—	—	229	161
432	46	73.6	83.5	—	—	222	156
421	45	73.1	83.0	—	—	215	151
409	44	72.5	82.5	—	—	208	146
400	43	72.0	82.0	—	—	201	141
390	42	71.5	81.5	—	—	194	136
381	41	70.9	80.9	—	—	188	132
371	40	70.4	80.4	—	—	181	127
362	39	69.9	79.9	—	—	176	124
353	38	69.4	79.4	—	—	171	120
344	37	68.9	78.8	—	—	167	117
336	36	68.4	78.3	—	—	162	114
327	35	67.9	77.7	—	—	157	110
319	34	67.4	77.2	—	—	153	108
311	33	66.8	76.6	—	—	149	105
301	32	66.3	76.1	—	—	145	102
294	31	65.8	75.6	—	—	142	100
286	30	65.3	75.0	—	—	138	97
279	29	64.7	74.5	—	—	135	95
271	28	64.3	73.9	—	—	132	93
264	27	63.8	73.3	—	—	128	90
258	26	63.3	72.8	—	—	125	88
253	25	62.8	72.2	—	—	122	86
247	24	62.4	71.6	—	—	120	84
243	23	62.0	71.0	—	—	117	82
240	—	—	—	100	82.0	116	82
234	—	—	—	99	81.5	112	79
222	—	—	—	97	80.5	106	75
210	—	—	—	95	79.0	101	71
200	—	—	—	93	78.0	96	67
195	—	—	—	92	77.5	93	65
185	—	—	—	90	76.0	89	63
176	—	—	—	88	75.0	85	60
169	—	—	—	86	74.0	81	57
162	—	—	—	84	73.0	78	55
156	—	—	—	82	71.5	75	53
150	—	—	—	80	70.0	72	51
144	—	—	—	78	69.0	—	—
139	—	—	—	76	67.5	—	—
135	—	—	—	74	66.0	—	—
130	—	—	—	72	65.0	—	—
125	—	—	—	70	63.5	—	—
121	—	—	—	68	62.0	—	—
117	—	—	—	66	60.5	—	—
114	—	—	—	64	59.5	—	—

DATA CHART 19
INCHES TO MILLIMETERS

Inches to Millimeters

1 Inch = 25.4 Millimeters   1 Millimeter = .0394 Inches

Inches	0"	1"	2"	3"	4"	5"	6"	7"	8"	9"	10"	11"	12"	Inches
	.000	25.400	50.800	76.200	101.600	127.000	152.400	177.800	203.200	228.600	254.000	279.400	304.800	
1/32	.794	26.194	51.594	76.994	102.394	127.794	153.194	178.594	203.994	229.394	254.794	280.194	305.594	1/32
1/16	1.588	26.988	52.388	77.788	103.188	128.588	153.988	179.388	204.788	230.188	255.588	280.988	306.388	1/16
3/32	2.381	27.781	53.181	78.581	103.981	129.381	154.781	180.181	205.581	230.981	256.381	281.781	307.181	3/32
1/8	3.175	28.575	53.975	79.375	104.775	130.175	155.575	180.975	206.375	231.775	257.175	282.575	307.975	1/8
5/32	3.969	29.369	54.769	80.169	105.569	130.969	156.369	181.769	207.169	232.569	257.969	283.369	308.769	5/32
3/16	4.763	30.163	55.563	80.963	106.363	131.763	157.163	182.563	207.963	233.363	258.763	284.163	309.563	3/16
7/32	5.556	30.956	56.356	81.756	107.156	132.556	157.956	183.356	208.756	234.156	259.556	284.956	310.356	7/32
1/4	6.350	31.750	57.150	82.550	107.950	133.350	158.750	184.150	209.550	234.950	260.350	285.750	311.150	1/4
9/32	7.144	32.544	57.944	83.344	108.744	134.144	159.544	184.944	210.344	235.744	261.144	286.544	311.944	9/32
5/16	7.938	33.338	58.738	84.138	109.538	134.938	160.338	185.738	211.138	236.538	261.938	287.338	312.738	5/16
11/32	8.731	34.131	59.531	84.931	110.331	135.731	161.131	186.531	211.931	237.331	262.731	288.131	313.531	11/32
3/8	9.525	34.925	60.325	85.725	111.125	136.525	161.925	187.325	212.725	238.125	263.525	288.925	314.325	3/8
13/32	10.319	35.719	61.119	86.519	111.919	137.319	162.719	188.119	213.519	238.919	264.319	289.719	315.119	13/32
7/16	11.113	36.513	61.913	87.313	112.713	138.113	163.513	188.913	214.313	239.713	265.113	290.513	315.913	7/16
15/32	11.906	37.306	62.706	88.106	113.506	138.906	164.306	189.706	215.106	240.506	265.906	291.306	316.706	15/32
1/2	12.700	38.100	63.500	88.900	114.300	139.700	165.100	190.500	215.900	241.300	266.700	292.100	317.500	1/2
17/32	13.494	38.894	64.294	89.694	115.094	140.494	165.894	191.294	216.694	242.094	267.494	292.894	318.294	17/32
9/16	14.288	39.688	65.088	90.488	115.888	141.288	166.688	192.088	217.488	242.888	268.288	293.688	319.088	9/16
19/32	15.081	40.481	65.881	91.281	116.681	142.081	167.481	192.881	218.281	243.681	269.081	294.481	319.881	19/32
5/8	15.875	41.275	66.675	92.075	117.475	142.875	168.275	193.675	219.075	244.475	269.875	295.275	320.675	5/8
21/32	16.669	42.069	67.469	92.869	118.269	143.669	169.069	194.469	219.869	245.263	270.663	296.063	321.469	21/32
11/16	17.463	42.863	68.263	93.663	119.063	144.463	169.863	195.263	220.663	246.063	271.463	296.863	322.263	11/16
23/32	18.256	43.656	69.056	94.456	119.856	145.256	170.656	196.056	221.456	246.856	272.256	297.656	323.056	23/32
3/4	19.050	44.450	69.850	95.250	120.650	146.050	171.450	196.850	222.250	247.650	273.050	298.450	323.850	3/4
25/32	19.844	45.244	70.644	96.044	121.444	146.844	172.244	197.644	223.044	248.444	273.844	299.244	324.644	25/32
13/16	20.638	46.038	71.438	96.838	122.238	147.638	173.038	198.438	223.838	249.238	274.638	300.038	325.438	13/16
27/32	21.431	46.831	72.231	97.631	123.031	148.431	173.831	199.231	224.631	250.031	275.431	300.831	326.231	27/32
7/8	22.225	47.625	73.025	98.425	123.825	149.225	174.625	200.025	225.425	250.825	276.225	301.625	327.025	7/8
29/32	23.019	48.419	73.819	99.219	124.619	150.019	175.419	200.819	226.219	251.619	277.019	302.419	327.819	29/32
15/16	23.813	49.213	74.613	100.013	125.413	150.813	176.213	201.613	227.013	252.413	277.813	303.213	328.613	15/16
31/32	24.606	50.006	75.406	100.806	126.206	151.606	177.006	202.406	227.806	253.206	278.606	304.006	329.406	31/32

DATA CHART 20
POUNDS TO KILOGRAMS
KILOGRAMS TO POUNDS

Pounds to Kilograms
1 Pound = 0.453592 Kilogram

Pounds	Kilograms	Pounds	Kilograms	Pounds	Kilograms	Pounds	Kilograms	Pounds	Kilograms
0.1	0.05	1	0.45	10	4.54	100	45.36	1000	453.59
0.2	0.09	2	0.91	20	9.07	200	90.72	2000	907.18
0.3	0.14	3	1.36	30	13.61	300	136.08	3000	1360.78
0.4	0.18	4	1.81	40	18.14	400	181.44	4000	1814.37
0.5	0.23	5	2.27	50	22.68	500	226.80	5000	2267.96
0.6	0.27	6	2.72	60	27.22	600	272.16	6000	2721.55
0.7	0.32	7	3.18	70	31.75	700	317.51	7000	3175.14
0.8	0.36	8	3.63	80	36.29	800	362.87	8000	3628.74
0.9	0.41	9	4.08	90	40.82	900	408.23	9000	4082.33
1.0	0.45	10	4.54	100	45.36	1000	453.59	10000	4535.92

Kilograms to Pounds
1 Kilogram = 2.204622 Pounds

Kilograms	Pounds	Kilograms	Pounds	Kilograms	Pounds	Kilograms	Pounds	Kilograms	Pounds
0.1	0.22	1	2.20	10	22.05	100	220.46	1000	2204.62
0.2	0.44	2	4.41	20	44.09	200	440.92	2000	4409.24
0.3	0.66	3	6.61	30	66.14	300	661.39	3000	6613.87
0.4	0.88	4	8.82	40	88.18	400	881.85	4000	8818.49
0.5	1.10	5	11.02	50	110.23	500	1102.31	5000	11023.11
0.6	1.32	6	13.23	60	132.28	600	1322.77	6000	13227.73
0.7	1.54	7	15.43	70	154.32	700	1543.24	7000	15432.35
0.8	1.76	8	17.64	80	176.37	800	1763.70	8000	17636.98
0.9	1.98	9	19.84	90	198.42	900	1984.16	9000	19841.60
1.0	2.20	10	22.05	100	220.46	1000	2204.62	10000	22046.22

INDEX

—A—

AC alternator design: 356
Acetylene: 3, 12, 14, 21
Air carbon-arc gouging: 335
  Electrodes, amperage range: 336
Allotropic: 81
Alloy addition: 85
Alloying elements, steels: 83, 85
  Manganese: 83, 85
  Phosphorous: 83, 85
  Sulphur: 83, 85
Alloys:
  Aluminum: 305, 307
  Carbon steel: 305
  Copper: 25
  Electrode wire: 307
  Low alloy steel: 75
Alternating current: 21
  Effective value: 38
  Frequency: 43
  Gas tungsten arc process: 250
  Impedance: 46
  Shielded metal arc welding: 147, 152
  Troubleshooting: 176
  Welding power sources: 152, 160
Alternating magnetic field: 126
Aluminum:
  Cleaning procedures: 347
  Conductor: 29
  DC component: 201, 251-253
Ampere: 22
Ampere-turn: 22
Amplitude: 26
Angstrom: 22
Anode: 22
Arc welding: 22
Arc blow: 22
Arc initiation:
  Force required for: 22
  Gas tungsten arc: 244
  High frequency: 46
Arc power ratio:
  Volt-ampere: 148, 315
Arc stabilization:
  High frequency: 264
Arc time: 22
Arc voltage: 23

Arc welding processes:
  Atomic-hydrogen: 23
  Gas metal arc: 281
  Gas tungsten arc: 236
  Shielded metal arc: 142
Argon: 239, 297
  Additions to:
    CO_2: 303
    Helium: 302
    Helium-CO_2: 303
    Oxygen: 301
Atomic-hydrogen process: 23
Atomic structure:
  Electron bond: 40
  Materials: 40
  Nucleus: 55
  Shielding gases: 239
Atomic weight: 101
AWS welding symbols: 386-387

—B—

Backfire: 24
Backhand welding technique: 24
Backing, Backup: 24
Bare electrode: 25
Base Metal:
  Alloy materials: 25
  Heat affected zone: 45
  Welding applications: 25
Basic metal structures: 78, 79
Bead:
  Top bead: 25
  Under bead: 25
Black iron: 1
Blacksmith: 1
Brazing: 18
  Filler rods: 19
Brush-commutator, MG: 181, 182
Buried arc transfer: 287
Buttering: 26, 82

—C—

Capacitance: 27
  Power factor: 137
Capacitor: 26
  High frequency: 268-269

Capillary attraction: 27
Carbide precipitation: 96
Carbon:
  Boiling point: 256, 257
  Iron alloy: 74
  Melting point: 256, 257
Carbon steel: 82
  High carbon: 82
  Low carbon: 82
  Medium carbon: 82
Cast iron: 82
  Buttering: 26, 82
Cathode: 27
CFH: 27
Charpy test: 27
Chemical cleaning:
  Aluminum: 347
Chipping: 27
Chlorine: 304
  Circuit: 28
  Diagram: 216
  High frequency: 267-268
Cleaning action:
  Gas metal arc: 297
  Gas tungsten arc: 250
Coil: 28
  Form: 28
  Primary: 61
  Reactor: 61
  Secondary: 65
Cold rolled steel: 84
  Grain size and control: 84
  Welding of: 84
Cold shuts: 300
Communications: 20
Conductor: 29, 102
  Circuit: 28
  Duty cycle: 36, 37
  Eddy currents: 130, 132
  Electrical resistance: 130
  Temperature: 131
Constant current power source: 29
Constant potential: 30
  Constant voltage: 30
  Power source: 30
  Slope: 325
  Voltages: 323
Contactor: 31
  Primary: 31
  Secondary: 31
Control circuit:
  Description of: 221
Cooling curve: 32
Cooling rates: 82

Cooling rates, steel: 86
  Bainite: 86
  Coarse pearlite: 86
  Fine pearlite: 86
  Martensite: 86
Copper: 25
  Alloy: 25
  Conductor: 29
Core: 32, 33, 131
  Basic design: 131
  Hysteresis: 133
  Insulation: 32
  Magnetic link: 32
  Permeability: 33
Coulombs Law: 379
Counter EMF: 34
  Lenz's Law: 379
Crystal formation: 78
Crystal structures: 78, 79
Current: 36
  Electrode melt rate: 146
  Methods of control: 172
  Movement: 36
Current density: 242-243, 342, 382
  Chart: 382
Current feedback control: 215
Current flow: 104
  High frequency: 268
Cycles: 34
  AC sine wave form: 35
  Electrical degrees: 39
  Frequency: 43
  Zero line: 39
Cylinder valves: 36

—D—

DC component: 251
  Formation of: 251, 252, 253
Defects, weld:
  Oxide inclusion: 251
Density, current: 242-243, 342, 382
Deoxidizer: 36
Deposition efficiency: 36
Deposition, weld metal: 339, 340, 341, 342
  Penetration: 343
  Rate: 341, 342
Dielectric: 36
Direct current: 36
  Arc blow: 22
  DCRP arc: 282
  DCSP arc: 283
  Gas metal arc: 282
  Gas tungsten arc: 274
  Polarity: 58

Ductility: 36
Duty cycle: 36
  NEMA Standards: 37
  Volt-ampere curves: 37
Dynamic electricity: 38

—E—

Eddy current:
  Direction of movement: 130
  Loss calculations: 132
  Minimized in conductors: 131
Effective value: 38
  RMS (root means square): 38
Elasticity: 38
Electrical charges: 38
Electrical conductivity: 39
Electrical conductor concepts: 107
Electrical conductors: 102
Electrical current:
  Definition: 102
  Measurement: 105
Electrical degrees: 39
Electrical efficiency:
  Conductors: 130
Electrical energy:
  Definition: 106
  For welding: 114, 115, 116
Electricity:
  Dynamic: 38
  Static: 67
Electrode: 39
  Holder: 40
Electrodes:
  Bare: 25, 144
  Flux coating: 144, 145, 146
  Gas metal arc: 305
  Gas tungsten arc: 242, 248
  Shielded metal arc: 145, 146
  Wire calculation chart: 380, 381
Electron: 40
  Flow: 104
  Ion: 48
Electron beam welding: 366
  Generation of EB: 366
  Process: 367
  Weldable metals: 367, 368
  Welding procedure: 367
Electron bond: 40
Electron movement concepts: 40, 110
Electron theory of current flow: 104
Electrostatic effect: 27
Embrittlement: 41
Energy:
  Arc: 147
  Electrical: 106

Induced: 47
Latent: 118
Mechanical: 181
Stored: 26
Engine driven generators: 349
Epilogue: 376
Equipment:
  Atomic-hydrogen: 23
  Gas metal arc: 284
  Gas tungsten arc: 236
  Safety: 159
  Shielded metal arc: 142

—F—

Farad: 27
Ferrous metals: 73, 74
Filler metal: 396
  Gas metal arc: 339
  Joint requirements: 339
  Losses: 340
Fixtures: 24
Flashback: 41
Flowmeter: 41
Flux, chemical: 41
Flux-cored electrodes: 308
  Benefits: 308
  Development: 308
  Manufacture: 308, 309
Flux, electrical: 41
  Magnetic lines of force: 41
Flux, electrode: 144-146
  Coating: 144-146
  Deoxidizers: 305
  Functions of: 145
Forehand welding: 42
Forge: 1
Frequency: 43
Fuse: 43
Fusion welding: 44

—G—

Gas flow rates: 242
Gas metal arc welding: 281
  Amperage control: 294
  Applications: 283, 344
    Aluminum: 347
    Steel: 345
  Chart: 396
  Electrode wire: 305
    Alloys: 305
    Deoxidizers: 305
    Flux-cored: 308
  Equipment: 284
  General: 281
  Open circuit voltage: 56, 323

— 404 —

Power sources for: 311
 Amperage control: 321
 Constant potential: 317
 Conventional: 313
 Inductor: 333
 Slope:
  Reactor: 327-333
 Transformer design: 318
 Voltages: 323
Shielding gases: 297-304
Slope: 325-333
Spot welding: 67, 333-334
Types of metal transfer:
 Basic: 286
 Buried arc: 287
 Globular: 291
 Pulsed current: 289
 Short circuit: 291
 Spray: 286
Gas pocket: 44
Gas pressure regulators: 8
 Acetylene: 8
 Oxygen: 9
 Single stage: 9
 Two stage: 9
Gas shielded welding: 44
 Gas metal arc: 281
 Gas tungsten arc: 236
Gas tungsten arc welding: 236
 AC welding power source parts: 276
  Derating for current: 278
 AC/DC welding power sources: 198
 Equipment: 236
 General: 236-237
 Shielding gases: 239
 Torch: 238
Generator, auxiliary power plant: 357, 361
Generator, DC, open circuit voltage: 349-350
Generator design:
 AC alternator: 356
 AC generator: 355
 DC generator: 353
Generator heat losses: 359
Generator motive power: 352
 Electric motor: 352
 Fuel powered engine: 353
Generator, welding:
 AC alternator: 356
 AC generator: 355
 DC generator: 353
Generator, welding output: 360
Groove angle: 44

—H—

Heat:
 Balance in arc: 147
 Gas metal arc: 344
 Gas tungsten arc: 236
  DC component: 251
  Energy input comparison: 238
  Thermal placement: 245
 Losses: 359
Heat affected zone: 45
 Macro examination: 150
 Profile: 90
Heat transfer rates: 17
Heat treatment: 74
Helium: 240, 298
Hertz: 45
High Frequency: 46, 264, 268
 Arc initiation: 46, 265
 Gas ionization: 264
 Oscillator unit: 266
  Circuit diagram: 267-268
 Problems with: 269
 Skin effect: 267
Holder, electrode: 40
 Gas metal arc: 40
 Gas tungsten arc: 40
 Shielded metal arc: 40
Hold time: 46
 Resistance welding: 64
 Weld nugget: 55
Horsepower: 46
 Electrical: 46
 Mechanical: 46
 Watt: 71
Hot start current: 233-234
 Hydrogen:
  Atomic: 23
 Embrittlement: 41, 86
 Hot cracking: 83
Hysteresis: 46, 133
 Definition: 133

—I—

Impedance: 46
 Reactance: 46
 Resistance: 46
Included angle: 44
Inclusions: 47
 Slag residue: 47
 Weave bead: 71
Induced current: 47
Induced voltage: 34
Inductance: 47
Inductor: 333

Inert gases: 47
  Argon: 239, 297
  Helium: 240, 298
Inert gas shielded arc cutting: 47
Inert gas shielded arc welding: 48
Insulator: 102
Ion: 48
  Negative ions: 48
  Positive ions: 48
Ionization potential: 48
  Gas metal arc: 297
  Gas tungsten arc: 243, 264
  Shielding gases: 297
I²R heating effect: 130
  Formula: 130
  Power losses: 130
Iron: i
  Alloy with carbon: ii, 74
  Melting point range: ii
  Transformer core: 131
Izod test: 48

—J—

Joint: 48
  Gas metal arc: 338
  Shielded metal arc: 338
Joint efficiency: 50
Joint penetration: 50
Joule: 50

—K—

Kilo: 136
Kilovars: 137
  Capacitive: 137
  Inductive: 137
Kilo volt-amperes:
  Definition: 120
  Rule for specific KVA: 120
Kilowatts:
  Primary: 122, 136
  Secondary: 122
Kinetic theory: 106

—L—

Laser welding: 373
Latent energy: 118
Lenz's Law: 379
  Applied to transformers: 168-169
  Principle of CEMF: 34, 168-169
Line voltage, primary: 136
Load voltage: 50
Locked rotor current: 51

—M—

Macrograph examination: 150
Magnetic arc blow: 22
Magnetic coupling: 51
  Iron core function: 133
  Transformer: 133
Magnetic field: 52
  Circuit: 125-126
  Flux: 125-126
  Hysteresis loss: 133
  Induced current: 126
  Reactor concept: 209
  Reluctance: 63
  Strength calculation: 209
  Transformer concept: 125
Malleability: 53
  Plastic deformation: 53
Martensite: 86-89
  Tempering and hardenability: 87-89
Matter, theory of: 99
  Atom: 100
  Atomic movement in: 106
  Atomic nucleus: 100
  Electrons: 100
  Molecule: 99
  Neutrons: 100
  Protons: 100
Mechanical properties of mild steel
at elevated temperatures: 390
Melting point:
  Carbon: 256-257
  Ice: i
  Refractory elements: 62-63
  Steel: ii
  Tungsten: 257
Metal:
  Conductors: 29, 102
  Ductility: 36
  Electron bond: 40
  Overheating: 56
  Properties .of: 384
  Removal, chipping: 27
  Weldability: 71
Metallurgist: 73
  Extractive: 73
  Process: 73
Metallurgy: 73
MIG: 53
Metric conversion table: 375
Micro-farad: 27
Mild steel: 53
Mixtures, shielding gases:
  Argon-CO₂: 303
  Argon-helium: 302
  Argon-oxygen: 301
  Helium-agron-CO₂: 303

Motor, electric: 53-54
  Design criteria: 201
  Energy conversion: 181
  Locked rotor current: 51
  Starting current: 184
Motor generator: 180
  Early models: 201
  Function and design: 180-182

—N—

Negative charge: 54
  Cathode: 27
NEMA: 54
  Duty cycle: 36
  Rating: 170
Newton's Law: 379
Nitrogen: 304
Non-ferrous metals: 73
  Hardening of: 73-74
Nucleus: 55
  Positive charge: 61
Nugget: 55
  Resistance welding: 64

—O—

Ohm: 55
  Ohm's Law: 379
  Unit of electrical resistance: 55
Open circuit voltage: 56
  Gas metal arc: 323
OSHA: 4
Overheating: 56
Oxide removal:
  Aluminum: 347
  Gas tungsten arc: 250
Oxy-acetylene torches:
  Torch body: 6
  Torch mixer: 7
  Torch tips: 8
Oxy-acetylene welding:
  Carburizing flame: 2
  Neutral flame: 2
  Oxidizing flame: 2
Oxygen: 56
Oxygen-fuel gas welding: 2
Hydrogen, propane, natural gas: 2

—P—

Penetration:
  Joint: 50
  Weld comparison: 343
Permeability: 56

Phase: 57
  Capacitive circuit: 57
  Inductive circuit: 57
  Resistive circuit: 57
Phase diagram, Iron-iron carbide: 81
Phase diagram, pure iron: 80
Phase transformation: 79
Plasma arc welding process: 368-373
Polarity: 58
  Connections: 58
Porosity: 58, 346
Positioned weld: 59
Positive:
  Charge: 59
  Pole: 22
Potential, constant: 30
Power: 117
  Line: 122
  Primary: 122
  Welding: 114
Power factor: 59
  Benefits of: 138
  Definition: 59
  Example: 138-139
Preparation, joint: 44
Primary coil: 61
  Transformers: 162
Primary contactor: 31
Protective lens, color chart: 397
Proton: 61
Pulsed current transfer: 289

—Q—

Quenching mediums: 87
Questions and answers:
  AC power sources: 177-178

—R—

Ratings, power sources:
  AC: 171
  NEMA: 170
Reactance, inductive: 61
Reactor: 61
  ANSI Standard symbol: 208
  Current control: 211-212
  Description of: 208
  Function of: 209-210
  Slope control: 327
  Tapped: 211
Rectified DC:
  Single and three phase: 202

Rectifier: 61
  AC/DC welding power sources: 198
  DC welding power sources: 201
  Function: 180
  Metallic types: 180
  Selenium: 185
  Silicon: 191
  Vapor-arc types: 180
Refractory materials: 62
  Known elements: 62
Relay: 63
Reluctance: 63
Residual magnetic field: 64
Resistance: 64
  Electrical conductor: 107-108
  Ohm's Law: 379
  Power loss equation: 130
Resistance heating: 116
  Power loss equation: 117
Resistance welding: 64
  Nugget: 55
Response time: 332
  Reverse polarity: 64
  Connections for: 65
  Gas metal arc: 311
  Gas tungsten arc: 246
RMS current: 38
  Effective value: 38
Root pass: 65
Rotating equipment: 349

—S—

Safety: 3
  Cylinders: 4
  Electrical: 142, 157
  Glasses: 5
  Oxy-acetylene: 3
  Protective clothing: 159
Saturable reactor control: 206
  Function: 211-214
Saturation curve: 213
Seal weld: 65
Secondary coil: 65-66
Secondary contactor: 31
Shielded metal arc welding: 142
  AC welding power sources: 152
  Basic equipment: 142
  DC welding power sources: 153
  Electrode losses: 340
Shielding gases:
  Gas metal arc: 284
  Applications chart: 383
    Argon: 297
    Argon-chlorine: 304
    Argon-CO_2: 303

Argon-helium: 302
Argon-nitrogen: 304
Argon-oxygen: 301
CO_2: 299
Helium-argon-CO_2: 303
Nitrogen: 304
Gas tungsten arc: 236
  Argon: 239
  Helium: 240
  Ionization of: 243
Short circuit transfer: 291
  Slope: 293
Single phase:
  AC welding power sources: 178
  AC/DC welding power sources: 198
Skin effect: 267
Slag:
  Shielded metal arc: 145
Slope: 66
  Function of: 325
Slope control:
  Reactor: 327
  Resistance: 325
Solid state:
  Metals: iii
Spatter: 66
Spot welding: 67
  Gas metal arc: 67
  Gas tungsten arc: 67
  Resistance: 67
Spray Transfer: 286
Squeeze time: 67
Stabilized AC:
  High frequency: 264
Stabilizer:
  AC/DC welding power sources: 198
Stainless steel, basic types:
  Martensitic: 94
  Ferritic: 94
  Austenitic: 95
Stainless steel filler metals: 95
Stainless steel, sigma phase: 95
Standard welding symbols, AWS: 386-387
Static electricity: 67, 98
Steel: 33
  Electrical: 33, 131
  Permeability: 33, 131
  Silicon: 33, 133
Straight polarity: 67
  Connections: 67-68
  Gas tungsten arc: 249
Stress relieving: 91-92
Stress relieving temperatures:
  Chart: 391

Stringer bead: 68
Submerged arc welding: 363
Switch: 68
  Contactor: 31

—T—

Technique, welding:
  Backhand: 24
  Forehand: 42
Temperature:
  Conversion chart: 388
Testing methods: 21
Thermal conductivity: 69
  Aluminum, copper, magnesium,
  carbon steel, nickel, nickel al-
  loys, stainless steel: 17
Three phase:
  AC wave form: 276
  Rectified: 276
TIG: 69
Timer: 69
Top bead: 25
Transfer:
  Buried arc: 287
  Globular: 291
  Pulsed current: 289
  Short circuit: 292
  Spray: 287
Transformation diagram, Isothermal
(TTT): 87
Transformation temperatures:
  Alloy metals: 76
  Pure metals: 76
Transformer:
  AC welding: 172-174
  Auto-transformer: 126
  Component relationships: 163
  Constant potential: 318
  Functions of: 125, 167
  Isolation: 128
  Power: 121
  Two winding: 126-127, 163
  Volt-ampere ratio: 126-127
  Volt-turns ratio: 126-127
  Welding power sources: 124
Transformer-rectifier power
sources:
  Constant potential: 30, 317
  Single phase AC/DC: 198
  Three phase DC: 201
Tungsten:
  Atomic-hydrogen: 23
  Boiling point: 256
  Centerless ground: 257-258
  Clean: 257
  Current carrying capacity: 257

Electrical pointing: 261
  Electrodes, chart: 248
  Emissive qualities: 257
  Gas shielding: 242
  Grinding: 258
  Manufacture: 255
  Melting point: 257
  Thoriated: 257
Turn:
  Ratio, volt-turn: 127

—U—

Ultimate strength: 70
Ultrasonic welding: 368
Underbead: 25
Undercutting: 301
Unit charge: 70

—V—

Vector:
  Definition: 137
  Diagram: 141
  Sum: 137
Volt: 70
  Meter: 224
  Volt-turn ratio: 127
Voltage:
  Arc: 23, 323
  Constant: 30
  High frequency: 267
  Load: 50, 323
  Open circuit: 323
  Potential: 109
  Primary: 226
    Terminal linkage: 226
Volt-ampere curve: 70
  AC/DC Power sources: 199
Voltmeter: 71

—W—

Watt: 71
Watt-hour: 71, 114
  As energy units (Joules): 114
Weave bead: 71
Welder: 71
Weld cleaning: 340
Weldability: 71
Weld deposit:
  Comparison of SMAW and
  GMAW: 344
Weld joint, metallurgy: 149
Weld metal deposits:
  Alloy steel: 92
  Carbon steel: 92

— 409 —

Welding:
  Cable size chart: 389
  Procedures: 72
Welding Metallurgy: 73
Welding power sources:
  AC: 160
  Constant potential: 29
  DC control panel: 228
    Start rheostat: 233
    Weld rheostat: 230
  DC Key circuits: 206
    Control circuit: 221
    Current feedback: 215
    Saturable reactor: 206-215
  DC Troubleshooting: 224
    Duty cycle: 36
  Gas metal arc: 311

  Constant potential: 317
  Conventional: 313
  Gas tungsten arc: 274
  Shielded metal arc: 150
Welding operator duty cycle: 22
Weldment: 72
Wire feeder equipment: 284
  General description: 284
Wire, flux cored electrode: 308

—Y—

Yield strength: 72

—Z—

Zero line: 39